Closing the Mind Gap

Making Smarter Decsions in a Hypercomplex World

Ted Cadsby

Foreword by Don Tapscott

Toronto and New York

Copyright © 2014 by Ted Cadsby

All rights reserved. No part of this publication may be reproduced or transmitted in any form or by any means, electronic or mechanical, including photocopying, recording, or any information storage and retrieval system, without permission in writing from the publisher.

Published in 2014 by
BPS Books
Toronto and New York
www.bpsbooks.com
A division of Bastian Publishing Services Ltd.

ISBN 978-1-927483-78-7 (paperback)
ISBN 978-1-927483-79-4 (ePDF)
ISBN 978-1-927483-80-0 (ePUB)

Cataloguing-in-Publication Data available from Library and Archives Canada.

Cover: Gnibel
Text design and typesetting: Daniel Crack, KD Books, www.kdbooks.ca
Index: Isabel Steurer

*For Jodie and Mackenzie,
inheritors of many choices
in an increasingly complex world*

Contents

FOREWORD xi
by Don Tapscott

PREFACE xvii

INTRODUCTION 1
The Mind Gap
An overview of the book and the main argument.

PART I   BRAINS COMING INTO THE WORLD
How Smart Are We Anyway?

1   Brain Evolution 13
    Two Big Bangs
    The circuitous journey from the first animal brains to human brains.

2   Bounded Brain 23
    The Cognitive Trade-Off
    Of the many constraints on our thinking, the most prominent is the trade-off between speed and accuracy.

PART II   BRAINS SORTING OUT THE WORLD
Construction Zone A Head

3   To Know Is to Construct 43
    Building a World
    Naïve realism is a useful operating assumption in the limiting case of straightforward challenges, but not in the case of complex problems.

4	**To Know Is to Simplify**	57
	Reducing Reality	
	A bounded brain interprets the world by simplifying it via three cognitive shortcuts.	

5	**When Simplifying Meets Complexity**	73
	Greedy Reductionism	
	Simplifying shortcuts work up to the point where we are too greedy in reducing reality.	

6	**To Know Is to Satisfice**	88
	Rushing to Certainty	
	We are addicted to certainty, courtesy of our physiology, so we rush to conclude by satisficing.	

7	**When Satisficing Meets Complexity**	106
	Missing Alternatives	
	Satisficing works only for problems that our intuitions are attuned to.	

PART III THE BRAIN–WORLD PROBLEM
When Intuition Meets Complexity

8	**Intuition as Expertise**	127
	Ten Years of Quality Practice	
	Expertise is impossible when we cannot accumulate practice with reliable cues and consistent feedback from which to learn.	

9	**A Tale of Two Worlds**	140
	Living in Complexity	
	We live in two worlds. World #2 consists of complex systems that lie between the simple systems of World #1 and complete randomness.	

10	**The Intuition–Complexity Gap**	158
	The Brain That Did Not Change Itself Fast Enough	
	The brain–world gap arises when we are overconfident in oversimplified interpretations. Complexity and cognitive sciences are the antidotes.	

PART IV THE BRAIN–WORLD SOLUTION
Complex Thinking

11 Two Types of Thinking — 171
Automatic and Effortful
Complex thinking begins with an understanding of the strengths and weaknesses of our two types of thinking.

12 Climbing the Cognitive Hierarchy — 185
Dual Thinking for Dual Worlds
World #2 requires a higher degree of Type 2 thinking: both more sophisticated modelling and more metacognitive intervention.

13 Rethinking Causality — 209
Systems Theory Reveals "It's Not So Simple"
Systems theory seeks to uncover the interactions among multiple causal factors, which are easy to miss and misinterpret.

14 Strategies for Complex Systems — 228
The Hunt for Signals
Specific strategies for assessing problems in World #2 originate from the insights of systems theory.

15 Rethinking Truth — 253
Provisional Truth Reveals "It's Not So Certain"
Provisional truth is the foundation of drawing conclusions in World #2, because it orients us toward probabilistic thinking.

16 Strategies for Provisional Truth — 270
Dogma Is for Dogs
Specific strategies for invoking higher forms of metacognitive thinking originate from the implications of truth as provisional.

PART V BRAINS AND PEOPLE
Human Complexity

17 The Complexity of Self — 305
The Depths of Our Hypocrisy
Each of us is a compilation of multiple selves, which puts bounds on our individual rationality, willpower and consistency.

18 The Complexity of Others — 328
Our Social Shortcuts
In assessing and interacting with others, we take shortcuts to reduce their complexity, usually in favour of preserving our own.

19 The Complexity of Being Human — 338
Coping With the Paradoxes
A complex brain gives rise to complex struggles.

EPILOGUE — 349
Two Games for Two Worlds
Two contrasting versions of the game of life.

APPENDIX 1 — 353
Sizing Up the Gap
A chart summarizing the cognitive strategies that suit each world.

APPENDIX 2 — 355
Humans Versus Other Animals: A Dialogue
A conversation highlighting the difference.

Endnotes — 363

Index — 377

FOREWORD

The first time Ted Cadsby told me about his book project, it was over coffee at the Café Doria in Toronto. He opened with a startling statement: "Two centuries ago most of us were farmers. Since then the world has changed profoundly, but our brains have not kept up. We have to acknowledge this brain–world gap if we are going to think more productively about our new world."

Flash forward two years and he has produced a fascinating tome that will disarm and enlighten many people. The book argues that the current human brain is struggling to deal with the hypercomplexity of modern society, and that the solution is to increase the complexity of our thinking.

True enough, if you go back a few hundred years, life was simple indeed. It was the agrarian age, and under feudalism — the economic and political system that surrounded most of the world — knowledge was concentrated in tiny oligopolies of the church and state. There was no concept of progress. You were born, you lived and you died.

But when Johannes Gutenberg introduced the printing press, parts of society began to acquire knowledge. New institutions emerged, and feudalism started to appear inadequate. It didn't make sense anymore, for example, for the church to be responsible for medicine. The new tool for disseminating information precipitated profound changes. It fuelled the growth of universities, new organizations, science, the Industrial

Revolution, the nation state and capitalism. It advanced our productive forces and our standard of living. It also eventually introduced mass media, mass production, mass marketing, mass education and mass democracy, not to mention the rise of knowledge work and industrial-age models of management.

Now, once again, another paradigm is emerging, and another technological genie is out of the bottle. This time it's the Internet and the digital revolution — but with an impact that is very different from that of the printing press. The printing press gave access to recorded knowledge. The Internet offers access to the intelligence contained in the brains of people around the world.

As I see it, we are not in an "information age." We are in the age of "networked intelligence." Rather than an economy based on brawn, we have one based on brains, characterized by collaboration and participation, and it offers huge promises and opportunities. As Cadsby writes, we are at an unprecedented point in our history, one in which we can respond stupidly or smartly.

These changes are leading to a much less ordered and therefore more complex world. In yesterday's corporation, information flowed vertically. People were separated into two groups: the governors and the governed. Most employees of large, vertically integrated companies contributed physical strength, not intelligence. Management invested in big factories with production processes and machinery that required little decision-making or operator skill. Employees were extensions of the machine. They were expected to follow orders and not to take much initiative — if any. Management was based on mistrust and command and control, and decision processes were totally opaque to employees.

And things weren't much better in the white-collar world. The goal was to climb the ladder and acquire more direct reports. The work goals were established higher up. This was the world of the "organization man." When a decision was required, a meeting or teleconference was convened. Each participant knew something about the problem at hand but shared only some information, not all of it. Each made assumptions about what everyone at the table should know, but these assumptions were often flawed. While it was convenient to make a decision by consensus, such decisions rarely tackled the toughest problems. It was a recipe for mediocrity.

It was fifty years ago that the management thinker Peter Drucker predicted the emerging force of the knowledge worker. He was the first to identify that the economy was moving from brawn to brain, a profound change that would enhance productivity. But productive work doesn't mean working harder or producing more ideas per hour. It means collaborating more effectively. The metabolism of work systems goes up when individuals engage with the rest of the world.

For example, if you are a Procter & Gamble chemist, the most productive thing may no longer be trying to invent some molecules by yourself but reaching out to other scientists in the ideagora of open markets to find the uniquely qualified minds, ideas and solutions that exist outside the boundaries of your company.

We must move beyond a world in which solutions are made and imposed from the top. Leaders can no longer learn for their organization as a whole, or make decisions for the enterprise. New networked models are critical for today's environment, which is unordered and complex. In this context, leaders curate conditions for the emergence of innovation through organizational learning.

As Cadsby argues, we need to rethink how we make decisions in this volatile and ambiguous environment. In an increasingly complex world, problems are less and less straightforward. To tackle them, leaders need input from different people, often in differing cultures and geographies, sometimes in different organizations, typically spanning multiple areas of expertise. Yesterday's yes/no decisions for managers have become multiple-choice questions.

Modern companies must be increasingly supple, dynamic and resilient, with many workers changing their processes often and continuously learning and adapting as they work. For example, a job like working in a call centre can generate considerable insights as every hour customers raise new questions and issues. The flexible corporate structure taps into this emerging knowledge.

A central theme of this book is that we need to be individually more humble, recognizing our own cognitive limitations, and need to supplement this humility with the strategy it implies: working together more proactively than ever before. Our maximum problem-solving capability is a function of how well we work together. Problems such as faltering education and soaring health-care costs are complex challenges that can be tackled only with a new approach that embraces information sharing and collaboration.

An informed society is one where citizens have the resources, education and skills to access and participate in the free flow of reliable and pertinent information. Allowed to flourish, new media technologies offer the promise to societies of being better informed, more open and more successful than their industrial-age counterparts. People in many parts of the world have unprecedented access to data, information and knowledge. They can inform themselves through collaboration like never before. People by the millions can contribute useful knowledge for everyone to share (as in the case of Wikipedia).

This creates many challenges for a modern society. How do we survive information overload? How do we sort through all the misinformation spewed forth when a billion people essentially have printing presses at their fingertips? How do we ensure quality news, investigative reporting and good journalism? How do we avoid a balkanization of news where we each simply follow our own point of view, placing each of us in a self-reinforcing echo chamber where the purpose of information is not to inform but to give comfort? How can schools and universities take advantage of the new tools and media to transform pedagogy and themselves?

Such questions are particularly challenging for those for whom the digital technology culture is not intuitive. These people, and I am one, are often referred to as "digital immigrants." For us, rising to the new challenges of the modern networked society is difficult, but we know it can be done. It takes resolve and creativity, and this book is a major contribution to the process.

I have long argued that, for the first time in history, children are more comfortable, knowledgeable and literate than their parents about an innovation central to society. I call them the "Net Generation." It is through the use of the digital media that young people will develop and eventually superimpose their culture on the rest of society.

Evidence is mounting that they can juggle multiple sensory inputs much more easily than older adults. Rather than our children having dysfunctional brains that can't focus, they are developing brains that are more appropriate for our fast-paced, complex world. The post-WWII generation spent many hours a week staring at a television screen, and that form of passive behaviour shaped the kind of brains they developed. Today, young people spend an equivalent amount of time with digital technologies, being the user, the actor, the collaborator, the initiator, the rememberer, the organizer — which is giving them a different kind of brain.

The interactive games that young people play today require both team-building and strategic skills. They collaborate constantly through online chats, multi-user video games and — more recently — text messages, Facebook and Twitter. For teenagers today, doing their homework is a social and collaborative event involving text messages, instant messages and Facebook walls to discuss problems while their iPods play in the background.

Youth are learning, playing, communicating, working and creating communities very differently than their parents did. They are a force for social transformation. The main interest of the Net Generation is not technology but what can be done with that technology. They are smart, have great values, know how to use collaborative tools and are well equipped to address many of the big challenges and problems that my generation is leaving them. Overall, their brains are more appropriate for the kinds of complex demands of the twenty-first century.

I'm of the school that the future is not something to be predicted but rather something to be achieved. Given that we live in a new world of interconnectivity, we have to adopt new ways of thinking and new approaches to solving problems. Rather than making decisions by blinking, we need to embrace new ways of thinking. *Closing the Mind Gap* is an important read because it explores the kinds of concepts that are crucial to "Thinking 2.0": how we both model complex problems and monitor our thinking about them.

DON TAPSCOTT

Don Tapscott is CEO of the think tank The Tapscott Group, Adjunct Professor at the Rotman School of Management at the University of Toronto, and Chancellor of Trent University. He is the best-selling author of fifteen books, including MacroWikinomics, *and most recently* Radical Openness.

PREFACE

For many years I have been fascinated by the haphazard way we humans think through problems and arrive at conclusions. At one point in my career, when I was immersed in the mutual fund industry, my interest in decision making led me to research and write about investor psychology (the investment arena being a playground for anyone curious about how people interpret and respond to complexity). As my career evolved, my fascination heightened — and broadened beyond the financial arena. It is clear to me that we do not think as constructively and reliably as we are able, and that closing this gap easily represents our greatest challenge in an increasingly complex world.

My initial objective was to write a one-hundred-page "airline read" — the kind of book you pick up at the gate and read on a flight. But the manuscript landed at over six hundred pages before being culled to its current length: any shorter and the ideas would not have had the space and detail they need to coalesce and reveal the bigger picture that they collectively produce. The ideas are drawn from academic research in multiple fields, and while they are not difficult to grasp, neither are they immediately intuitive. So I have erred on the side of including more rather than less: examples and anecdotes, contrasting perspectives of a variety of experts, some repetition of key themes in diverse contexts and a gradual building of a larger, deeper story.

One of the arguments in this book is that our physiology and psychology are not designed for patience, yet patience is precisely what a brain needs to tackle complex problems. My hope for the patient reader is that the journey through this book, albeit longer than a short flight, will be as stimulating and rewarding as the research and writing were for me (and nowhere near as harrowing as the actual flight described in chapter 10).

INTRODUCTION
The Mind Gap

Compared with what we ought to be,
we are only half awake.
William James*

We all struggle: as individuals creating our lives, as communities interacting with one another and as a species surviving on the planet.

A large part of this struggle originates in how we think; in particular, how we think about complex things.

The premise of this book is simple enough to be captured in one sentence:

> The complexity of our thinking has not kept pace with the complexity of the world we have created, and we need to catch ourselves up.

But a single sentence cannot convey the all-encompassing significance of this insight. In fact, a whole library of books would barely do justice to the density of ideas that underpin it. What this book attempts to do is explore just some of the main concepts involved. It does so on the basis that catching up to complexity is a two-step process: understanding the problem, then fixing it. First, we have to understand how the gap arises so we know what *not to do*. Then, we have to explore how to close the gap so we know what *to do*. Our survival and the quality of our survival both depend on smarter decision making in the hypercomplex world that we live in today.

---
* *Energies of Men*, Kessenger Publishing, 1998.

Two Worlds

Human-generated complexity is evolutionarily new. Brains have been evolving for tens of millions of years. Only in the past hundred thousand, however, did the human brain evolve to a sufficient level of consciousness and creativity to give birth to complexity. Even then, it was only ten thousand years ago, with the advent of the Agricultural Revolution when we started living in larger communities instead of nomadic tribes, that complexity became significant in our lives. Complexity steepened during the Industrial Revolution and has jumped to hypercomplexity in the Age of Information.

So here we are today — all seven billion of us — trying to make sense of things.

Making sense is a struggle because we no longer live in just one world: we straddle two. As this book explores, we have one foot in our ancestors' world and the other in a new world that snuck up on us, gradually at first, then at a rapidly accelerating pace over the past few hundred years.

First, there is straightforward World #1, in which countless generations of our ancestors lived and in which we continue to spend much of our time. We have evolved intuitive expertise in handling this world because it allows us to. The signals we need to decipher it are unambiguous because cause and effect are tightly linked in time and space and easy to access. These causal relationships create patterns that are consistent from one situation to another. And when we interact with this world, the feedback we receive is direct, timely and "clean" (i.e., not mixed with distracting noise). In this world, learning is easy and predictions are reliable; the automatic way that we think is very productive since our errors are infrequent, and the mistakes are typically not that severe.

Second, there is complex World #2, in which cause and effect are not as closely connected so the cues that we need to make sense of things are buried and ambiguous. The patterns vary since no two situations are identical, and the feedback is delayed, indirect and "dirty." Learning is difficult and predictions are not reliable. Here, our expertise is underdeveloped: our automatic intuitions that serve us so well in the first world do not reflect the operating structure of new-world complexity. Today's challenges are very different from foraging for food and fleeing from predators — for example, navigating fulfilling career paths; raising well-adjusted kids in a world dominated by social media; mapping out corporate strategies; negotiating with menacing dictators; reducing global warming; and just separating good information from bad in the Age of Big Data. Not to mention creating personal meaning for ourselves.

We are experts in World #1, but novices in World #2. *Whereas in World #1 we know more than we can say, in World #2 we know less than we think.* Our problem is that we are typically oblivious to the difference between the two worlds and how underdeveloped our expertise is in World #2. We think that we are playing a game that we have mastered. Like an overconfident amateur tennis player smashing the ball into the net, we overconfidently force fit our basic intuitive models onto complexity, which causes us to misinterpret complex situations and make bad decisions. This is the gap we face: the chasm between complexity and the intuitions that we rely on to interpret it.

Mind — the Gap

There is no polite way to put it: we are not as smart as we think we are. We are irrational, illogical, innumerate, unreasonable and overreactive. But, as the following chapters reveal, our two biggest problems — the most prominent (and interesting) of our cognitive frailties — are oversimplification and overconfidence. We simplify everything we think about, and we treat our simple interpretations as final. Given the urgency with which we need to respond to the world, and our limited brain-processing power, we are forced to simplify the barrage of sensory data that we are exposed to. Then we "satisfice": we lock on to the first reasonable interpretation we come up with.

Our brains evolved over an unimaginably long period in a setting that was harsh but comparatively simple to figure out: there is nothing confusing or complex about a charging tiger, and a high level of cognitive sophistication is not required to interpret the situation and start running. During those millions of years, our simple view of the world, which we rarely second-guessed, was well matched to the straightforward threats and opportunities that confronted us, many of which demanded fast, decisive responses. It served our purpose to rush to certainty, and it still does, much of the time: the simplifying and satisficing shortcuts that work for hunting, finding shelter and avoiding predators are equally useful for crossing busy intersections and avoiding dangerous neighbourhoods.

But the two shortcuts are no longer sufficient: applied to the complex problems of a modern society, their efficiency is offset by a reduction in effectiveness. Rushing to certainty works only when it is based on expert intuitions about how things work: only then does the rush not force a significant trade-off in accuracy. But the sacrifice in accuracy from speedy thinking rises quickly when we confront a level of complexity for which our intuitive shortcuts were not designed. Our lack of expertise results in *oversimplified* interpretations and *overconfident* conclusions.

The acceleration of complexity has outpaced the way we interpret things. H.G. Wells noted this gap in the mid-1940s when he wrote that "hard imaginative thinking has not increased so as to keep pace with the expansion and complications of human societies and organizations."[1] More recently, Nassim Nicholas Taleb has pointed out that the world we live in is different from the world we think we live in: we misunderstand complexity, because human knowledge developed in a world that "does not transfer properly to the complex domain."[2]

Everything is different now. We need a more sophisticated way of thinking in World #2: new models for interpreting that are not as vulnerable to oversimplification and new ways of concluding that are not as vulnerable to overconfidence.

So the premise of this book can be expanded:

When intuition that works in a straightforward world is applied to a complex one, a brain–world gap arises. Closing the gap depends on replacing our automatically invoked shortcuts with more sophisticated ways of interpreting and deciding.

We get into trouble when we treat the two worlds as though they are the same (they are not), as if our expertise in one is fully transferable to the other (it is not). As this book will demonstrate, the intuitions that work in World #1 need to be extended.

Extending Our Basic Intuitions

Science typically progresses not by rejecting previous theories outright but by extending them: new theories do not necessarily contradict older ones; rather, they reconceive the older ones as limited to certain conditions. The new theories extend the older ones to different domains so that more conditions can be explained. For example, Galileo's theory about falling objects (that they all fall at the same speed) works in a vacuum, but Newton's laws work in vacuums and extend to non-vacuums. It became clear in the twentieth century that Newtonian mechanics could not adequately describe objects that are very fast moving (approaching the speed of light) or massive (like black holes). Newtonian physics generated significant errors in these cases, so broader theories were needed that extended to these conditions, which is exactly what Einstein's special relativity did for high speeds and his general relativity did for gravity. But even Einstein's theories did not adequately account for the behaviour of subatomic particles, for which quantum mechanics had to be developed. Scientists are still working to extend general relativity and quantum mechanics (which are incompatible) to a generalized theory that works under all conditions (various string theories are such an attempt).

Just as Newtonian mechanics is limited to working for slow speeds, our intuitive models are limited to working for the straightforward problems of World #1. Just as we need relativity for fast speeds, we need more sophisticated thinking for tackling complexity. And just as broader scientific theories extend narrow ones, complexity demands broader models that extend our automatically invoked, intuitive ones.

Closing the Mind Gap

With the right coaching and sufficient high-quality practice, amateurs can become more expert, gradually developing more productive intuitions about the dynamics of whatever game they are mastering. So it is for all of us with respect to complexity: we have the opportunity to develop skill in avoiding common playing errors, putting the odds of success more in our favour. To be effective in a complex environment — to achieve our goals and minimize our failures and frustrations — we can no longer rely exclusively on our basic evolutionary intuitions. We need more sophisticated models and decision-making criteria than our default simplifying and satisficing shortcuts. We have to meet increasing complexity in the world with greater complexity in our thinking.

The insights of cognitive and social psychology, anthropology, biology, neuroscience, physics, philosophy and many other fields collectively describe how complexity operates differently from the way our brains interpret it, how this gap undermines our effectiveness and how we can work to close the gap. We need the insights of complexity science to assess World #2. And we need the insights of cognitive science to draw conclusions about this world.

The following diagram maps out the central argument of this book, moving from the problem of the gap to the solution for closing it.

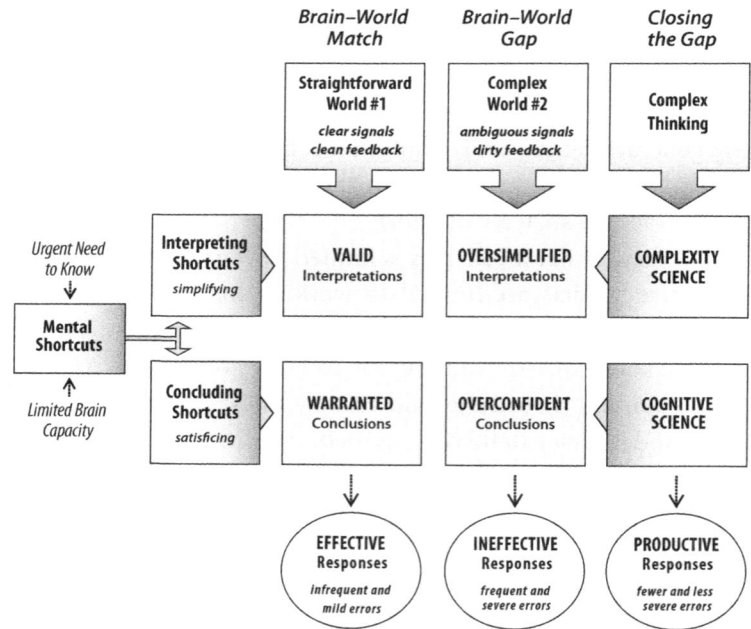

New thinking models give us more productive ways to understand and manage the underlying complexities of our modern challenges; they explain how:

- cause and effect are rarely as straightforward as we assume;

- randomness, unpredictability and probabilities impact our lives;
- relationships of every kind are defined by systems in which we unknowingly participate;
- our psychological disposition is not conducive to assessing and coping with ambiguity.

These complexities are hidden beneath the simple view that we routinely apply to our world. They are the "difficulties of the game" that we are vulnerable to in World #2; they are the conditions for which our intuitions must be extended so we can close the gap. Without understanding them, we are not nearly as effective as we could be, but understanding them requires a greater degree of self-reflection, or mindfulness, than we traditionally employ.

An Urgent Need for Mindfulness

Our struggle is unique in the animal kingdom because we are the only species living in an environment that is very different from the one our brain developed in and was designed for. At the same time, our capacity for meaningful self-awareness enables us to overcome our limitations in a way that no other animal can. Our distinctive form of human consciousness allows us to identify and correct the flawed ways in which we think, to fight against the part of us that does not work as well as we need it to. This is the fight *of* our lives, because it is that big a challenge, and it is the fight *for* our lives, because we are now capable of annihilating most of the planet.

The ability to overcome ourselves, by thinking about how we think, is what makes us human. Our mindfulness is perhaps the most sublime aspect of the natural world. Unlike anything else in nature, we can shape our individual and collective destinies, an ability that gives us choice and therefore power. But choice and power are available only to the extent that our beliefs about the world, other people and ourselves match how things really work. When our beliefs are poorly calibrated to the world, our responses are ineffective and our power is undermined.

To avoid the oversimplification and overconfidence that our intuitive shortcuts generate in World #2, we must invoke less intuitive, more sophisticated models. To make this substitution we have to rebalance our thinking by giving up the efficiency of our speedy, automatic intuitions, opting instead for the greater effectiveness of slower, mindful deliberation — the unique human form of cognitive self-monitoring.

The Next Few Hundred Pages

This book begins, in part I, by tracking the evolution of our brains, which reveals the limits of our thinking. From there, part II explores how we construct a view of reality for ourselves — in particular, how the cognitive shortcuts we rely on most of the time are ill suited to tackling complex problems. Part III delves into the differences between Worlds #1 and #2, and why we develop intuitive expertise for the former but not the latter, giving rise to the brain–world gap. Spoiler alert: complexity is defined as much by what we *cannot* see as by what we can, because it lurks in the second of the two worlds we occupy. Part IV focuses on how scientists across various domains, as well as philosophers from diverse areas, have developed more sophisticated ways of interpreting complexity and solving complex problems. Part V applies these newer models to the personal complexity that we come up against in World #2, taking a deeper look at ourselves, other people and the challenge of being human.

Quick Fixes

The brain–world gap does not accommodate quick fixes: they are the source of the gap in the first place. But there are better ways to interpret and respond to complexity, many of which are covered in parts IV and V. But before exploring concrete solutions, it is necessary to understand the gap, which is the first step in closing it: understanding our mistakes gets us a long way toward playing a more productive game. Managing complexity requires cognitive flexibility, the kind that is only possible with an underlying base of humility. After all, it was only a handful of generations ago that doctors did not think of washing their hands before operating, the age of sexual consent in most countries was between ten and twelve and slavery was the norm, and only four decades ago that homosexuality was considered a psychiatric illness. Our great-grandchildren almost certainly will shake their heads in disbelief when they reflect on the thinking of today. We can lessen our future embarrassment by pulling back the curtain to reveal the error-prone thinking that hides behind our self-assuredness. It is arguably our obligation to do this: we owe this to ourselves, to one another, to our children and to all human beings who have yet to be born, not to mention the other creatures with whom we share the planet.

Complexity in the world has outpaced the complexity of our thinking, and the gap is widening. We need to catch up. As it happens, thinking about how we think is one of the most fascinating endeavours human beings are capable of undertaking.

Part I

Brains Coming Into the World
How Smart Are We Anyway?

1

Brain Evolution
Two Big Bangs

With all his noble qualities ... Man still bears ...
the indelible stamp of his lowly origin.

Charles Darwin*

In order to move, a body needs a brain. Unlike living things that are immobile, such as plants, sponges (which attach to rocks) and jellyfish (which float in ocean currents), mobile organisms evolved brains to navigate their environments. Human cognition has the same general purpose as the brain activity of all mobile animals, and the building blocks of the human brain are virtually identical to those of other animals. Our brain, however, is estimated to be six times larger than needed for orchestrating a mammalian body of our size. Fossil evidence reveals that, in the last two million years, our brain tripled in volume. While all primates have bigger-than-necessary brains, ours is extra large, but it is not just the size of our brains that makes them unusual. While our brains have remained about the same size over the past one hundred thousand years, the neuronal connectivity within them has dramatically increased over this same period.

How did our brains get so big and complex? The answer lies in long time lines that the human mind is nearly incapable of comprehending. From a wormlike creature with a few hundred neurons arose the modern human brain containing about one hundred billion neurons, each with about ten thousand connections to other neurons, producing trillions

---

* *The Descent of Man*, Penguin, 2004.

of neural networks. This incredible system of networks gives rise to the complexity of human consciousness, characterized by our distinct form of self-awareness. Brain evolution, which is the focus of this chapter, is an important starting point to understand how imperfect the human brain is, and therefore how vulnerable to error our thinking can be.

Since its first full publication by Charles Darwin in 1859, the story of evolution has been modified and expanded as new discoveries force adjustments to the details. But common descent — the idea that every living thing descends from a common ancestor — is a scientific fact based on observable evidence from genetics, carbon dating of fossils, comparative anatomical biology, molecular time-dating, paleontology, molecular biology, embryology and biogeography. Natural selection is one of the dominant drivers of evolution: genetic traits that enhance survival and reproduction in a given environment are gradually "selected" for preservation, by virtue of the fitness of the organisms that have these traits; the fitter organisms are better able to survive and pass the favourable traits on to successive generations. Natural selection is the best explanation that we have come up with to account for the fact of common descent, just as general relativity is the best that we have come up with to explain the fact of gravity.

What follows is a preposterously short journey through fourteen billion years of the brain's history.

1.1 Getting From There and Then to Here and Now

After exploding from a point of pure energy 13.7 billion years ago, a sea of hot gas cooled to form the universe. Because the gas was not perfectly uniform, the gravitational pull of the denser parts attracted nearby matter, creating galaxies, one of which housed a tiny planet that formed five billion years ago. A billion years after that, on this speck of planetary matter, a source of energy like lightning or solar radiation triggered a series of chemical reactions between some simple molecules, giving rise to the first living cells. These single-celled organisms exhibited the first forms of decision making: they were able to sense and respond to different concentrations of chemicals in their environment.

Over the next few billion years, these single-celled microbes began to coalesce, leading to more complex organisms with specialized cells, like neurons that processed sensory information. The first nervous systems consisted of a string of neurons that enabled organisms to ingest food by enveloping it.

First Brains

As organisms began evolving a mouth, their neurons began to congregate closer to where food was ingested. The first worms, five hundred and fifty million years ago (mya), had rudimentary nervous systems operated by a clump of neurons at the top of their head: this ganglion was the precursor to the first rudimentary brain.

Some worms evolved into primitive fish five hundred mya, and their neurons began to specialize and form differentiated clumps, each performing distinct functions. One neural clump specialized in the sense of smell, while another specialized in visual stimulation. These primitive fish had brains that resembled the three-part structure of all animal brains today: hindbrain, midbrain and forebrain.

Some of the early fish developed throat pouches that took in oxygen in times of drought; they were able to squirm from swamp to swamp using these primitive lungs. By four hundred mya, lizardlike creatures that could survive both in and out of the ocean became the first amphibians. Their larger brains enabled them to orchestrate the additional balance and flexibility required for them to move around on land.

Over the next fifty million years, some amphibians developed hard skins to resist heat and dryness, which enabled them to lay hard-shelled eggs on land. These were the first reptiles, whose bigger forebrains allowed them to analyse and store more information, especially once the outer layer expanded, becoming the cerebral cortex.

By three hundred mya, some reptiles were evolving into dinosaurs, while others were becoming mammalian, with teeth that allowed more efficient digestion, which in turn allowed more energy to be directed to the growth of their brains. About two hundred mya, some of the mammalian reptiles developed into the first tiny mammals, and by seventy mya, some of these had shifted location, becoming tree-bound. These primates did not thrive until sixty-five mya, when their predators, including dinosaurs, were wiped out along with 70 percent of all plant and animal life, the likely result of climatic changes triggered by an asteroid colliding with what is now northeastern Mexico.

Mammalian Brains

The brain was becoming more complicated. The first tiny mammals had significantly bigger forebrains and cerebral cortexes, which consisted of the original reptilian cortex (paleocortex), as well as a newer growth (neocortex), a six-layered construct that was more complex than the simple cortexes of reptiles. The mammalian brain also developed structures (like the hippocampus and amygdala) that allowed for the formation and storage of memories, especially those associated with danger. The combination of more intricate cortexes and new memory structures enabled these mammals to process and respond to sensory data in ways that were not exclusively automatic. The expanding neocortex differentiated the brains of early primates — in particular, the prefrontal cortex at the front of the brain, which was unique to primates and the site where higher-order thinking originated. To expand within the confined space of the skull, the neocortex became grooved and wrinkled; its surface area was much larger compared with the smooth cortex of earlier mammals.

By forty mya, monkey primates had evolved. Their bigger brains were becoming visually dominant; this was an evolutionary twist since the sense of smell is the dominant sensory input for most mammals. A tree-living primate, jumping between branches in a forest, needs to be able to judge depth, height and texture: there was pressure on their brains to reduce their olfactory bulbs to make room for their expanded vision centres, eventually giving rise to stereoscopic and colour sight. Of the monkey primates, a species evolved by about fifteen mya that had lost its tail and developed brachiation, the ability to swing from tree to tree. Flexible shoulders and strong arms led to more erect posture in these early apes — a precursor to upright walking.

At about this point, fifteen mya, the earth was entering an ice age, which reduced rainfall as oceans and lakes froze. The African jungles where the primates lived began to recede, and vast plains expanded in their place. In their search for food and shelter, some of the tree-loving apes slowly adapted to living on the ground, and by around six mya, one type of small African ape had evolved to walk upright on two legs. Bipedalism was a tipping point for the species because of its multiple benefits: it enabled these apes to see farther into the horizon, wade deeper into the water, have less of their body exposed to the hot sun and use their hands more resourcefully, as in carrying food while moving quickly. To top it off, a fully erect posture allowed for a lower larynx and freer vocal tract, conditions for the development of verbal language.

Early Human Brains

The ice expanded, receded and expanded again, putting significant environmental pressure on all animals. Some of the bipedal apes perished while others adapted; by four mya, one branch of the species had evolved into four distinct great apes: orangutans, gorillas, chimpanzees and humanlike primates (*Australopithecines*, the protohumans). While the latter were not meaningfully different in their habits from the other African great apes, these humanlike primates had larger brains, by virtue of a neocortex that enveloped the older brain parts and pushed forward in the skull as a larger prefrontal cortex. This species eventually branched into different species, including the first *Homo* (human) primates, which entered the scene about 2.4 mya. The first humans were apelike and still spent much of their time in trees, until *Homo erectus* appeared around 1.8 mya, a species that led a fully terrestrial lifestyle.

While six to twelve different *Homo* species evolved, it was two hundred thousand years ago that one species had evolved into modern humans: *Homo sapiens*. These primates ended up being the only species of the *Homo* genus to survive, possibly because they had developed body types that adapted to both extreme cold and heat.

The early humans did not have the kind of consciousness that we have today. Like all mammals, they would have had domain-specific brain structures that were largely isolated from each other; the parts of the brain that regulated communication, nurturing, eating and so on were not as integrated as they are now. It took thousands of years for these modules to become intricately interconnected, largely by the neocortex, which is linked by neural pathways to most other brain areas. We do not know what the semiconscious minds of our ancestors felt like, but the human brain probably underwent what scientists call a phase transition, a threshold of neuronal activity in which a fuller form of consciousness began to gel in humans.

Modern Human Brains

And then there was a "cultural big bang," between sixty thousand and thirty thousand years ago, the period when art, sewing needles and musical instruments were all developed, as well as early kinds of religious rituals, like burial rites. Human consciousness began to take on its contemporary form, with higher levels of conscious awareness becoming available, as well as expanded working memory, the mental space where we hold ideas temporarily while we consciously work on them. While other animals are largely imprisoned in the present moment, reacting to immediate stimuli, we developed the ability to hold and manipulate our ideas to create hypothetical or imaginary situations, which enabled us to escape the present by reflecting on the past and imagining the future.

Higher-level consciousness enabled our ancestors to mimic and learn from one another. While some animals, like chimps, can pass on learned traditions, only humans can accumulate their innovations by building on prior ideas and sharing them with a broad audience. This ability to learn by mimicking facilitated the rapid cultural spread of ideas, which sped up even further when rudimentary language developed to replace the primitive sounds and protolanguage of earlier *Homo* species.

The cultural big bang was the starting point of accelerated change, fuelled by the human ability to communicate complex ideas and generalize learning by applying insight from one task to different ones. The complexity of the human living environment increased substantially when bands of hunter-gatherers became farmer-herders, living in larger communities. In these larger groupings, we developed more nuanced social interactions and cultural traditions; government and religion arose, presumably as ways to instill social harmony within our more intricate social arrangements. Not only did social complexity explode, so did innovation. As population density increased, more ideas were generated by the larger number of people; living in closer proximity meant higher connectivity between people so that new technologies were shared in a reinforcing feedback loop of accelerated creativity. New civilizations expanded by conquest to become empires. Farming innovations increased food production while medical science increased resistance to disease: population growth exploded, as did technology, ramping up environmental complexity even further.

1.2 The Uniquely Large Human Brain

The rudimentary gene sequence that determines the structure of our bodies and our brains was present in the first vertebrates five hundred million years ago. We are off-the-rack primates with special-order brains: our brains are generally cut from the same cloth as all primate brains, but ours are set apart by the quantity of material used and how it is structured. It is the intricacy of our brain's design that distinguishes it: the breadth and depth of neural networks that constitute human consciousness. Chimps may use tools to dig, but only humans use tools to make other tools. Chimps may mimic one another, but only humans create lasting cultural rituals. Chimps may attack to protect their territory, but only humans engage in elaborate religious warfare.

The regulatory gene that determines how many times a neural cell divides is coded for many replications in humans; the same gene is coded for less neural duplication in other primates, and less still in other animals. There are some compelling theories about how our brains got so big, and the explanation likely lies in a combination of them. These theories can be classified into two broad categories based on two distinct questions: "What external environmental pressures encouraged the evolution of bigger brains?" and "What internal biological factors enabled that growth to occur so dramatically in human primates?"

Environmental Pressures

The brains of any organism are extremely energy demanding, so there has to be sufficient motivation to grow them. For humans, there were two motivating pressures: climate change and social complexity.

Extreme climate changes swung temperatures and moisture in different directions throughout various ice ages and forced the extinction of many species, including a dramatic change just 74,000 years ago that wiped out many species and most humans (likely the result of a climate-altering volcanic eruption in Indonesia). Those that survived were able to adapt in part through their ingenuity (for example, by constructing protective shelters). The survivors then passed their smarter brains on to their offspring.

Harsh environments also put pressure on our ancestors to become more social: living in the open savannah grasslands that were created by post–ice age drought meant that they needed to develop the social skills necessary for survival, like the co-operation required to find food and fight off predators. Social living requires far more intellectual sophistication than the challenges posed by nature. To be social, for instance, requires at least a rudimentary understanding of how other people think (known as "theory of mind"). There is a strong, positive correlation between the size of an animal's neocortex and the size (and therefore complexity) of the animal's social group.[3]

But these environmental pressures are exerted on many animals, so why are we the only ones with big and complex brains?

Biological Enablers

All the external pressures in the world will not create a big brain if it does not have the fuel it needs to grow and the space to expand. The first and probably most profound milestone in human brain evolution was getting more meat into our diets. Meat has a high concentration of protein and fat, both of which are superlative sources of sustained energy. Once we began to experiment with stone tools, we could cut open the carcasses of large dead animals and access the meat within, no longer relying on small rodents for our protein. When stone tools evolved into hunting gear, the quantity of meat we ingested increased significantly yet again. Compared with the predominantly vegetarian diets of other great apes, the human diet is much more protein oriented (less than 5 percent of a chimp's diet is meat, for example).

Stone tools contributed a second advantage: they allowed us to cut food into small pieces, which enabled our bodies to divert energy from digestion toward brain development. The rule in the animal kingdom is that the more intricate the digestive system, the dumber the animal, since there is less metabolic energy left over for thinking. Once we developed the controlled use of fire, making it possible to cook meat, even more of our bodily energy could be diverted from digestion to brain growth.

Even with sufficient fuel, we had to overcome the bodily constraints of bigger brains. Having evolved to be competent upright walkers, we had narrow pelvises, allowing us to generate a high amount of forward momentum. But a narrow pelvis requires a narrow birth canal. Whereas a female ape's birth canal is a straight tube, the birth canal of human primates was not uniform: it was narrower in some spots to accommodate closer pelvic bones. This configuration presents a serious obstacle for a growing human brain to fit through, on its way out and into the world. Natural selection solved this problem over the course of hundreds of thousands of years by orchestrating the premature delivery of newborn humans. Hormones that trigger maturation in the wombs of other mammals do not appear in large quantities in humans until we are out of the womb. We pop out about six months early, relative to typical mammalian delivery, with our brains only 25 percent developed; compare this with chimps at 50 percent, other primates at 75 percent and other vertebrates with close to 100 percent brain formation at birth.

Because we are born with undeveloped brains, our parental dependence extends beyond that of other animals. While most of our neural circuitry is in place by fifteen months after birth, it is not until age six that our brains reach 95 percent of their ultimate size, which they finally achieve at age thirteen. Even then, significant neural restructuring must occur before the final circuitry is in place, which is not complete until the late teens. (This is when most neuronal axons are fully myelinated, myelin being the fatty substance that lines the axons to speed up transmission of nerve impulses.) Even with the ingenious solution of protracted brain growth, our huge brains still make for a very difficult labour: until the advent of modern obstetrics, many women died giving birth, at rates much higher than in other animals.

The opportunity for a brain to grow without the constraints of a womb was a gold mine for human cognitive development. But our ever-expanding neocortex caused another constraint, even after the narrow pelvis problem was solved: bipedalism puts a limit on how big of a head a human can have and still balance on two legs. Natural selection had a solution to this problem, as well. The neocortex folded itself again and again, to the point that two-thirds of its surface area is in grooves.

Without this folding mutation, the human neocortex would be about the size of a stovetop (compared with a chimp's unfolded neocortex, which would be the size of a sheet of paper, or a monkey's, which would be the size of a dollar bill).

Given enough time and enough mutations, Mother Nature's trial-and-error process, a.k.a. natural selection, is a competent, albeit imperfect, engineer. The impact of our incredible evolutionary journey on who we are today is so profound that it is easy to underestimate — and, typically, we do just that. We have astoundingly impressive brains, but one of the most important implications of the evolution narrative is how fundamentally imperfect they are. Each of us has a single three-pound organ with limited energy supplies and limited processing power, which we rely on to figure things out quickly. To say that it does not always work as well as it could is a colossal understatement.

2

Bounded Brain
The Cognitive Trade-Off

No man can achieve the greatness of which he is capable until
he has allowed himself to see his own littleness.

Bertrand Russell*

Little in the natural world inspires more awe than the insight and inventiveness of the human mind. Yet equally astounding is how feeble and misguided that mind can be, and how oblivious we can be to its shortcomings. Our art, our industry, our science: we create in a way that is nothing less than miraculous. Our wars, our pollution, our cruelty: we destroy in a way that is nothing less than shameful. We may be getting closer to unifying the laws of physics, but we are also closer to nuclear annihilation.

These contradictions are probably what Blaise Pascal had in mind when he wrote that humans are "the glory and the shame of the universe,"[4] and they raise the question, addressed in this chapter, why such an incredible instrument, the human brain, entails such failings.

2.1 Biological Imperfection

Natural selection is as miraculous as time and mutation allow it to be. After all, it can select only from building blocks that already exist, and useful mutations are vastly outnumbered by useless, even harmful ones.

---

* "Dreams and Facts," *Skeptical Essays*, Routledge, 1996.

Millions of years are not nearly enough to create perfection — or even near perfection — from this protracted, hit-and-miss process. We are the product of the imperfect creation that arises from the evolutionary process. The various forms of biological imperfection fit into the two categories of circuitous design and environmental mismatch.

Circuitous Design

This category of imperfection originates from evolution's tendency to incorporate changes into a pre-existing framework. Like all living things, we are a patchwork of bits and pieces that were shaped, reshaped and added to over time. For example, our throats have two passages: one leading to the lungs and one to the stomach. While this particular engineering works most of the time, it makes us extremely vulnerable to choking when food goes down the wrong passage, especially since we also use our mouths and tongues to talk while we eat. Our chronic lower back problems originate from spines that were not initially designed for constant upright posture; our knees are weak and vulnerable, in part because our feet are insufficiently cushioned to absorb stress. We are also vulnerable to autoimmune diseases, because our immune systems occasionally attack organ tissue by mistaking healthy cells for malignant ones. The list of biological imperfections goes on.

Environmental Mismatch

The second category of imperfection results from a design that served us well at one time but has become ill suited to our new environment. Consider our modern-day diet. We spent most of our history hunting and scavenging for complex carbohydrates and protein, driven by a desire for sugar (in fruits) and salty fat (in meat). However, at precisely the time that our lives have become increasingly sedentary, we have generated through technology an abundance of sugary, fatty, simple-carb foods. Our instincts and cravings are still useful today, but the *degree* or *intensity* of many of them is mismatched with our new world. The mismatch between our powerful craving for sweetness and the modern abundance of sugar is causing no end of medical problems.

Even before non-nutritious food became abundant, the invention of cooking, and then farming, led us to eat food that was much softer than the kind we had ingested for millions of years. We chew our food far less than other primates, and we chew it much less intensely. As a result, our jaws have become smaller. But the number and size of our teeth have shrunk only modestly, leaving our mouths overcrowded with too many teeth, all of which are prone to premature decay. And the speed with which we can ingest soft, easy-to-digest food is poorly matched with the fifteen- to twenty-minute lag it takes our bodies to recognize that we are full, leading to chronic overeating. To make matters worse, we are tempted to overconsume high-glycemic carbohydrates like sugary drinks and snacks, in a desperate attempt to keep ourselves alert: the invention of electricity has enabled us to stay up past our ancestors' bedtime, contributing to an epidemic of sleep deprivation.

2.2 Cognitive Imperfection

What are the cognitive equivalents of our dual-purpose throats and our antiquated sugar cravings? These will be explored in subsequent chapters, although some of them get a mention later in this chapter in a discussion of the bounds of our thinking. These mental weaknesses fall under the same two categories of biological imperfection: the design of our brain is flawed by the circuitous journey the brain has taken over hundreds of millions of years, and our thinking is often ill suited to a world so different from the one it was designed for.

Circuitous Design

The human brain is, figuratively speaking, two brains: the "old" one, which operates largely below our conscious awareness, and the "new" one, which evolved over the past one hundred thousand years, and which we think of as human consciousness. The old one is our ancient automatic thinking, the basis of subconscious intuition. The new one is our modern effortful thinking, the basis of conscious deliberation.

But the two are not equal partners: evolution layered the new on top of the old, and although they work together, our newer, conscious deliberation has limited access to its ancient, subconscious brethren, so it has a hard time understanding and influencing it. We are largely captive to our ancient brain because it orchestrates the vast majority of our thinking and behaviour, but its workings are largely opaque to our conscious awareness. The clumsy marriage of subconscious and conscious thinking gives birth to one of our biggest challenges as human beings. Because our brains are a mishmash of old and new, our more basic urges (like lust) clash with newer, more complicated urges (like monogamy). These conflicts are difficult to sort out and prioritize because it is difficult to manage our motivations when they are largely invisible to our conscious awareness. One part of our brain is not fully transparent to the other: this a brain-to-brain gap, or **brain–brain gap**.

One of the primary reasons for this gap is energy conservation. Neurons need about two times more energy than other cells since they are always in a state of metabolic activity, creating the enzymes and neurotransmitters that are the bases of bioelectric signaling, which is how they communicate with one another. Neural activity has to compete for energy with the digestive system, among others. The human gastrointestinal system uses 70 percent of the energy we take in. Another 15 percent goes to movement, repair, reproduction and other activities. This leaves a mere 15 percent for the brain (which uses about 20 percent of the oxygen we take in and about 10 percent of our glucose stores). Our ability to absorb fuel and convert it to the energy needed for maximum brain functioning is limited. Our bodies are not equipped to provide our brains with a constant, uninterrupted supply of just the right kind and quantity of energy, which is why our thinking and ability to concentrate deteriorate precipitously when we are tired or hungry. So we must contend not only with a brain–brain gap, but also with a **brain–body gap**.

For the same reason that we instinctively preserve our energy in movement (we do not constantly run from place to place or dance on the spot while we are waiting), we also preserve our mental energy by being stingy with the effort we put into hard, concentrated thinking, like the kind required for complex problems. Concentration takes a lot more energy than automatic, intuitive thinking: it requires energy-intensive mental attention, which literally chews up a lot more glucose than automatic thinking does. Our brains prefer to run on a kind of autopilot the vast majority of the time, to burn less fuel.

Environmental Mismatch

How we think is not always matched to our modern world, which is riddled with much more complexity than our automatic intuition was designed to accommodate. The mental models we use to make sense of the world were developed over a long period in an environment that bears little resemblance to our environment today. The challenges we confront are so fundamentally different from those of our ancestors that we also have another gap, a **brain–world gap**.

Our brains have been caught off guard by the relative speed with which modern complexity arose. We have been mammals for two hundred million years, primates for eighty million, great apes for fifteen million, humans for two million, *Homo sapiens* for two hundred thousand, and language-using, culture-creating modern humans for fewer than fifty thousand years. If your outstretched arms represented the time line of human evolution starting from the first mammals, all you would need to do to eliminate the entire history of *Homo sapiens* would be to file off the tip of your fingernail.

For the vast majority of the long history leading up to our "sudden" modern incarnation, the challenges that we faced were generally consistent. Granted, we had to cope with multiple ice ages. But other than the occasional asteroid colliding with the earth, these changes were very gradual, and our ancestors' objectives remained straightforward: to survive another day by securing nourishment and protection from predators and the weather, and to reproduce. Then, almost instantly, we converted our world into something completely unrecognizable: a highly populated, frenetic place with an unimaginably large number of interconnections between individuals, their groupings and their environments.

More than one hundred thousand generations of our ancestors faced challenges that were virtually identical. But only about ten generations have had to cope with the kinds of problems that originated with the Industrial Revolution. Our more distant ancestors had to deal with a world over which they had no control, but which operated in fairly straightforward causal ways. We face challenges that are far less straightforward than food, shelter and sex, but have not had much evolutionary time to develop reliable intuitions about how complexity works. Paradoxically, we have to deal with complex systems that we ourselves created and therefore have some control over, but that are extremely difficult to understand because the causal relations that characterize them are intricate and ambiguous.

Over millions of years of relying on simple visual cues, we have become accustomed to using a mental model of cause and effect that could be easily ascertained by basic observation. This same mental model is not nearly as reliable in a world where cause-effect relationships hide beneath the immediately visible surface and are far more ambiguous than those that dominate a straightforward life of surviving on the savannahs. That our evolved mental models are often mismatched to the complex world that we have created is especially problematic because these ancient mental models are deeply embedded in our subconscious, making them difficult to evaluate and modify. Our default ways of conceptualizing give rise to misunderstandings of how complexity works: our evolved mental models lead to conceptual illusions in the same way that they cause optical illusions.

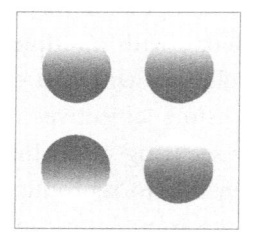

For example, the object above appears to contain one concave cup, like an empty muffin cup, and three convex domes, like pitchers' mounds. But if you turn the page upside down, you will see three cups and one dome. Psychologists use this illusion to demonstrate that our brain interprets the image differently when it is rotated, even though we know that the drawing has not changed. Why? Over millions of years, our brains were programmed to assume that light comes from above — from the sun. This assumption became embedded in how we view the world, because it was a reliable constant, a mental shortcut that consistently helped us to interpret things. When the page is upright, we see one three-dimensional concave cup because the shading in the circle simulates the light source we are predisposed to see. When we turn the page upside down, the same modelling by the brain, relying on the same assumption about the light source, creates three cups based on the inverted shading.

The assumption that light originates from above works most of the time in our modern environment, so it is not a mental model that we should abandon. But when the mental model does not match the real world, an optical illusion arises. The same applies to conceptual illusions. Just as we are programmed to assume a light source from above and can be tricked into seeing something that does not exist, our brains rely on numerous mental shortcuts that make us vulnerable to misinterpretation when our models do not jibe with the way things work: illusions in how we view events, other people, ourselves and even the meaning of life. Misconceptions arise similar to the way misperceptions do: our automatic way of looking and thinking can mislead us in certain circumstances. Here is where the parallel ends: optical illusions are infrequent and typically innocuous, whereas conceptual illusions are common and can have severe consequences.

2.3 Bounded Brains

The Nobel prize–winning economist Herbert Simon coined the term "bounded rationality" to describe how our reasoning is constrained by the limited computational power and storage capacity of our brains, and how we use shortcuts in our decision making to get around these constraints.[5] One of the outgrowths of Simon's work is the field of behavioural economics, which arose to challenge the assumption of classical economics that people behave in consistently rational ways. Behavioural economics focuses on the particular shortcuts we take and the biases in our decision making that arise from these shortcuts. In fact, this field goes beyond Simon's suggestion that there are bounds to our rationality, demonstrating that we are patently irrational.

"Bounded" connotes a line beyond which we cannot travel, which is perhaps an overly pessimistic view of human cognition. Highlighting our constraints does not detract from the fact that we are getting smarter.

In the 1950s, only half of the world's population was literate; today it is over 80 percent and still rising. Research suggests that the average IQ is substantially higher today than that of our ancestors a century ago. Some of the deepest research in this field was conducted by psychologist James Flynn, who identified this upward trend in IQ (known as the "Flynn effect").[6] According to Flynn, our basic math capabilities are comparable with those of our recent ancestors, but our abstract reasoning skills have improved dramatically. Flynn's explanation is that, whereas a century ago our ancestors applied learned procedures to perform basic jobs, our challenges today force us to engage greater conceptualization, including the contemplation of hypothetical outcomes to possible scenarios. So while there are limits to our mental capacities, they are not necessarily fixed. Natural selection has gifted us with a mind that can transcend itself and extend its boundaries. Both types of cognitive imperfection — circuitous design and environmental mismatch — can be mitigated, and improved thinking is precisely the focus of this book.

First, we have the ability to overcome some of our automatic intuition's inaccessibility by understanding how thinking works — specifically, how automatic thinking operates and how effortful, deliberative thinking can influence it. This is trickier work than dealing with the design imperfection of choking when we breathe, talk and eat at the same time. But reducing the brain–brain gap is doable.

Second, we can address the mismatch between our mental models and our complex world by examining the ways in which automatic, intuitive models have become obsolete in particular circumstances and then adopting new models that are more effective in these situations. It is trickier work than overcoming the environmental mismatch created by a sugary, mushy diet that leads to cavities. But reducing the brain–world gap is also doable. And there is urgency to close this gap, given the unrelenting escalation of complexity in our time.

Our challenge is somewhat parallel to the awkward and protracted transition we underwent between knuckle walking and bipedalism, when we roamed the earth using some version of both. We are in just such a "between state" today, one in which we are slowly and clumsily evolving to use our minds more effectively, having not yet mastered the full potential of our complex brains. Once we understand where our minds fail us, we can work to close the brain gaps. Without diminishing the power and creativity of human thinking, we have to start with a genuine humility about how our minds work: we cannot solve a problem if we do not first acknowledge that there is one.

2.4 How Smart Are We, Anyway?

As reasonable, cogent, level headed and insightful as we are much of the time, we can also be illogical, emotional, irrational, absent-minded and outrageously hypocritical. Behavioural economics has shone a bright light on our cognitive weaknesses, including how misleading our intuitions can be and how oblivious we can be to our limitations. In the spirit of breaking us down to build us up, below are some demonstrations of how our thinking can lead us astray: examples of how bounded our thinking can be.

Bounded Awareness (a.k.a. Selective Attention)

Bill loves to read and tends to be reserved. From everything you know, is he more likely to be a librarian or a salesperson, or is it impossible to guess which?

Most of us will assume that Bill is more likely to be a librarian, because we tend to be very selective in what we attend to. We focus almost exclusively on information that is immediately available, like Bill's love of reading and reserved nature. We overlook the undeniable fact that there are hundreds of times more salespeople in the world than librarians, including many reserved, book-loving ones. This basic fact trumps whatever intuitive stereotypes we have about librarians; Bill, therefore, is much more likely to be a salesperson. In our rush to conclude, we tend to gravitate to facts that are immediately available, at the expense of facts that we know but are not top-of-mind.

Are there more words ending in "ing," like "running," or words whose second last letter is "n," as in "friend"?

Because we know so many words that end with "ing," most of us assume that there are more of them. The list of "ing" words is more cognitively available, but this availability obscures the fact that all "ing" words have "n" as their second last letter. The answer, then, is that there are more words with "n" as the second last letter, since they include all the "ing" words.

Our attention to detail takes energy and focus, so it is often easier just to go with what is immediately available and obvious — a shortcut that will be explored in 6.4.

Bounded Probabilities

There is a test that can detect a certain disease in 90 percent of people who have it. You are tested, and the results indicate presence of the disease. What are the odds that you have it? Ninety percent, higher, lower or cannot be determined?

Our understanding and use of probability is pretty rudimentary (and that is being charitable). It is surprising how innumerate — mathematically illiterate — we are when it comes to assessing probability. Without sufficient training in probability and statistics, most of us think the answer is 90 percent. In fact, we do not have enough information to support that conclusion. We need to factor in what portion of the population has the disease before we can assess our own test results. For example, if only 1 percent of people have the disease, then the probability of our having it if the test results are positive is only 8 percent. This likelihood is substantially lower than almost everyone guesses, because most of us are oblivious to the missing information, without which a meaningful estimate of probability cannot be established.

Unfortunately, having all the information does not necessarily make the problem easy to solve: when it comes to probability, we often do not know what to do with the information we have, as the next question demonstrates.

You're out for a walk and run into an old friend. She mentions that she has two children. What are the odds that she has one of each gender? A moment later, a girl runs up behind her, whom your friend introduces to you as her daughter. What are the odds that your friend's other child is a boy? Do the odds change if you are told that her daughter is the elder sibling?

There is a 50 percent chance that a two-child family has a child of each gender. If you know that one is a girl, the odds that the other is a boy rise to 67 percent. However, if you also know that the girl is older than the other child, then the odds that the sibling is a boy drop back to 50 percent. Our intuitions fail to accurately calculate probabilities that are conditional on other probabilities; our probabilistic thinking is not as sophisticated as we need it to be for assessing complexity, as will be explored in 15.3 and 16.7.

Bounded Randomness

If Jill is better at shooting hoops than Jack, should Jack care whether they play an eleven-point game or a twenty-one-point game?

Luck, which is a form of randomness, is more pronounced in a shorter game, but takes a back seat to skill in a longer game. That is because, in the short term, many unexpected things can happen: Jill could trip, or Jack could get in some lucky shots. In the long term, the randomness of lucky and unlucky things balances out, and Jill's skill would dominate. Jack should angle for the shorter game, since it offers higher odds of unexpected luck.

In an office of twenty-three people, what is the probability that two of them share a birthday? Is it 6 percent, 22 percent or greater than 50 percent?

We are pattern-seeing creatures; our intuitions about probability typically underestimate the expected frequency of coincidental, random events, such as in this case: the answer is 51 percent. Our bias for patterns, key though it is to our survival, causes us to see them where none exist. We are chronically inclined to overinterpret by ascribing meaning to events or patterns that have none, such as the five-year return of a mutual fund. Our misinterpretation of randomness is addressed in 13.4 and 14.6.

Our most challenging confrontation with randomness may be the fact of our own existence. We *Homo sapiens* are only one of many human species, all of which are extinct except us. You are here only because at one particular romantic moment, one of a few hundred million of your father's sperm joined with one of the four hundred eggs that your mother ovulated over her lifetime. This happened years after your grandparents joined forces at some random point, following another random point of interaction between your great-grandparents, and so on. It is difficult to appreciate the inherent randomness of that long line of fertilizations, going back in time for hundreds of millions of years, creating you instead of someone else — or nobody at all.

Bounded Causality

Research indicates that most successful companies have well-articulated mission statements. Is it safe, then, to conclude that a clearly articulated mission is key to corporate success? If a company is successful, can we assume it has a clear mission?

We are extremely prone to misinterpreting causal relationships. The research cited above is useless, because there is no evidence to support a causal connection between a clear mission and a company's success. Our intuition makes a case for causality, but it does so outside the bounds of cogent proof. Even if 100 percent of successful companies had mission statements, that does not preclude a missionless company from being successful for other reasons, which means that a mission statement is not necessarily "key." What is more, the research above does not tell us how many *un*successful companies have clear missions: if most companies — whether successful or bankrupt — have missions, the research says nothing interesting whatsoever.

Studies suggest that people who have good work relationships are happier in their jobs than people who do not. Should companies promote social activities to increase job satisfaction?

There may well be another factor causing the correlation between work relationships and job satisfaction; in fact, the two may not be directly related at all. Extroverted individuals, for example, tend to have happier and more optimistic temperaments in general; the research may confirm only that extroverted people, who are more inclined to seek out relationships at work, are happier in their jobs, as they are in all areas of their lives. More social activity is unlikely to change their already positive attitude. Introverts, meanwhile, are not likely to enjoy participating in work-related social activities, because they prefer one-on-one interactions. Furthermore, introverts are unlikely to convey as much job satisfaction anyway, since they are usually less upbeat in general.

Our proclivity for ignoring the possibility of a third factor, which generates the illusion that two things are related, reveals itself in countless ways. One example comes from studies purporting to show that people who meditate have lower blood pressure and more relaxed demeanours than people who do not. The simplicity of this connection is appealing: meditate to lower blood pressure. It could just as easily be the case, however, that people with a relaxed temperament are better able to sit still, meditate and enjoy the experience. This "truth" proves nothing except that Type A personalities are not inclined to meditate and Type B personalities enjoy being still. The way causal complexity trips us up is featured in 13.2 and 14.3.

Bounded Rationality (a.k.a. Irrationality)

You're about to buy a toaster for $45 when your neighbour spots you and tells you the same toaster is on sale for only $20 at a store that is a five-minute walk away. Are you likely to take the trip to buy the toaster on sale? What if you're about to buy a $2,545 television and you discover the same one is on sale for $2,520 at a store five minutes away?

Most people would take the trip in the first scenario, but not the second, even though the impact to their net worth is identical. There is no rational reason to go out of your way in one case and not the other. It is a perfect example of the irrational motivation generated by the different starting prices of each product (a phenomenon called "anchoring"). But our irrational thinking and behaviour do not stop there.

You decide to buy a stunning sweater at an upscale second-hand clothing store. As you get your money out to pay for the item, the checkout clerk tells you that the sweater's previous owner was a convicted mass murderer. Do you still buy it?

Most people recoil at the thought of wearing a sweater previously owned by a murderer, even though the sweater itself is just wool and thread: it contains no evil residue. We are a very superstitious species.

Who would you rather be, the 100th customer to walk into a new store where you are awarded $100, or the person right behind the 1,000th customer to walk into a new mall, where the person in front of you wins $10,000, and you win $125 for being the 1,001st?

Most people prefer to take the lesser prize to avoid the painful regret of narrowly missing out on the big prize. Regret is something we go great lengths to avoid, as will be explored in 16.3.

Bounded Conceptualization

If a two-hour movie represents the time line from when the earth formed to the present moment, how far into the film would humans appear?

In geological time, the human species would appear in the last four seconds, and modern humans would be in the last third of the last second — you would miss us entirely if you blinked. It is virtually impossible for us to conceptualize deep time.

It is no easier for us to conceptualize deep space. If the earth were the size of a pea, the sun would be a beach ball, lying about ninety meters away on the eight-kilometre-long beach that is our solar system. The sun is only one star of more than one hundred billion in the Milky Way galaxy, which is itself only one of more than one hundred billion galaxies in the universe — all moving away from each other at an increasing rate.

Not only are space and time deeper than we can imagine, but they are also far more malleable than our intuitions suggest.

If you take a plane around the world and return two days later, will you have aged the same as the people you left behind, less or more?

The relativity of time is completely foreign to our intuitions. The answer is that time unfolds more slowly for a moving body than a stationary one, so you would be younger (imperceptibly so, but younger nonetheless). The aging difference would be partially offset by the fact that time also moves more slowly the farther out you are in the earth's gravitational field. Time and space are constructs that change with the motion of the observer. This aspect of physics is anathema to our intuitions about how time and space work.

We do not need to delve into physics, however, to reveal our conceptual bounds; we can get ourselves into mental contortions working out even simple questions.

Paul is single and is being watched by Mary, and Mary is being watched by Peter, who is married. Is a married person looking at a single person?

This question manages to be as confusing as it is straightforward. It does not matter whether Mary is married or not, since, either way, a married person *is* looking at a single one. If she is married, then the answer is yes, because she is looking at single Paul; if she is not married, Peter is looking at her, so it is still yes. Our working memory — what we can hold in our mind's eye while we work out a problem — is severely limited. It is difficult for us to envision multiple relationships simultaneously, even if there are just three people and the relationships are straightforward.

We have not even scratched the surface of the myriad ways in which our thinking is bounded. Our memory is hopelessly inaccurate, our willpower is more like whim-power (think dieting and exercise), our emotions are overreactions that reduce our reasonableness to that of an angry monkey, and we are hopelessly hypocritical yet aggressively self-righteous. We are supremely overconfident in our own abilities and highly adept at excusing our own bad behaviours but quick to be outraged at the perceived injustices perpetrated by others. Our thinking is patently imperfect, and more so than we typically appreciate.

2.5 The Cognitive Trade-Off

"All life is problem solving," according to philosopher Karl Popper.[7] The function of the brain is to orchestrate action that facilitates the survival (and reproduction) of the organism. Since humans aspire for more than mere survival, our goal is to thrive. The brain interprets its environment so it can motivate actions that are conducive to thriving.

The catch with constrained cognitive power is that it forces a trade-off between speed and accuracy, between efficiency and effectiveness. It is impossible for a bounded brain to maximize speed and accuracy simultaneously, because the two are not complementary — they are in opposition. Speed forces a sacrifice in accuracy: the faster we draw conclusions, the less time there is to test them against alternatives that also explain the data our brains take in.

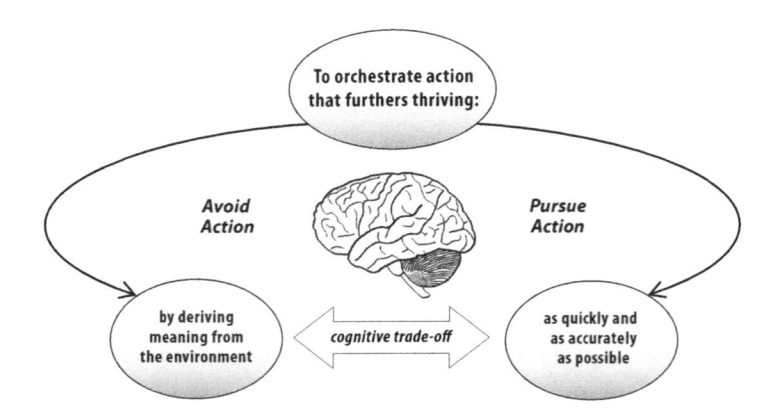

We are not programmed to optimize accuracy exclusively but to optimize the balance between speed and accuracy. Natural selection had to find the balance between these two that maximized survival and reproduction. Animals that survived to reproduce and pass their genes on had brains that attained this balance, generating interpretations of the world that were quick and accurate more often than not.

The balance that served our ability to thrive evolved to be highly skewed toward speed. There were two principal reasons for this. First, taking time to second-guess yourself in a threatening environment can be the difference between eating and being eaten: an indecisive caveman is a dead one. Second, jumping to conclusions does not force a significant sacrifice in accuracy most of the time. Speed of response served our survival needs for the vast majority of our time on this planet, while invoking few life-threatening errors. Speedy decision making is effective for problems that can be assessed on the basis of simple visual cues, when all we need to know is what we are looking at. A charging tiger does not require us to do a lot of compare-and-contrast processing to arrive at the best possible conclusion; nor does a bush of lush berries that other animals are eating; nor does an attractive member of the opposite sex; nor does a storm cloud. Extra time does not generate much of an improvement in accuracy for straightforward challenges.

Evolution furnished us with cognitive shortcuts that facilitate speedy interpreting and concluding. One of our dominant shortcuts, for example, is to relate our present experience to what most resembles it in our previous experience: we interpret situations on the basis of beliefs that we already have. Like the behind-the-scenes algorithms that drive Google searches, our minds have filters that narrow the possibilities according to what we already know. This operating principle works perfectly in a simple world in which threats and opportunities resemble one another quite obviously. Once we get into a complex environment in which challenges are not so clear-cut, the same cognitive shortcuts can have a large, detrimental effect on our understanding, and when we do not slow down to consider alternative interpretations of complex phenomena, the reliability of our conclusions falls dramatically.

A Different Trade-Off for Complexity

The above examples of bounded thinking demonstrate how our mental shortcuts can lead us to wrong conclusions. Harvard academic David Perkins notes that "the question is not so much *whether* human intelligence works well as *when*."[8] He quotes author Jeremy Campbell's pithy observation about how a human brain operates: "It needs to be as bad as it sometimes is in order to be as good as it usually is."[9] The point is that our brains work well precisely because our thinking shortcuts are so effective much of the time, at the cost of misleading us some of the time. But the rise of complexity means that the situations for which our shortcuts are ill suited are on the rise. Our default efficiency–effectiveness balance, skewing as it does toward speed, works well in a straightforward world. It is not suited to complex problems, however, because we have not yet developed expert intuitions about complexity: the same shortcuts generate a high rate of error, pushing us toward grossly oversimplified interpretations of new-world complexity. As complexity continues to accelerate, we are vulnerable to an increasing frequency of ever-bigger errors.

We need a different balance for complex problems: one that deprioritizes speed in favour of greater accuracy; one that gives up efficiency for gains in effectiveness; one that reduces reliance on default shortcuts and invokes more sophisticated ways of interpreting the world.

There is no quick fix to closing the brain–world gap, but attempting to do so is well worth the effort, because the payoff is huge and the downside of not getting smarter about complexity is even greater. The story of our evolutionary history and the demonstration of our cognitive imperfection are crucial starting points. The next step is to surface and evaluate the main assumption that underlies our thinking, one that does not serve us when we confront complexity: "naïve realism," the belief that the world works exactly the way that we perceive it. To truly understand when and how our thinking goes wrong, we have to appreciate how we construct our view of reality.

Part II

Brains Sorting Out the World
Construction Zone A Head

3

To Know Is to Construct
Building a World

*Physical concepts are free creations of the human mind,
and are not, however it may seem, uniquely determined by the external world.*
Albert Einstein*

We may feel bad for colour-blind people and their inability to differentiate green from red, shut out as they are from seeing the whole of reality, but our sympathy is misplaced. All of us, including those of us with "full" vision, do not see reality "as it really is." Consider how other animals could feel bad for humans:

- A cardinal could not imagine how dull the world must look to us, since we lack its ability to see the ultraviolet that makes flowers more dazzling than we could ever appreciate.

- A dog would pity us because we cannot smell the subtle odours that are picked up so vividly by its two hundred million scent receptors (compared with our measly six million) and processed by its larger olfactory bulb, which interprets all these scents.

- A bat would be stunned to learn how blind we are at night without the benefit of sonar-like hearing.

- A rattlesnake would wonder why we have not yet evolved the heat-sensitive equipment necessary to pick up the long infrared wavelengths that make it possible to track prey in the dark.

---
* *The Evolution of Physics: From Early Concepts to Relativity and Quanta*, Touchstone, 1967.

All of us — birds, dogs, bats, snakes and humans — have our own way of representing reality: in fact, one of the profoundest insights that humans are capable of is that reality is *constructed* by us to a larger extent than it is *received* by us. This insight, which is the focus of this chapter, contradicts our intuitive assumption that the world works the way we perceive it. The latter assumption serves us well because much of the time we have little reason to think otherwise. But when it does not serve us, it can be extremely unproductive and even dangerous.

3.1 Naïve Versus Constructive Realism

We may not find it surprising that we play a role in constructing our version of reality when we consider the "he said / she said" nature of most arguments, in which two different perspectives generate two sometimes radically different interpretations of the same event. Religious clashes, border disputes and class warfare are examples of this problem played out on a larger, more complicated stage. We should be surprised, however, by the extent to which we are not more cognizant of how these disagreements arise, and how difficult and unnatural it is for us to exert the effort to view things from someone else's perspective. Why is it so difficult? In part because of our operating assumption that the world is delivered to each of us in an unaltered form, and our inference, based on this assumption, that our individual perceptions about it must therefore be true.

Naïve Realism

It is logical that human experience is available only through human cognitive faculties. It is not necessarily intuitive, however, because our minds default to naïve realism: the belief that the external world works exactly as we perceive it. We come into the world with this assumption and most of us carry it to our graves. Naïve realism ignores any difference between the world we perceive and the world that lies beneath our perceptions. It neglects the profoundly important insight that what we perceive is both a small piece of reality and one that is shaped by the particular cognitive apparatus of our species.

Naïve realism is a shortcut that expedites the rush to conclude because it obviates the need to second-guess our conclusions about the world and other people. If what we see is what there is, then we are inclined to treat our interpretations as truths and move on, rather than as mere possibilities that may require further examination. Second-guessing is not conducive to survival in a straightforward environment; in fact, it can be distractingly detrimental in dangerous situations in which survival requires fast and decisive action.

Our view of the world, however, is not given to us from the outside in a pure, objective form; it is shaped by our mental faculties, our shared cultural perspectives and our unique values and beliefs. This is not to say that there is no reality outside our minds or that the world is just an illusion. It is to say that our version of reality is precisely that: *our* version, not *the* version. There is no single, universal or authoritative version that makes sense, other than as a theoretical construct. We can see the world only as it appears to us, not "as it truly is," because there is no "as it truly is" without a perspective to give it form. Philosopher Thomas Nagel argued that there is no "view from nowhere," since we cannot see the world except from a particular perspective, and that perspective influences what we see.[10] We can experience the world only through the human lenses that make it intelligible to us. We do not experience things "as they really are"; we experience things "as *we* really are."

Constructive Realism

An alternative assumption to naïve realism is constructive realism, which differentiates the external world ("world out there") from the representations we have of the world in our minds ("world in here"). To be sure, the world out there influences the world that we represent in our heads, but so does our cognitive apparatus. Constructive realism explicitly recognizes that knowledge is the product of both the outside world and the equipment we use to interpret it. As philosopher Paul Thagard describes it, our experience of reality is based on both bottom-up and top-down processes, and these processes interact: our bottom-up sensory information, which originates from the physical world, interacts with our top-down internal beliefs and expectations, which shape our interpretations of the sensory information.[11]

We cannot possibly be effective in responding to the world if we do not appreciate the influence that the assembly plant in our heads has on our ultimate perceptions. And we risk serious conflict with others if we assume that we all share one unalloyed version of reality: worst-case scenario, we go to war. Unlike naïve realism, constructive realism is based on an understanding that we are each restricted to flawed perspectives of reality.

The various versions of constructive realism have long histories, beginning in the East with the Hindu sacred texts, the Upanishads, most of which were written between 800 and 400 BCE. These texts distinguish the transitory world that we experience from the permanent world that lies behind it. Indian culture influenced the thinking of Siddhartha Gautama (the Buddha), born around 560 BCE, who based his philosophy on the examination of our fleeting sensory experience.

In roughly the same period, Greek philosopher Parmenides distinguished unchanging, unified reality from the varied, shifting opinions that we have about it. Fewer than two centuries later, around 400 BCE, Plato built on this division by distinguishing between non-material "Forms" that we cannot directly perceive and the material world of our experience, which mimics reality as a shadow mimics the object that casts it.[12]

As the scientists of that age, these philosophers shared a conviction that our sensory experience deceives us, and that an ultimate reality is accessible to us only if we take the right steps to circumvent our senses (Buddha, through enlightenment; Parmenides and Plato through pure reason). Fifty years after Plato's writing, Aristotle entered the fray by declaring that Forms had no reality of their own: they were, instead, categories that gave structure to things and could exist only as represented in those things, not independently of them. Aristotle's contribution to our conception of reality was significant: he argued that matter without form was unintelligible, since matter and form are inextricably linked.[13]

The notion that it is impossible to access a "formless" state of reality has been explored by many intervening philosophers, finding its fullest expression in the work of eighteenth-century philosopher Immanuel Kant. Kant worked out a very detailed description of the distinction between things as they appear to us and things as they are in themselves.[14] Optical illusions demonstrate this difference quite nicely, but Kant's claim encompassed far more than just the tricks our eyes can play on us. His separation of the world we experience and the world that underlies our experience was not original, but his *explanation* of the separation was.

Kant suggested that our experience of the outside world is shaped by our uniquely human cognitive structures. In his view, we perceive external reality through our sensory and mental faculties, which employ specific forms, like time, space and causality, to structure and order the world. We thereby create the world that we experience, a world that is a function of the forms we impart to it. The properties that we associate with the world are features of our cognitive apparatus, not of "things-in-themselves." If pink lenses were implanted over our eyeballs at birth, the world would appear to us with a pink tinge, and we would have no way of envisioning reality without this pink overlay. Similarly, we cannot see reality without the influence of how our eyes and brains are constructed to view things.

According to Kant, when we attribute properties like causality, space and time to the world outside our experience we run into conceptual confusion and create contradictions, because these properties are conceptual structures, not structures of things-in-themselves. These contradictions are known as Kant's antinomies of pure reason, and they reveal the limits of our knowledge: we are restricted to things as they appear to us; we cannot know the world as it exists without the form of these appearances. Kant did not deny the existence of objects outside us; rather, he asserted that we perceive them in a form that is determined by the way the human brain works.

Leading Up to Kant

Many of Kant's ideas had their origins in earlier thinkers, but he packaged their thinking in a more coherent and detailed way. If each of these thinkers had had the opportunity to talk through their ideas, as a stream of thought leading to Kant, it might have looked something like the following hypothetical conversation.

> **René Descartes:** What can I know for certain? That there are two distinct ontological substances: that which is mental and that which is material. I know the material world through my mental ability to reason.[15]

John Locke: We have to acknowledge that although the external world underlies our sensory experience, we cannot access it directly: we know it only as it is mediated by our senses and our ideas. What lies beneath our mediated experience is a reality that we have no access to; it remains unknown; it is "I know not what." Although we cannot know reality directly, we can know its properties. Some qualities, like an object's size, are independent of observation, while other qualities, like the object's colour, exist only because of our interaction with the object.[16]

George Berkeley: We need to go further. The notion of a mysterious external world that you label as "I know not what" is unintelligible for the very reasons that you suggest: all we can know is the world as it appears to us; all we ever have access to are the contents of our minds. We cannot step outside ourselves and escape our perspective to identify an external world that lies beneath our experiences. Absolutely nothing in our experience warrants the assumption that something external underlies our perceptions. It makes no sense to talk about something that cannot be conceptualized as being the source of what we can conceptualize![17]

David Hume: In fact, we are not even in a position to assume the existence of a "self" that houses experience. For the exact same reasons that you argue against an external world, Berkeley, I argue against a distinct, conscious self; all we can know is a bundle of sensations that feels like a self. We are inescapably restricted in our ability to be certain about anything, because we are locked inside our habits of mind and have no ability to escape them to examine the "real" world. Having said that, we do have to conduct our lives, so for purely practical reasons, we can follow the habits of our mind in thinking that there is an external world organized by cause-effect relationships. But these relationships are merely expectations of how things work, not provable facts.[18]

Immanuel Kant: Actually, we can be certain of some things. Your description of the mind's habits is a good starting point, Hume, but it is not just our habits that create expectations of how things work. The structure that our minds give to the external world entitles us to have some certainty about it. Our cognitive apparatus interprets the world in specific and consistent ways, giving order and predictability to our experience. We perceive objects in space and time because space and time are forms of knowing that we impose on the external world. It follows, then, that we cannot have any experience without the forms of space and time, because they are the lenses through which we are able to perceive. Because these forms are standardized, we can use geometry to describe space with certainty and arithmetic to describe time with certainty. Similarly, our minds use the form of causality to establish relationships between events.

You are right that cause and effect are not properties of things-in-themselves, because they are a form of understanding that we use to conceive the external world. But the constancy and consistency of these forms of knowing are what allow us certainty. While certainty is available about things-as-they-appear, you are correct, Hume, to propose that we can have no certainty about the external world in its "unstructured" state — before our cognitive structuring of it. The only aspects of the external world that we can apprehend are the ones that conform to our cognitive structures for sensing and understanding. Beyond the world as we perceive it, all we can say about things-in-themselves is Locke's "I know not what."

Kant's insight that our experience conforms to our cognitive structures is even more compelling today, given the recent findings of cognitive science. Studies that examine brain-damaged patients reveal that the human brain binds information from various modules, each of which processes a component of the overall impression. So motion-reporting neuronal microcircuits are joined with shape-reporting modules and colour-reporting modules, which themselves are combined and compared with memories of experiences to determine if we recognize the car, flower or person in front of us.

Even Kant, though, would be surprised by psychologists' discoveries of how actively we construct reality: we are far more involved than he ever imagined. It is not just that things-in-themselves conform to our cognitive structures; it is also that we play a surprisingly interventionist role in structuring the external world, creating a version of reality that suits our idiosyncratic purposes. From the sensory tsunami that rolls over us, moment by moment, we actively *select* what we pay attention to, picking out what we think is relevant. And we *add* to these data, striving to create interpretations that are coherent to us personally.

Typically, we are unaware of this active process of construction, because our intuition encourages us to believe that we are passive recipients of an external world that impinges on our senses. G.W.F. Hegel, writing thirty years after Kant, observed that this intuition comes very naturally to us: mind mistakenly thinks reality is "out there" and independent of it, with little recognition that reality is entirely its own creation.[19] While Hegel overstated the case by denying that consciousness was separate from the external world, there does remain an insurmountable chasm between the external world and our perceptions of it — a gap that typically we are oblivious to. The relationship between the world outside and the one in our heads is much looser than our intuitions suggest; the correspondence between the real world and our perception of it often is astonishingly low.

Since Kant

How can we have certainty about anything if we are locked into our own constructions with no direct access to the "real world"? Many thinkers have weighed in on that question. Albert Einstein credited Hume with inspiring him to develop the theory of relativity (which ended up disproving Kant's thesis that space and time are independent of each other and fixed for all observers). Physicists are still struggling to come up with one all-encompassing theory of how everything works, at both the level of massive objects (where general relativity applies) and the level of the submicroscopic (where quantum mechanics applies). Maybe Hume had a point about the fundamental lack of certainty that is available to us.

There is virtually no controversy, however, over the proposition that there is a fundamental difference between our view of reality and the external world that features in our view. If we asked some of the best contemporary minds to comment on Kant's distinction between things-as-they-appear and things-in-themselves, their responses might look a lot like the Einstein quote at the beginning of this chapter.

> **Oliver Sacks** (neurologist): I think of our version of reality as being a "statistically plausible representation" of the real thing. We come up with what we think is a good approximation of what we observe.[20]

Stephen Hawking (theoretical physicist): We perceive the world through the particular lens of the human mind's interpretive structure as a "model-dependent reality," distinct from the real thing that we have no direct access to.[21]

Richard Dawkins (evolutionary biologist): I call our conceptual construct of reality the "middle world": the world that is between subatomic particles and celestial bodies, neither of which is visible to the naked eye.[22]

3.2 A Uniquely Human Version of Reality

It is worth noting a point that Kant did not emphasize. Not only do we model reality, but our models are largely contingent: they are a function of our particular cognitive structures, which could be other than they are. Humans model reality very differently from, for example, the way squirrels do.

Consider our perception of colour, which is a function of: (i) the three kinds of photoreceptor cone cells in our retinas, which enable us to distinguish blue, red and green (and the colours generated by their combinations), and (ii) the particular wavelength of light that is not absorbed by an object but reflected off it. A red tomato appears red to us because the molecular structure of a tomato absorbs all visible wavelengths of electromagnetic radiation except the long ones that bounce off it. When this reflected light reaches our eyes, the colour receptors in our eyes, which are sensitive to this wavelength, work in conjunction with the visual cortex of our brain to translate the data into the colour red. If you shine a pure blue light on the tomato, it will all be absorbed and nothing will be reflected, so the tomato will appear black.

Some fish and birds have four kinds of colour receptors, enabling them to detect ultraviolet light that is completely invisible to us (although we experience its effects when we are in the sun too long). Squirrels, along with most mammals, are dichromatic, having only two colour receptors in their eyes: red and green are indistinguishable for them. Squirrels do not have to worry about traffic signals, so the red-green distinction is not much of a problem. They are completely oblivious to what they are missing, as are the small percentage of humans who are also dichromatic.

Each animal has its own world of representations that are particular to it, what German biologist Jakob von Uexküll refers to as each species' subjective world (*umwelt*) in distinction to the universe-in-itself (*umgebung*).[23] An animal's *umwelt* is determined by the way its particular body interacts with the world; in other words, all cognition is embodied, so the way we perceive the world is shaped by the particular kind of bodies we have.

Brain Versus Mind

Descartes made a clear distinction between the corporeal and the mental as two fundamentally different substances (a distinction referred to as "dualism."). Diverging from Cartesian dualism, other philosophers argued that there is only one substance. Berkeley and Hegel, for example, championed the idealist view that all is mental. Others took up the materialist view that there is nothing except matter, of which thinking is just a fancy by-product. Twentieth-century philosopher Gilbert Ryle insisted that Descartes had committed a "category mistake" by describing a physical brain with non-material qualities; Ryle argued that mental activity does not require a new category of substance.[24]

The latter view is shared by many (probably most) neuroscientists and many (probably most) contemporary philosophers, who believe that mind and brain are distinguishable, but not as radically as Descartes suggested — not as two different substances. Because mind does not exist without brain, some scientists refer to the brain as "brain/mind," to enforce the conception that the two are so tightly linked that we might as well call them the same thing. Others insist on using "mind" as a verb, on the basis that mind is what the brain does.

There is consensus (although by no means unanimity) that brain and mind refer to the same material system, with the difference between them being based on the perspective that one takes of the system: "brain" captures its physical properties, whereas "mind" captures how it processes information. The two are fundamentally related: the brain is embodied in a corporeal framework and embedded in a specific cultural framework, so it processes information only by interacting with the world in a way that we call mind.

Brain and mind are inseparable in the way that hardware and software are inseparable in a functioning computer. Consider the corporeal brain to be the hardware, and mind the software that constitutes the operating system and runs application programs (such as a language program for speaking English). A computer's software has no physical presence: it cannot be viewed by removing the computer's casing, because the coding that constitutes the software is embedded in the hardware. The hardware and software are mutually dependent: the hardware is inert without the instructions supplied by the software, and the software coding is useless without the hardware to run on.

The analogy implodes when pushed too far because software does not rise from the complexity of the computer: it is installed separately after the computer is built. Thinking, though, does arise from the complexity of our brains. How exactly does mind arise from brain? Neuroscientists do not know. Some contend that we will never figure the mystery out, but many are optimistic that we will, just as we are likely to eventually unlock many of the still-to-be-solved mysteries of science.

"Emergence" is the concept in complexity science that describes how new properties arise from complex systems, such as how mind emerges from brain. When flour, butter, water, egg and baking powder are mixed and heated, a cake emerges with properties that are very different from the ingredients. You would never predict a cake by just examining the ingredients, until you have witnessed the baking process. When one hundred billion neurons interact, human thinking somehow emerges. The cake analogy does not detract from the mystery of mind, especially the special form of human thinking that allows us to reflect on how we think. But it does reinforce the notion that, while we currently lack a complete understanding of the "mental baking process," mind emerges from brain in a way that does not necessitate Cartesian dualism.

Not only is our world different from a squirrel's, but it is also different from the world that we cannot see at all, a world inhabited by atomic and subatomic particles. In this world, there is no solidity and therefore no touching of objects. Even though it feels to us that we are continually touching solid objects, they are mostly space: the space between an atom's nucleus and the negatively charged electrons that swirl around it. Objects feel solid because of the electromagnetic repulsion that resists our hands when they approach. Negatively charged electrons in the atoms of approaching objects repel one another in the same way that two negatively charged ends of different magnets do. Even though what is happening at the atomic level is electromagnetic repulsion, our brains create a model that we call solidity, which is a useful concept that works in our version of reality — in what Dawkins calls "the middle world." In fact, electrons themselves are not "things" that can be seen with a powerful enough microscope; they are just theoretical concepts used to explain the behaviour of subatomic parts.

Then we come to the world of fast speeds, in which time is different for observers moving at different speeds. We model time as being fixed, because that is the way that it appears in the middle world. Any discrepancies are imperceptible, because we are all moving at around the same speed, which is nowhere close to the speed of light. At faster speeds, time slows down. A watch worn by someone travelling in a very fast rocket would show less time elapsed than one on an observer on earth.

The worlds we cannot see — both the ones of other observers and the ones outside our middle-world perceptions — operate very differently from our intuitions. The recurring theme of this book is that ignoring this difference gets us into trouble in situations where the difference matters.

3.3 Our View as a Limiting Case

We construct our realities with a limited amount of data about the external world: we see only a narrow range of the sun's electromagnetic radiation, hear only a fraction of the sound vibrations around us and smell only a limited number of odours. Even within this limited set of data, our brains still cannot handle all that is available, because it is simply too much for in-depth processing. This is why we select what we pay attention to: we work with a small portion of data that we select from the small portion of data that are available to human cognition.

Our version of reality is what could be called a "limiting case" of the real thing.

Physicists use this term to describe the conditions of a theory that make it consistent with another more expansive theory. For example, while Newton's theory of motion is contradicted by Einstein's theory of relativity, Newton's theory remains consistent with relativity if we make the simplifying assumption that objects move much more slowly than the speed of light. The errors generated by Newton's laws are imperceptibly small at slow speed, so his laws are useful in the *limiting case* of our middle world of slow-moving objects. The error rate of Newtonian mechanics grows exponentially, however, as bodies accelerate to very high speeds. This problem is what Einstein corrected for: special relativity works for all speeds, including the speed of light. Put another way, the limiting case of low speeds masks the nature of space and time, which operate differently at very high speeds. (Einstein's general relativity did the same thing for Newton's law of gravitation, which works only in the limiting case of weak gravitational fields where objects do not warp space because they do not have significant mass.)

As scientists develop new theories, the prior theories often still stand up as limiting cases of the new ones, which extend to a broader set of conditions. The concept of a limiting case is a useful analogy for our perspective of the world and how it works. That our view of reality is a limiting case of the external world is obvious when we consider how unintuitive many of science's insights are: how large-mass objects warp space (general relativity), how fast speeds warp time (special relativity) and how basic cause-and-effect does not work for subatomic particles (quantum mechanics). Our day-to-day, intuitive view of how things work is meaningfully incomplete: it is severely limited and subsumed by larger views, which are themselves subsumed by even larger views, ad infinitum, until we get to the theoretical "view from nowhere."

What are the limiting conditions that make our intuitive view of the world useful to us? We have already seen that things cannot be too big, too small or too fast for our intuitions to be approximately correct. But it is another limiting condition that informs the focus of this book: *Our intuitive view of the world only works well when the problems that we are dealing with are straightforward.* Once we enter the realm of complex problems, our intuitions get things wrong, just as Newton's laws generate errors when the limiting conditions are removed. Complex problems are usually beyond the scope of our limiting-case perspective. Our middle-world view is limited not only to slow speeds and objects that can be seen without microscopes or telescopes but also to the straightforward problems that we are accustomed to solving — the kind that dominated our long evolutionary past and much of our present daily lives.

Our cognitive faculties have been designed to deal with a straightforward but harsh and unforgiving world in which fast interpretations and responses are necessary for survival. They work in these circumstances, as well as for basic quotidian challenges. We are well equipped, in the middle world, to handle the straightforward problems of hunting for food and driving to work. But the same intuitions are prone to failure when we expand the set of problems to the non-limiting world that includes new-world complexity. Our familiar and predictable middle world is a limiting case of the bigger set of challenges, which includes complex problems.

Away from the centre of our middle world problems are more complicated challenges like navigating interpersonal relationships, choosing and building a career and raising well-adjusted children. Just beyond the fuzzy, outer edge of the middle world are the crises often referred to by decision theorists as "wicked problems": religious wars, global warming, personal happiness. Far beyond the border of the middle world are problems that may be unsolvable, like how consciousness arises from a bunch of neurons and how a point of pure energy with no dimension suddenly exploded into an ever-expanding universe.

To deal with complexity, we need the problem-solving equivalent of relativity and quantum mechanics. In other words, we need more sophisticated thinking: complex thinking for complex problems. Only by acknowledging and understanding our construction of reality can we open up the possibility of interpreting the world differently: in more sophisticated ways than our intuitions offer; in more effective ways for coping with complexity.

The first step is to acknowledge the gap between our constructions of reality and reality itself: to acknowledge *that* we construct. The second step is to examine the specific ways in which we create our view: to understand *how* we construct.

4

To Know Is to Simplify
Reducing Reality

The human mind is a story processor, not a logic processor.
Jonathan Haidt*

How do we construct reality? We select data to pay attention to, we model these data to interpret them and we weave the modelled data into a story that makes sense to us. We reduce the world to something that fits into our minds. In a word, we *simplify*.

Consider the amazing lurches forward and backward in nutritional science as just one example of how captive we are to constructing simple stories about complex phenomena: stories that are only partially true, if at all.

> The history of nutrition is a parade of stumbles and trips, jumping from one "definitive discovery" to another, promoted by books and media that purport to reveal one-size-fits-all solutions to weight loss and healthful eating. In the 1970s and early 1980s, reports linked dietary fat to poor health, and in 1984, the American National Institute of Health recommended a reduction in dietary fat intake; the food industry responded by marketing fat-reduced and fat-free products. But reducing fat reduces flavour, so food producers dumped enormous amounts of salt and sugar into their products (salt and sugar counterbalance each other in a flavour-enhancing way). Fat consumption declined, but sugar intake increased and obesity

---

* *The Righteous Mind*, Pantheon Books, 2012.

levels climbed. Eventually, the pendulum swung to another simple solution: attention shifted to the glycemic index, an invention in the early 1980s that rates foods on the basis of how quickly they increase blood-sugar levels. High-glycemic carbohydrates were linked to fat storage, giving rise to the explosive popularity of low-carb diets in the 1990s, which have recently morphed into an antigluten craze.

Then there is the vitamin and mineral supplement industry, based on a complete disregard for the difference between how we metabolize and absorb nutrients when they are contained in foods versus when they are separated out. Scientists are only beginning to probe these differences and are discovering that they are very significant. Most supplements that were once considered important, if not essential, have since been discovered to have little benefit when taken on their own, and in some cases to cause harm in high dosages, as in the case of vitamins C and D, as well as fish oils. Research scientist Edward Archer and his colleagues studied forty years of nutritional research and concluded that "the results and conclusions of the vast majority of federally funded nutrition studies are invalid."[25]

We have no choice but to simplify: during any given millisecond, far too many sensory data are available to our limited brains for us to sort out. We are forced to choose what to pay attention to. We also choose how to put the pieces together and how to create a story that gives meaning to the sensory data we select. We transform the world into mentally digestible pieces that eliminate ambiguity, ostracize complexity and banish uncertainty. At the end of it all, we are left with a highly simplified account of reality: a rarified, reduced, relegated representation that often bears some resemblance to the real thing.

Simplifying makes perfect biological sense: our cognitive processes are dominated by the urgent need to figure things out (after all, charging tigers and menacing strangers wait for no one). Combine the need for speed with a brain that has limited processing power, and what do you get? Survival by way of cognitive shortcuts. However, to understand the mismatch between our simplifying strategies and the complexity that lurks beneath the surface of our experience, we have to understand *how* we simplify — that is, how we use shortcuts to make sense of the world by:

- selecting what we pay attention to,
- connecting the selected pieces,
- creating coherent stories about the pieces.

4.1 To Simplify Is to Select

The essence of selection is separating signal from noise. Our minds constantly sift through the mass of sensory data that we are exposed to, in search of potentially useful signals. This process involves picking out relevant cues from irrelevant data. The direction and speed of an approaching tiger are highly relevant cues, easy to pick out from the noise of other data, like how clean its fur is. Our brain's task is to detect signals; it is our brains, not our sensory organs, that constitute the bottlenecks that force selection choices on us. We can take information in much faster than we can process it. Scientists have estimated, based on the number of sensory receptor cells connected to our brains, that our brain processes about eleven million bits per second — ten million of them from our optic nerve and the rest from other sensory organs.

Of the eleven million bits processed, only forty make it to conscious awareness. The vast majority of them remain hidden in the depths of subconscious processing. So the spotlight of our conscious attention casts a strong but very narrow light. The spotlight metaphor is particularly apropos because, unlike the many animals that rely largely on smell or touch, humans rely predominantly on vision. The limits of our attentional spotlight are a function of two things: how narrow its beam is and how easily its focus can be manipulated.

Narrow-Beamed Spotlight

The narrowness of our conscious attention is so extreme that we can be completely oblivious to things that are right in front of us, like the ambient background noise you did not hear as you were reading this sentence — until now, when it was pointed out.

The most famous demonstration of this oddity is the gorilla experiment undertaken in 1999 by professors Daniel Simons and Christopher Chabris.[26] Subjects were asked to watch a video of basketball players passing a ball between them and count the number of passes made by the players in white shirts. The concentration required to count was sufficient to make half of the subjects blind to the fact that a person in a gorilla costume traipses through the proceedings on the video. These subjects seemed to be partially blind and had no recollection of seeing the gorilla. The ones who acknowledged it were usually wrong in their pass counting, which indicates that their attention was not focused on the task.

As surprising as the experiment is, it is not that unusual. Think of how often we discover that we have failed to hear what someone was saying to us because we were momentarily distracted, often by nothing more intrusive than our own ruminations. Our subconscious can manage a multitude of activities simultaneously, but our conscious attention is much more constrained. For example, we can only handle challenges that require conscious attention serially: we may feel that we are proficient multitaskers, but actually we are good at doing single tasks quickly by switching from one thing to another and back again: we cannot do them in tandem. We cannot converse with someone while we read, or work on two problems at the exact same time. Studies have revealed that texting while driving has the same impact on our driving skills as intoxication, and that even carrying on a simple conversation is a significant distraction for us when we are at the wheel.

Our spotlight of attention is narrow in part because our subconscious is adept at suppressing the processing of information that it deems irrelevant. When we focus our attention, we activate inhibitory neurons that suppress the firing of neurons in our visual field that might otherwise cause distraction. This is why many people miss the gorilla altogether, and why, for example, our brain subtracts blurs of movement in our visual experience. If you look in a mirror and switch your focus from one eye to the next, your eyes will move, but you will not see that movement in the mirror: in what is referred to as "saccadic suppression," your brain blocks out the motion.

Magicians rely heavily on "attentional blindness" for the vast majority of their tricks: by diverting our attention with their hands and other tools of their trade, they render us oblivious to all kinds of things that they are doing right in front of us, things that are stunning when revealed. In the same vein, "change blindness" is a well-documented form of attentional blindness in which significant changes can be made in a scene or event without the awareness of observers. For example, the colour of background drapes can be switched without notice. In one of the most celebrated experiments, a stranger asking for directions was replaced with a different stranger after a brief interruption, without any acknowledgment whatsoever by the observer who was asked the questions.

Not only is our attentional spotlight itself narrow in scope, but its direction is also highly influenced by factors that we are usually oblivious to.

Easily Influenced Spotlight

Where we choose to shine the light of conscious attention is highly influenced by our subconscious expectations in the moment. We will be much more attentive to a growl-like sound if we are walking in a jungle than if we are sitting in a coffee shop; we will be much more attentive to the facial expression and tone of our boss if we are expecting to get fired than if we are chatting at a cocktail party. Our attention is also highly influenced by our own emotional state. Psychologist Lisa Feldman Barrett has demonstrated that what we pay attention to is partly determined by our mood: if we are in a lousy mood, we will be more acutely aware of frowning or threatening people than of cheery people. Our emotional state determines our conscious perception.[27]

The biggest influence on where we direct our spotlight is our past experience: we tend to focus on what is familiar to us because our survival depends on recognizing signals that were previously useful to us. The problem is that when we are confronted with new situations, we gravitate to earlier learning even though it may not be applicable. We focus our attention on what we think are useful signals, even though they may be useless or deceiving in certain situations, especially novel and complex ones. As will be reviewed in chapters 13 and 14, our minds are not adept at separating the noise from the signals embedded in complex phenomena.

Selecting what to pay attention to is just the beginning of our simplifying process; we also rely on simple models to construct relationships between things.

4.2 To Simplify Is to Model

Mental models are tools for organizing information. They enable us to determine patterns, and patterns are what we relentlessly seek — regularities that are invariant or stable from one situation to another. If we could not pick out patterns, we would not be able to learn, predict and adapt. Regularities are the signals we use to understand and respond to the world.

Our minds are meaning-making machines. The route we take to construct meaning is quick predictions based on recognizable patterns that we pick out courtesy of our models. Models allow us to relate what we experience to what we already know. Our brain constantly compares incoming sensory data with what it has already stored as knowledge. More than that, we interpret the world largely by our expectations of it, based on what our prior knowledge suggests that we are likely to see. So instinctive is this process that we barely miss a beat in making sense of something that is, on the face of it, unintelligible: *Froexpmle, tihs sntnece istn toodffcutl frus todcphr*. We experience the letters and words largely as we expect to experience them.

In fact, our proclivity for prediction is so ingrained and relentless that we can be tricked into reversing a sequence of events. Neuroscientist David Eagleman had subjects become accustomed to seeing a flash of light following a brief delay after they pushed a button. Once the delay was eliminated, however, so that the flash immediately followed the button, the subjects actually perceived the flash as preceding the button; their brains had been trained to expect a delay.[28]

Our brains engage in a constantly interactive, iterative loop with the world, making sense of it, checking the validity of what we sense and updating interpretations as necessary. Selection is simultaneous with modelling in this circular loop: our detection of signals is largely based on extracting bits of data that fit our mental models; models determine what is relevant and worth attending to as they organize the information. The process is so fast that it appears nearly instantaneous to us, which makes it difficult for us to intervene and challenge our own conclusions.

Just as we use a prey–predator model to assess a tiger, or a danger–safety model to assess a stranger walking behind us, so too do we use different models to assess our relationships, our careers and our self-worth, to name just a few of our bigger challenges. Where do our models come from? Some are innate and common to most of us, and some are developed after birth and more personal to each individual. The innate models are proclivities that become fully operational when combined with experience, like the basic cause-effect model we all use. When Kant wrote about the models that humans use to represent reality, he emphasized these innate ones, like space and time, which are fundamental to all human cognition. He neglected, however, the importance of personal models, which are also fundamental even though they vary among individuals and cultures.

Personal Models

Each of us evaluates the world through a huge web of interconnected beliefs, a web that is as distinctive as our fingerprints. Unbundling this web to differentiate the innate models from the learned ones is difficult, if not impossible in some cases. Nature and nurture are misleadingly simple concepts that belie the complex intermingling of both; they defy easy separation in the human species. But we do know that environmental conditions play a significant role. Because we are born "prematurely," our developing brain is exposed to circumstances that are much more complex than those of a womb, where most of the cognitive maturation of other animals occurs. The complexity of the human brain and its protracted development in a complex environment are what make individual humans more diverse than any other species.

In our early years, experiences that are unique to each of us create neural pathways that are reinforced by repetition, ingraining idiosyncratic views of the world. So deep are these views that we typically make no distinction between the two lenses — shared and personal mental models — through which we view the world. Each of us collapses these into a unified perspective of how the world works.

Our individuality is reflected in the highly personal models each of us holds: one of us may have a deep-seated belief that a benevolent god rewards good deeds, while another may have a difficult-to-dislodge intuition that the only way to get ahead is by deceiving others. Whereas common models exist because they have been evolutionarily useful to large numbers of us for a long period of time, based on a large sample size representing our ancestors' collective experience, personal models are based on a small sample of individual experience. There is a high risk, therefore, in generalizing personal models beyond whatever use they have in a narrow range of circumstances. But generalize is precisely what we do. For example, a small number of negative experiences early in life may cement a personal view that nobody can be trusted or that we are not worthy of love.

The focus of this book is common models, which are by no means foolproof, either. Just as we can overgeneralize our personal models, common models can be overextended to situations in which they do not apply.

Common Models

Today, cognitive psychologists are virtually unanimous that we are not born as blank slates, as John Locke had argued.[29] We come into the world with some basic models already hardwired into our brains, ready to be applied. In Kantian terms, our brains are preformatted to interpret the world in certain ways.

We are instantly able to recognize faces (which has been demonstrated in babies only one hour after their birth) and pre-equipped to learn language. An overwhelming percentage of human beings around the world are instinctually afraid of spiders, snakes, rats and heights — fears that are evident long before anything that could give rise to them has been experienced. We also have a very early instinct for fairness: psychologists have observed that toddlers instinctually punish perpetrators who take their toys by taking the perpetrators' toys. Psychologist Elizabeth Spelke has shown that five-month-olds will convey surprise when they are shown impossible things, such as simple magic tricks.[30] Even this young, they already have a rudimentary sense of how physical things work.

These models are innate because versions of them helped our ancestors survive, and, for the same reason, other animals share many of these models, like the innate model that identifies light as coming from above, as illustrated in the optical illusion demonstrated in 2.2. A baby chick born in a research lab in which light is shone only from below will still peck only at photographs of grains that are shadowed as if lit from above. Of course, many models are species unique. Even if they are born in a room with no wood, beavers will automatically simulate dam-building activities. Lab-raised chickens do not react to the projection of a cardboard cutout of a goose but will panic at a cutout of a hawk. These species-specific models give rise to the phenomenon of "supernormal stimuli": birds will ignore their own eggs to sit on fake ones that are bigger and brighter, or even attempt to mate with more colourful but fake partners.

The fake hawk, egg and mate experiments may sound amusing, but they hit closer to home when we consider the myriad ways a human will overreact to a mouse, a garden snake, a perceived insult or a slow driver in the passing lane. We may think that other animals are driven by instinct whereas we are "free" by virtue of our incredible reasoning power, but this belief belies how captive we are to our own deeply embedded, automatic motivations. Psychologist William James insisted that instincts were the building blocks of every animal's behaviour, and that human uniqueness lies in our having more instincts than any other animal.[31] James' view understates the unusual human ability to assess and transcend instinct, which is truly *sui generis*. But his view does place a suitable emphasis on just how prevalent and powerful subconscious motivations are in our daily lives.

One way we model is by using tried-and-true rules of thumb, referred to collectively as "heuristics," which enable us to solve problems quickly and efficiently. Heuristics are subconscious strategies that rely on easily accessed cues to facilitate quick decision making without expending unnecessary time and effort. They are mental shortcuts, or what psychologist Gerd Gigerenzer calls "fast and frugal" mental rules.[32] We use all kinds of heuristics to navigate various challenges. We have, for example, a heuristic for catching a ball based on observing its arc (the gaze heuristic), another one for picking a hotel based on a familiar name (recognition heuristic), another for picking a movie based on a sample of people who liked it (representative heuristic) and yet another for assessing and responding to the co-operativeness of others by treating them as they treat us (tit-for-tat heuristic).

Heuristics are developed and enforced by repeated experience, so they are largely automatic. We do not contemplate them or consciously evaluate their appropriateness to a given situation; we merely invoke them in our haste to evaluate and decide.

Resemblance and Causality

Underlying all heuristics are two central models that form the bases of how we interpret everything: resemblance (applying our understanding of one thing to another thing) and causality (identifying the relationship between things). Resemblance allows us to group similar things so we can generalize our understanding from one situation to another. Causality allows us to relate objects and events to one another so we can understand how they work.

The models of resemblance and causality are Kantian cognitive mechanisms for making sense of the world: we come into the world looking for and relying on resemblance and causality, and a years-long process of applying these models reinforces their use to the point of our overapplying them, as will be explored in 5.1. If adult subjects are shown a video of a large ball following a small ball and then a video of the reverse (a small ball following a large ball), they will report that, in the first video, the large ball is chasing and in the second video it is leading. We automatically implant into the scenarios a cause of the movement, without any context or cues other than the size of the balls. If you hear the sentence, "He was rushed and he tripped," you will likely interpret that to mean he tripped *because* he was rushed, even though the two facts could easily be unrelated. We constantly invoke causality to relate events, whether there is a real connection or not, and our use of language reinforces this habit.

Language

Resemblance is the cornerstone of our ability to categorize; categorizing enables us to chunk the world into pieces for easier assessment and remembering. Language is another tool that helps us categorize: we organize things by describing them. Language distills reality into discrete parcels that we use to communicate with one another. It serves as the scaffold for our thoughts, starting around the age of two, when the brain areas responsible for creating language come online.

Just as models help us organize data but also influence what we pay attention to, language describes what we perceive but also influences how we perceive. The "linguistic relativity hypothesis" is based on a well-researched description of how language shapes our experience of reality. The Russian language, for example, has more words for blue than English, and Russians are better at discerning subtle differences in shades of blue. The link between language and the real world is often more tenuous than we assume: language, resemblance, causality and other basic versions of our common mental models are all vulnerable to overextension, as will be explored in the next chapter.

4.3 To Simplify Is to Storytell

It is not just the models we use to simplify but also the stories we concoct that help us make sense of the world. As Haidt suggests in the epigraph at the beginning of this chapter, we look for a satisfying story, not a logically airtight explanation.

We start by selecting signals to pay attention to, searching for familiar patterns by using models to organize data into relationships. We end by creating stories about the modelled data. Where the key to selection is perceived relevance, and the key to modelling is pattern recognition, the key to storytelling is coherence: we create stories that cohere with everything we know, stories that make sense to us.

But we are not just leisurely looking for stories that make sense. We are desperate for stories: we will do whatever it takes to formulate coherent explanations of our experiences. Our brains actively shape the data we take in to eliminate any ambiguity that gets in the way of quickly establishing stories, making predictions and responding to the world. Our brains add in, take out and distort; in short, they confabulate.

Confabulating

To storytell is to confabulate: to "concoct fables," to manufacture, to make stuff up. We do this all the time, automatically and unknowingly. Confabulating is the fastest way to simplify.

Neuroscientist Michael Gazzaniga developed a number of experiments that demonstrate just how far we will go to come up with a coherent story that explains our experience, and just how oblivious we are to our own inventions. He uses the descriptor "the interpreter" to identify the mental system in the brain's left hemisphere where the work of confabulation is done: the processing module that infers causality, searches for patterns and works hard to come up with a story. Unlike our right brain, which operates on impressions and looking at the big picture, our left brain is analytic and language-based. Left has no tolerance whatsoever for ambiguity: its job is to interpret, not to equivocate. It is very stubborn: it will not abandon a story it has created unless competing evidence is sufficiently compelling.[33]

Much of Gazzaniga's research is based on split-brain patients who have a damaged or surgically severed corpus callosum, the pathway that joins the two hemispheres in the human brain. In one of his better-known experiments, the left eye of a subject was shown an image of a snow-covered house and the right eye a chicken claw. Since the left eye is routed to the right hemisphere and the right eye to the left hemisphere, the right hemisphere saw the snow and the left hemisphere saw the claw. Asked to pick out a drawing that related to the two images, the subject picked a snow shovel with his left hand (corresponding to the right hemisphere perceiving snow) and a chicken with his right hand (corresponding to the left hemisphere perceiving the claw). When asked why he picked a snow shovel, the subject responded that one needs a shovel to clean out a chicken shed. The key to understanding this odd response, which fails to acknowledge the snow, is to appreciate that the right hemisphere (which saw snow) does not have a language capability or a strong sense of causality, so it is up to the left hemisphere to articulate the rationale for picking the shovel image after the fact. The left hemisphere has not seen the snow so it has to confabulate a reason for choosing the shovel: it makes up a story that has internal coherence, which is that the shovel goes with the chicken claw, to clean out the chicken coop. The left hemisphere refuses to be stumped, so it invents a satisfactory explanation.

What is interesting is that this form of confabulation is an extension of what normal brains do all the time. *Because we do not like uncertainty, our brains do not tolerate gaps in understanding; they confabulate to close them.*

Confabulating to Close Sensory Gaps

The classic example used by neuroscientists to demonstrate how our brains fill sensory holes is the blind spot we have in each eye. The optic nerve of each eye is inconveniently positioned right in the retina, creating a small spot in our field of vision in which we can see nothing.

To fill this hole, our brain simply puts in whatever it thinks makes sense, usually whatever pattern surrounds the spot, so that we are oblivious to the small gap in our sight. Vision is not the only area where we fill in gaps. If someone coughs as they are speaking to us and a couple of their words are muffled, we still manage to hear complete sentences, in a process called "phonemic restoration," which fills in the auditory holes.

A good example of our brains' compulsion to fill holes is chronostasis, the backdating of time by the brain to eliminate gaps in our perception. Have you ever noticed that, when you first look at the second hand on a clock, it appears to languish for a second longer than usual before proceeding to move? The same phenomenon occurs when we phone someone and pull the phone away from our ear for a second; when we bring it back, we experience an extra delay before we hear the next ring. These illusory delays are the result of the brain constructing a pause, typically around 120 milliseconds, to cover the time that is otherwise filled with a short blur of sensory input. It backdates the image of the second hand and the silence on the phone to cover the time that the eyes and ears are moving; rather than registering a sensory blur, the brain creates continuity in our experience by filling these gaps.

Our brains do an enormous amount of smoothing to make our sensory experience continuous and coherent, all without our awareness. Because our eyes have a high concentration of photoreceptors only in the middle of our retinas (the fovea), we have detail only in a very tiny part of our total field of vision: about four centimeters in front of us at arm's length. Everything around that tiny field is our peripheral vision, which is fuzzy. But we do not see the world as a fuzzy blur, because our pupils constantly move around so our brains can take in the detail of a bigger picture. What our eyes actually take in, as a result of this constant movement, is a very shaky series of images, but our brain smoothes them out for us by eliminating the movement, constructing a more continuous and coherent picture than the bits our retinas take in every millisecond. This kind of active brain involvement has led Oliver Sacks to point out that we see with our brains, not our eyes: the discrete impulses that our retinas pick up are transmitted to our brains, where they are consolidated and massaged into the images we perceive.[34]

Confabulating to Close Emotional Gaps

Much of our emotional life is reasonably transparent to us: we know why we feel fear when we are alone in a dark alley and anger when we feel betrayed. But the causes of some feelings are less obvious, because their roots are buried deeper in our subconscious and are not as clearcut. We are always in some state of feeling (our "core affect") that we are often unaware of unless someone points it out to us: "Why are you so irritable?" or "What's put you in such a good mood?" We often struggle to answer these questions when our emotions have no immediate or obvious cause. In these cases, we search for reasons that seem plausible: "I must be stressed from work" or "I don't know — it's a beautiful day." In the absence of clear and distinct causal explanations, we confabulate to eliminate the ambiguity of our feelings.

Even when our feelings reflect purely physical states, like fatigue or hunger, we are still inclined to construct a story to explain or justify our emotions. When subjects in experiments are injected with adrenalin but not told what the substance is, they concoct explanations for why they suddenly feel jumpy. Psychotherapy — psychoanalysis in particular, with its emphasis on childhood experiences — is viewed by some academics as an exercise in confabulation, in which patients create stories that explain why they feel the way they do. This perspective does not rule out the potential value of talk therapy; rather, it argues that we have no reliable way to ascertain the validity of the connections between our present selves and our childhood experiences. The research of psychiatrist Joel Paris, for example, argues that childhood experience is not the dominant determinant of an individual's current problems; rather, he says, it is only one of many factors, including the temperament that one is born with and the social relationships one experiences at school.[35] Similarly, psychologist Bruce Wampold's clinical research suggests that the value of therapy lies largely in the quality of the patient–therapist relationship, not in uncovering the truth about the patient's early influences.[36]

Our confabulating tendencies are not constrained to the present or the past. Cognitive psychologist Daniel Gilbert has revealed that we confabulate even when it comes to projecting our feelings into the future. We are quite sure that we know how our future selves will feel under certain circumstances, but typically we are way off in our expectations of the kinds of future emotions that we will experience, as well as their duration. Gilbert's work shows that we are highly unlikely to feel ecstatic for as long as we might predict after winning a lottery, or despondent for as long as we might expect after being disabled in an accident.[37]

Confabulating to Close Conceptual Gaps

It does not take a severed corpus callosum to make us vulnerable to imaginative storytelling. The term "choice blindness" is used to describe how we justify our choices, even if they are not our real choices. Cognitive psychologist Petter Johansson had subjects choose, from a group of photos, the people whom they felt were the most attractive. The selected photos were then swapped, unbeknownst to the subjects, and the swapped photos "returned" to them for discussion. When asked to explain why they chose the particular person in their hands, most not only were oblivious to the switching of the photos but also were able to concoct explanations for the choice that they thought they had made. The same experiment was performed with samples of jam. Subjects were asked to explain why they chose a particular kind of jam, even though the jams were swapped such that the subjects were explaining their preferences for jams that they had actually rejected. They confabulated explanations to bridge the conceptual gap between their perceived preference and their actual preference.[38]

Confabulating to Close Memory Gaps

We have a surprisingly strong proclivity for closing gaps in our memory of things. Just as we are selective in what we pay attention to, so we are selective in what we choose to remember, because there are bounds to our memory storage capacity. We can hold a limited amount of information in short-term memory (around seven items, give or take); beyond this, our only way of retaining information is to transfer it to long-term memory for storage. Because we cannot possibly store every moment-by-moment sensation and experience, we are forced to select what gets stored long term. We do not store whole scenes in long-term memory — just pieces of the events. Cognitive scientists have demonstrated that our memories are not perfect replications of past events that can be easily restored. We do not store things in our brains the way a computer or filing cabinet stores files for easy retrievability. Instead, we store *parts* of an experience and then reconstruct a narrative from these parts. When we extract experience from long-term memory, we are not *replaying* an event but *re-creating* it, using the cues that we have previously encoded in storage, to which we add a lot of information to make a coherent narrative. Our memories are subject to both storage and retrieval problems: we store cues that are influenced by our perspective at the time of storage, and we retrieve and reconstruct the memories in ways that are highly influenced by our perspective at the time of retrieval. Because remembering is based on retrieving bits and filling in gaps, our memories are extremely malleable and surprisingly unreliable.

Despite overwhelming evidence to the contrary, we persist in believing that we remember things pretty much exactly as they happened, and false memories feel as authentic to us as accurate ones, as will be revealed in the next chapter, under 5.1. This illusion is just another simplifying shortcut that circumvents our cognitive boundaries: a shortcut that, like all the others, serves us well much of the time but can be very misleading when we interact with complexity.

5

When Simplifying Meets Complexity
Greedy Reductionism

The world is satisfied with words. Few care to dive beneath the surface.
Blaise Pascal*

Reductionism is a fundamental principle of science. Scientists *reduce* phenomena to smaller components to understand how they work: an animal can be reduced to its distinct biological systems to understand how it functions as a whole. It is not just science that depends on reductionism, however; we are all naturally reductionist in the sense that we construct our understanding by simplifying — by selecting what we pay attention to, relating things with models and compressing reality into coherent stories. We reduce reality to the simplest possible terms to manage our cognitive inbox, which is constantly overflowing.

Reductionism, of course, can be taken too far, as this chapter explores. Consider the evolution of psychiatry over the past few decades: while not quite as tumultuous as that of nutrition, it does have a unique history of its own.

> Prior to Freud, psychiatry was the domain of neurologists who based their work on the assumption that mental illness was a nerve disease (hence "neurosis"), but Freud (who was also a neurologist) maintained that many mental illnesses were psychological rather than purely biological. He brought psychiatry out of hospitals and into outpatient offices where people who were not severely sick could be

---
* *Pensées*, Penguin, 2003.

treated with his new technique of psychoanalysis (the "talking cure," which was created by Freud's mentor Josef Breuer). The revolution that Freud unleashed was the notion that all people suffer a degree of mental illness and can therefore benefit from psychoanalysis to help them understand their neuroses.[39]

Sadly, a substantive portion of people with severe mental illness remain undiagnosed and untreated, many because they do not have the financial wherewithal to seek treatment. Ironically, at the same time, the field of psychiatry is prone to what psychiatrist Allen Frances calls "diagnostic inflation": an explosion of doctor-diagnosed mental disorders.[40] Meaningful advances have been made in the diagnosis and treatment of mental illness since the 1980 publication of DSM-III (the third edition of the *Diagnostic and Statistical Manual of Mental Disorders*, which classifies mental illnesses). But the number of people diagnosed with a mental illness has far exceeded what many medical experts believe is a realistic reflection of incidence within the population.

The problem is that, while the dividing line between severe mental illness and good mental health is clear, the distinction between the latter and mild-to-moderate conditions is very fuzzy. The line dividing normality from mental illness has never been fixed (which led philosopher Michel Foucault to argue in 1961 that definitions of mental illness are social constructions that reflect the desires and morality that prevail in a society at any particular time[41]). This fuzziness opens the door for excessive diagnoses: unlike, for example, cancer or diabetes, which can be detected with a fair degree of certainty, mental illness lies on a broad spectrum where the feelings and behaviours that are typical of everyday life can too easily be interpreted as symptoms of a psychopathology.

The fuzziness in assessing non-severe cases has been aggravated by the lowering of the threshold for disorders (e.g., attention deficit disorder), at the same time that new disorders have been defined (e.g., binge eating disorder). With the 2013 publication of DSM-5 (the fifth edition), psychiatry has, in the words of Frances, moved toward "the medicalization of ordinary life." By his estimate, the expanded definitions of mental illness mean that 50 percent of the population would now be considered to have some form of mental illness in their lives; a very small minority of these cases, according to Frances, should justify pharmacological intervention as opposed to non-drug psychotherapy or just a basic "wait and see" approach where the condition dissipates on its own. But as definitions of mental illness have expanded, doctors have prescribed medication for conditions previously considered normal, especially antianxiety and antidepressant medications, the vast majority of which are prescribed by family doctors. The result, in addition to all kinds of known and unknown side effects, is that more people die from prescription overdoses than from illegal street drugs.[42]

Many academics point the finger to pharmaceutical companies, which aggressively market new drugs to both doctors (through their extensive sales rep forces) and the public at large (through advertising). In 1997, the American Government's FDA opened the door to direct-to-consumer advertising by drug companies (ads that make direct product claims are still prohibited in Canada and Europe). And to capture an even larger customer base, the drug companies began to target mental illness in children by promoting awareness of attention deficit disorder and the drugs marketed to solve it. The companies' profit model is driven by convincing patients and doctors that everyday problems are psychiatric disorders, for which there happen to be myriad chemical solutions. So while a number of severely affected people are ignored by the system, the "worried well" are at risk of being overtreated.

The complexities of mental disturbance, both in the form of disorder (such as severe depression) and normal emotional volatility (such as coping with a lost job or loved one), have been reduced in many cases to simplistic labels and treatments, resulting in an epidemic of pill-popping. We are a long, long way from understanding and treating the various forms and degrees of mental illness, and, given the complexity of the human brain, probably farther away than we are from curing cancer. Many patients diagnosed with a non-severe mental illness are likely better off not ingesting massive amounts of chemical compounds. Philosopher Daniel Dennett coined the expression "greedy reductionism" to describe the not infrequent attempt by scientists and philosophers to reduce complex phenomena to overly simple explanations.[43]

5.1 Greedy Reductionism

Taking Dennett's notion one step further, it could be argued that all of us are greedy reductionists, all the time. We constantly reduce reality in order to squeeze some meaning out of it for ourselves, and we are greedy in wanting that meaning to be tidy and easy to intellectually digest. We are greedy in what we expect from the limited data that we select to pay attention to. We are greedy in what we expect from the basic interpretive models that we use to relate the data. And we are especially greedy in expecting the simple stories that we weave to capture reality accurately and comprehensively.

Greedy Selectivity

We are selectively greedy in how we grab anything that resembles a signal and ignore anything that resembles noise. Where complexity is concerned, the two are not so easily differentiated. As psychologist Philip Tetlock notes, "Picking up useful cues from noisy data requires identifying fragile associations between subtle combinations of antecedents and consequences." This is a skill that "probabilistic-cue learning indicates people do poorly."[44] His description captures the essence of the problem that we face with complexity: we are hard-pressed to make sense of the hidden, ambiguous and shifting causal connections that characterize it. Our facility with probabilistic thinking is rudimentary, further limiting our ability to manage these connections.

Greedy Modelling

Our simplifying models are reductively greedy when we overapply them by forcing them onto complex situations. It is not surprising that we overapply shortcuts in our struggle to make sense of complexity: not only are the signals and noise difficult to differentiate, but complexity rarely offers consistent, reliable, real-time feedback. Complexity can be very difficult to learn from: unlike the case of optical illusions where we can easily discover the error of our perceptions, conceptual illusions are more insidious because they are often not as obviously contradicted by real-world experience.

If you are driving on a hot, sunny day down a long stretch of highway, you will see puddles on the road that instantly vanish when you get closer to them. Hot air, warmed by hot asphalt, meets the cooler air above it; where the two air masses meet, the border acts like a lens, bending light such that, from a distance, it appears to be reflected by the asphalt, the way a puddle reflects light. Knowing the mechanics of this illusion does not matter; we learn early in our driving careers that these reflections are illusory.

But complexity does not offer the same quick feedback: we cannot just get closer to it to make it disappear. When we make mistakes in interpreting complexity, we are hard-pressed to understand what went wrong, because unbundling all of the relevant factors lies somewhere on the spectrum between difficult and impossible. Even our interactions with one another are a function of multiple causal factors (the particular moods of each person, their individual motivations and immediate desires, the history and character of the relationship, etc.) as well as the feedback loops that spin out of the various causal factors when they interconnect (the reaction to another's facial expression, tone of voice, choice of words, etc.). Reducing interpersonal conflict to "He is just self-centred," or "They are right-wing nutcases," or "She is overreactive" belies the complexity of the interacting factors that constitute any relationship.

Greedy Resemblance and Causality

We may chuckle at a four-year-old who insists that her stomach is upset because she drank water from a kitchen tap, "which is the same water that is in the toilet." But her error is simply a more obvious version of the resemblance and causal mistakes that adults constantly make.

Resemblance is a powerful shortcut, as long as knowledge of one thing can be legitimately applied to another. Behavioural economists use the term "representative bias" for the broad tendency to draw conclusions about one thing based on the larger group that represents it. It makes sense, for example, to avoid hooded men in dark alleys. But like all our shortcuts, when it is overextended, it becomes an unproductive bias. If two things or events are significantly different but have surface similarities, the temptation arises to be greedy by overextending the similarities. For example, Nobel laureate psychologist Daniel Kahneman uses the example of unreasonably concluding that someone with a lot of tattoos is unlikely to be successful in an academic career, scholars being a group of people who typically have no tattoos.[45]

Representativeness is often facilitated by "attribute substitution," the mental shortcut of assessing something that is difficult to decipher by substituting something easier and more accessible: skin colour for intelligence, height or attractiveness for credibility, today's emotional mood for overall quality of life assessment, or Apple's current innovations for the company's future success (a typical stock picker's mistake). Representativeness can take the form of assuming, for example, that prior foreign policy exploits contain important learnings about current ones, when the similarities are only superficial at best.

For example, pundits are often inclined to proclaim that certain international interventions, like the involvement of the United States in Afghanistan, demonstrate an example of "not having learned from the war in Vietnam." Many credible geopoliticians, however, are convinced that if the U.S. had not intervened in Vietnam, both Russian and Chinese Communism would have spread much farther into South and East Asia, and that Russia likely would have been emboldened to invade the Middle East and control its oil supplies. This is not to say that going to war in Vietnam was the definitively correct policy decision, nor does it undermine the argument that the war was poorly executed. It is to say that we can contemplate a reasonable counterfactual history in which the U.S. did not intervene, resulting in an escalated Cold War that may have led to a significant confrontation between the NATO partners and Communist regimes. We have to be very cautious about drawing definitive lessons from complex events, and even more cautious about assuming that they represent lessons that can be applied to other equally complex scenarios that are only superficially similar.

The overextension of representativeness is often expedited by an oversimplified version of cause-effect: a basic cause-effect explanation of one thing makes it tempting to assume that the same causal relationship explains something else. We are tempted to believe that we understand one scenario because we think its causal structure is represented by another scenario that we are familiar with. It is too easy for us to apply one-size-fits-all solutions to business strategies and interpersonal challenges, among other situations. "That idea won't work because our competitor tried it and failed" and "She needs to be stricter with her kids if she wants them to behave" are examples of a superficial assessment of causal intricacies. If we really understood the complex mix of causal relationships underlying a situation, we would be less inclined to assume resemblance between it and other complex phenomena.

Cause-effect relationships are much more complicated and nuanced than our intuitions suggest, and dealing with complexity requires a much more sophisticated notion of cause-effect than the one that works for all the straightforward problems we tackle. Chapters 13 and 14 will delve deeper into this problem of cause-effect; for now, it is worth noting that "because" is probably the most overused word in the English language. "Because" suggests that we have figured things out, when, much of the time, we have not.

Greedy Dichotomizing

Just as "because" can reduce an event to a deceptively simple causal explanation, so "either/or" enables us to effortlessly turn gray into black or white. Either the governing party is effective or it is incompetent. Either religion is beneficial or it is detrimental. Either a person is extroverted or introverted.

One of the more interesting examples of either/or simplification got its start in 1874 when scientist Francis Galton coined the dichotomy nature versus nurture and launched an investigation into which of the two has more influence on human personality.[46] Galton's work was largely based on studying twins. He was convinced that nature was the exclusive contributor to intelligence, talent and everything else about us, just as it is for eye colour. Today, we know that Galton was wrong on two counts: his emphasis on the exclusive role of genes is greatly oversimplified, and a simple dichotomy pitting nature against nurture is facile because of how they interact.

In the past two decades, rigorous twin studies have analysed the personalities of fraternal twins (similar environments, different genes) and identical twins who have been separated (different environments, same genes). Behavioural geneticists compare these two types of twins with non-twin siblings, identical twins raised together and people who are unrelated. The results converge to the conclusion that our personalities appear to be a mix of both factors. Genes are more probabilistic than deterministic.

Identical twins raised apart in different homes, for example, are much more similar in personality than any two strangers or two siblings would be, but their temperaments, interests and beliefs can be surprisingly dissimilar. Twin studies tend to attribute the variation in personality traits to genetic factors in the 40 to 60 percent range, depending on the particular trait. The difference between any two people is on average, therefore, around 50 percent genetic and 50 percent non-genetic. Interestingly, most of the environmental influence on personality is attributable to factors outside the home: childhood peer groups appear to have a much greater influence than parenting styles. But none of the research means that we can make any prediction whatsoever about a particular individual; we can only make generalizations about the differences between people within a sample.

Epigenetics: The Missing Link?

We know the issue is not either/or, but neither is it as simple as both-and. Genes and the environment interact through intricate feedback loops that include the influence of non-genetic factors that are the focus of the field of epigenetics (meaning "in addition to genes"). Epigenetic factors are attached to DNA; many genes are expressed only when triggered by the epigenetic factors that surround them. While we inherit fairly stable genes and our DNA reproduces quite accurately, epigenetic factors are more variable and more prone to random copying errors when reproduced. Epigenetic variability leads to variation in gene expression. For example, epigenetic variation explains some of the differences observed in disease susceptibility between two genetically identical twins.

The relative instability of epigenetic factors is interesting for another reason: it is likely the "missing link" in the nature–nurture story because it helps explain how genes and environment interact. The environment can have a significant influence on epigenetic factors, which in turn affect gene expression. So, via epigenetic factors, the outside world can influence the thousands of genes in a strand of human DNA that initiate the process of producing the proteins that are the building blocks for every part of us, including the neurotransmitters that influence our mood.

> To complicate things further, epigenetic modifications are, at least in part, heritable. If the environment influences your epigenetic factors such that they activate a particular gene that otherwise would be inactive, the gene modification may be passed on to your children, meaning that the environmental influence on your genes can be passed on in some circumstances. Genes are propensities that we are born with, the expression of which is influenced by the environment that affects them, largely via epigenetic mechanisms that trigger their activation.

Nature–nurture interactions are complicated further by the fact that the two can reinforce each other, making the difference between them more difficult to detect. If a person is born with a proclivity for being reserved and cautious, he is less likely to engage the interest and affection of other people and more likely to be picked on or bullied at school. These environmental factors will reinforce and perhaps exaggerate his temperament as he "learns" that the world is hostile.

The upshot? "Nature versus nurture" is a preposterously oversimplified either/or dichotomy that belies the complexity of personality development. Asking which has more influence in generating personality traits — genes or experience — is like asking which has more influence on a meal — the ingredients or the cooking process. Genes, like ingredients, are the starting point that limits the range of possible outcomes. While we can make some rough estimates of the influence of inputs, we cannot predict the final product from the ingredients or genes alone. Two bakers will create very different cakes with the same ingredients, just as history created very different results in Gandhi and Hitler, who had nothing in common despite sharing 99 percent of the same genes.

Greedy Language

Language promotes naïve realism, our default operating assumption, as discussed in 3.1. We assume that language is the conduit by which our thoughts connect us with "things in themselves": we forget how loose the relationship is between the two, and that language is often far too restrictive to do reality justice.

The philosopher Martin Heidegger insisted that language is *in* the world, not *of* it: language is so entwined with experience that we cannot use it to take us out of the world to capture something real about the world; language cannot transcend our thoughts and experience, because it is part of them.[47] Philosopher Ludwig Wittgenstein proposed in his later writing that language is a tool with many uses: words do not have fixed meanings but can vary with social and historical contexts because they derive their meaning from the sentences they are placed in.[48] Words like "good," "beautiful," "justice," "freedom," "capitalism," "reality," "truth" have no fundamental essence since they do not map onto the world in a consistent, unambiguous way. Words derive their meanings from the way people use them, not from a mystical power that ensures perfect representation of the way the world actually works. In fact, Wittgenstein argued that language as a whole is like a game: as a social activity, it has rules but also flexibility in how we use and interpret it. There is no immutable link between words and the world. For example, we can give an example of "freedom," but it does not have an identical meaning in each usage: "liberals" typically think of it as the right to the same base level of security and comfort, whereas "conservatives" typically think of it as the right to the same level of base opportunity to achieve those outcomes. Author Michael Shermer refers to the sloppy use of vague words like "free will" or "mind" as "linguistic placeholders substituting for our ignorance."[49]

Language makes complexity seem more straightforward than it really is, so language ends up, according to philosopher Friedrich Nietzsche, falsifying nature.[50] We end up, as the ancient philosopher Heraclitus declared, imposing order on fluidity. Language flattens out reality by tricking us into believing that it is always as simple as the way we describe it, forcing the world into tidy verbal packages. The routine vocabulary of most people is only about three hundred words; although we recognize many more than that, typically we compartmentalize our experience of reality within this finite number, giving us the sense that reality is just as straightforward as we describe it.

When we use words to disambiguate complex phenomena, we lose important meaning. Just as we cut up the rainbow into discrete colours, even though it is a spectrum with no dividing lines, we mistakenly think of animals as being separated by clear and distinct lines — hence nonsensical questions like "Where is the missing link?" The evolution from ape to human is a gradient with very fuzzy transitions. We cannot see this gradient, because so many of the intermediary species did not leave reliable fossils behind. To the uneducated observer, evolution appears to jump from ape to human with a few distinct stops along the way, and language reinforces this idea by giving discrete names to the intermediary ancestors whose remains we have discovered. But language fails to convey how gradual the transition actually was.

> ### The Ambiguous Second Amendment
>
> The ambiguity of language makes it vulnerable to misuse. The second amendment to the U.S. Constitution reads: "A well regulated Militia, being necessary to the security of a free State, the right of the people to keep and bear Arms, shall not be infringed." The amendment was composed and adopted near the end of the eighteenth century when a common fear was the tyranny of the federal government — an unsurprising concern in a newly formed nation in which each state had its own citizen militia that could, if necessary, stand up to a national army.
>
> This concern, and the wording that addresses it, does not easily translate into a twenty-first-century context. The introductory clause of the amendment has caused no end of controversy: Are the first thirteen words a condition or an example? Is the right of individuals to possess firearms conditional on their use by well-regulated militia protecting the rights of a State from tyranny? Or is militia activity just one of many examples that justify that right?
>
> Obviously, the National Rifle Association subscribes to the last interpretation. And Supreme Court decisions have supported this interpretation of individual rights in its rulings, writing that the first clause does not qualify the clause that follows it. But there are many Americans, perhaps a majority, who interpret the right to bear arms in much narrower, more restrictive terms.

Greedy Storytelling

We are compulsive meaning-makers because that is precisely the job that the brain was designed to do. On the one hand, we are parsimonious with the amount of concentration that we are willing to expend on revisiting problems we think we have figured out, no matter how superficial our analysis is. On the other hand, our minds are always on, looking for information to process and problems to solve (and even creating problems to solve). This is one of the many paradoxes of being human, a topic that will resurface in chapter 19. As meaning-making machines, our brains churn away on a countless number of problems that have not been solved to our satisfaction. Nature abhors a vacuum, and a mind that is not churning is a vacuum that thoughts rush to fill. Our relentless rumination is a product of involuntary mental systems over which we have limited control; it takes years of training and practice in meditation to stop the conscious mind from trying to figure things out, an achievement that very few of us accomplish.

This constant churning means that, with very few exceptions, our brains refuse to be stumped. We confabulate to fill in gaps, in order to construct coherent narratives. We do whatever it takes to satisfy our left-hemisphere interpreter. "Apophenia" is the finding of patterns in meaningless data: the image of Mary in a slice of toast, a lucky streak in a card game, a bad-luck streak of three mishaps in a row, or a premonition that happens to come true. Taken to the extreme, apophenia is a symptom of schizophrenia in which patients perceive random noises and sightings as evidence that they are being followed and persecuted. But compulsive pattern seeking and consequential storytelling are not limited to those afflicted with mental illness. Our tolerance for ambiguity — incomplete stories or unsolved mysteries — is very low, which is why we are attracted to and create simple narratives. The very structure of politics and media, for example, leads to the production of simple bite-sized stories. Politicians cannot get elected by opening campaigns with, "It's complicated." The media cannot attract viewership and sell advertising with long headlines that capture multiple perspectives on an issue, as opposed to attention grabbers like, "President Fumbles Ball" or "Carbs Are Your Enemy."

Our proficiency in filling conceptual holes and banishing ambiguity comes to the fore in our memory-based stories. The gaps in our sketchy memory are no challenge for our well-developed ability to make the past appear valid and vivid by confabulating it.

Psychologist Elizabeth Loftus and others have demonstrated, in a variety of experiments, that we can be made to remember things that never happened, and that we can be very confident of the authenticity of these false memories.[51] Loftus has revealed how unreliable eyewitness testimony can be in legal cases. Well-known examples include false memory syndrome, by which some adults are convinced that they have been abused in their childhood when evidence overwhelmingly proves them wrong, and false confession, by which suspected criminals, subjected to long and aggressive interrogation, start to believe that they have committed the crimes that DNA testing later confirms they did not. The Reid technique was designed to include a harsh form of interrogation, largely based on the assumption that innocent people would never confess to a crime that they had not committed, no matter how sleep deprived and harassed they were. We now know this to be a faulty assumption, because people's memories can be manipulated with active intervention. The complicating factor of harsh interrogation is that innocent people tend not to ask for lawyers to be present for their interrogation; they naturally think that they have nothing to hide, which makes them particularly vulnerable to the power of suggestion.

We also are expert at compartmentalizing the past into discrete events, by ignoring their connectivity. Just as we cannot witness evolution occurring right in front of us and are therefore prone to viewing ourselves as completely separate from non-human animals, so are we unaware of the numerous twists and turns religions have taken over periods of time. We tend to view religions as fundamentally distinct and unrelated, even though many of today's religions have evolved over thousands of years from common roots. The evolution of Christianity from Judaism is one example.

Early Christians worshipped in both Jewish synagogues and newly constructed churches well into the third century. The term "Christian" to denote followers of "Jesus the Christ" does not appear in writing until the beginning of the second century, nearly one hundred years after Jesus' death. Christianity had no standard form at that point: it was not until the fourth century, about three hundred years after Jesus' death, that the Roman Emperor Constantine convened Christian bishops to establish a common creed that confirmed agreement on a number of Christian doctrines, including moving the Sabbath to Sunday instead of the traditional Jewish Sabbath of Friday night and Saturday; separating Easter from Jewish Passover celebrations; and confirming that Jesus was the Son of God but not subordinate to Him (many bishops disagreed with this "homoousian" doctrine, until Constantine threatened to excommunicate them). We think of Judaism and Christianity as being fundamentally different, yet history reveals that Christianity could have evolved into a distinct sect of Judaism, or, as many scholars have speculated, could have disappeared if Constantine had not embraced it.

> ### The Evolution of a Messiah
>
> The patriarchs of Judaism — Abraham, his son and his grandson — were likely not monotheists, since Judaism arose from a polytheistic region where people believed in many deities, including Yahweh (which is why the first commandment is to worship only Yahweh). Jesus was a Jew: his name was translated from the Hebrew "Yeshua" by the authors of the Gospels, and by Paul, who introduced the term "Christos," which was a Greek translation of the Hebrew "Messiah," meaning "The Anointed One." Jesus was a nomadic Jewish teacher whose ambition, many scholars believe, was not to start a new religion but to refine some of the practices of Judaism.
>
> The gospels were written forty to seventy years after Jesus' crucifixion by people who never met him but relied on word-of-mouth stories, which they shaped in ways they thought appropriate, in most cases to match the prophecies of the Old Testament. Paul's epistles were written twenty to thirty years after Jesus' death. Paul preached that Jesus' followers did not need to follow the Jewish laws: his intention was not specifically to replace Judaism but to spread the teachings of Jesus to Jews and gentiles. Scholars like Karen Armstrong and Donald Harmon Akenson argue that neither Paul nor the first three gospel writers would have considered Jesus to be the literal son of God: "Son of God" had a traditional Jewish and Roman usage to indicate closeness to God, like the ancient kings of Israel and the Roman emperors.[52] Only gradually did disagreement arise about Jesus' status as the Messiah and as the incarnation of the Word of God, which is how John, writing later than the other gospel writers, described him. Only gradually did Christianity form as a religion distinct from Judaism. Historian Carlos M. N. Eire speculates that if Jesus had died a natural death rather than being crucified, he likely would have been perceived as a prophet and not granted the status of divine saviour.[53]

We weave stories as instinctually as spiders weave webs, and many of our stories are just as precarious as their dainty constructions. When simplifying meets complexity, the result is oversimplified interpretations. But greedy reductionism is only the first of two problems: not only do we construct by simplifying, but we typically are oblivious to the construction process we are engaged in. So greedy reductionism is compounded by our unawareness of what we are doing, which means that we do not have the opportunity to change our approach when it is not working. Worse, we often do not even realize that it is not working, just that we are not getting the outcomes we want or expect.

We operate efficiently and effectively in a certain kind of world with certain kinds of problems. In this world, there is no significant benefit to being aware of the construction process, because it works most of the time, and second-guessing it would just slow us down and waste energy. In a world where complexity features more prominently, our lack of awareness compounds the problem of reductionism: rushing to conclude with confidence is the last thing we should do if our starting point is based on oversimplified interpretations.

6

To Know Is to Satisfice
Rushing to Certainty

It's not that I'm so smart. It's just that I stay with problems longer.
Albert Einstein*

The first half of the brain–world gap story is that we construct by simplifying, and we simplify by invoking shortcuts to interpret. We reduce the world to pieces that our bounded brains can handle. Much of the time simplifying works for us, but increasingly it does not.

The other half of the story — and the focus of this chapter — is that we are too confident in the accuracy of our simplified interpretations, too ready to feel that we have figured things out. Much of the time this confidence is warranted, but increasingly it is not. When our certainty is premature, it precludes us from deeper exploration, the kind that complexity demands from a bounded brain. It is crucial to understand the mechanisms that give rise to cognitive confidence; only then can we counter them when they are unproductive.

Confidence arises when we feel that we do not need to expend further energy on figuring things out; mental effort is, after all, something that we expend very selectively. Faced with the choice between an escalator and a staircase that runs parallel to it, the vast majority of us opt for the lazy solution. The same goes for sitting rather than standing on a subway or bus and watching TV rather than reading a book. When we handwrite, most of us opt for barely legible scribbling instead of writing as neatly as possible. All of these examples reveal a

---
* *Bite-Size Einstein*, edited by J. Mayer and J. Holmes, St. Martin's Press, 1966.

general truth about biological creatures: we conserve energy, because we do not have an unlimited supply of it. The same applies to how we think. Our conscious thinking takes effort. We are willing to expend the effort to concentrate, but we are selective and need a strong incentive. Why take the stairs when we can take an escalator? Why concentrate when we can just rely on our automatic intuitions?

The cognitive equivalent of taking an escalator is to "satisfice," a term coined by Herbert Simon that combines "satisfactory" and "sufficient."[54] We satisfice when we draw a conclusion that is both satisfactory and sufficient to explain what we are experiencing or make a decision on that basis. Satisficing is the mental algorithm that we rely on to draw conclusions, in the fastest and most reliable way possible within the constraints of human cognition, given the need for speed and accuracy. The algorithm of satisficing is *to accept the first satisfactory explanation*.

What constitutes a satisfactory explanation? Our test is coherence, the degree to which our explanations fit with everything else we know. When this test is met, we accept an explanation with confidence, the feeling of knowing that closes down the search for alternative interpretations. Just as we take shortcuts to select, model and storytell, so we take shortcuts to conclude — to achieve coherence and confidence in our beliefs. Just as our interpreting shortcuts can lead us astray, so too can our concluding shortcuts.

When it comes to straightforward problems, the first reasonable interpretation is often the best, or at least good enough. When it comes to complexity, the first is rarely the best, and usually not good enough. It is far too tempting for us to gloss over the gaps in our thinking, however, because there are a lot of pressures on us to conclude quickly. First, evolution installed in us the survival-enhancing habit of being decisive. Second, our lack of expertise in solving complex problems is not obvious to us. Whereas an amateur tennis player is acutely aware of her weaknesses, the errors we are prone to when handling complexity are less visible and more readily rationalized. Third, we are disinclined to spend unnecessary cognitive energy on problems that we think we have solved, just as we are disinclined to spend unnecessary energy in all aspects of our lives where the benefit is not clear (sometimes referred to as the "principle of least effort"). Fourth, and most prominently, we have a deeply ingrained need to feel that we know: *we are addicted to certainty.* The last two motivations for rushing to conclusions are worth exploring in more detail.

6.1 We Conserve Cognitive Energy

"Cognitive miserliness" is sometimes used by psychologists to refer to how stingy we can be with the energy that we are willing to expend on figuring things out. We are especially stingy when it comes to conscious thinking that takes concentration: we are selective about what we deliberate on, and once we have arrived at a solution that feels right, there is insufficient motivation for us to expend further conscious energy. Subconscious thinking does not strain us nearly as much because it is automatic, like a heartbeat. We know this because scientists can observe the energy demands on our brains of various cognitive challenges, because the more active our neurons are, the more oxygen they need; oxygen is brought to them via the bloodstream, which can be tracked by magnetic resonance imaging (blood cells contain hemoglobin, which is magnetically resonant because of its iron component). Conscious thinking takes more energy to execute: it is evolutionarily new and does not run as efficiently.

Numerous experiments show our cognitive miserliness in different ways. One of the most revealing, A Tale of Two Pizzas, was conceived by four psychologists.[55] Subjects were divided randomly into two groups; each participant was asked to pick the toppings that he would like on his pizza, with the choice of twelve possibilities that cost fifty cents each. The difference between the two groups was simple: the first group started with only a crust and cheese and had to build their individual pizzas from that base (paying fifty cents for each topping); the second group started with a fully loaded pizza with all the toppings and had to remove the toppings that they did not want (deducting fifty cents for each removal).

People who made their pizza by adding ingredients ended up with fewer toppings than those who subtracted ingredients: of people in the first group, 70 percent chose between one and three toppings, whereas 64 percent in the second group chose between four and eleven. Psychologists explain this asymmetry in decision making on the basis that adding options to something takes a lot more cognitive effort than subtracting from something. This experiment has been repeated with options to choose for a car, computer, treadmill and stock portfolio: the results are always the same.

Behavioural economists have a field day with this peculiar habit: they point out that we tend to resist parting with something we already have (referred to as the "endowment effect"), because we feel that we are losing something ("loss aversion"); we also tend to wed ourselves to our starting positions ("anchoring"). These psychological quirks are hurdles that require conscious effort to overcome, which is why it requires more cognitive exertion to remove than it does to add. So we end up taking the easier cognitive path of acquiescing (and paying for) a lot of extra pizza ingredients.

Whereas the stairs and escalator both take us to the same destination, the easy cognitive path often takes us to the wrong place. In a different form of experiment, psychologist Shane Frederick illustrated cognitive miserliness by exposing how many people fail to check their thinking when given a simple math problem.[56] A bat and ball together cost $1.10; on its own, the bat costs $1.00 *more* than the ball. How much does the ball cost? Most people quickly land on ten cents as the answer. But if the ball costs ten cents and the bat costs $1.00 *more* than the ball, then the ball and bat together add up to $1.20 (ten cents plus $1.10), which is ten cents too high. The right answer to the question is five cents: if the ball costs five cents, and the bat is $1.00 more than the ball, then five cents plus $1.05 adds up to the $1.10 total.

What is interesting about this question is not that most people get it wrong, but that the people who get it wrong rarely check their math to see if it adds up to the total $1.10. People who respond quickly (and incorrectly) tend to think that the question is easier than those who take their time and answer correctly. Why do people who get this wrong not take the five seconds necessary to double-check their work? For the same reason that the vast majority of people choose the escalator: the alternative is too much effort.

Energy conservation is a good general operating strategy to cope with limited resources, both physical and mental. But its usefulness depends on an ability to recognize the situations that require extra effort. The problem of cognitive miserliness is that we are not attuned to the modern-day scenarios that autopilot thinking does not work for: complexity is often not as obvious a mental trigger for increased attention and probing. We skimp in these scenarios because we assume that our intuitions are reliable and there is therefore no need to ramp up our cognitive effort. Cognitive miserliness means that once we arrive at a reasonable explanation that satisfies us, we consider our work to be done.

In the words of cognitive psychologist Keith Stanovich, we are "strongly disposed to deal only with the most easily constructed cognitive model."[57] Cognitive miserliness translates into a strong disposition to default to the simplest model we can construct to explain what we perceive, and then leave it at that — full stop, without any motivation to revisit our conclusions unless there is clear and sufficient evidence that our conclusions are patently false. The challenge with complexity is that the contradictory evidence is rarely clear, so we are inclined to stubbornly stick to our early conclusions and enjoy the reassuring calm that certainty affords us.

Firm conclusions are enticing because we have a deeply embedded physiological motivation to avoid stress — in this case, the stress of not knowing. Make no mistake: we do suffer stress when we do not understand. Which is precisely why …

6.2 We Need to Know

Just as profound as the two insights that we construct our version of reality and that we do this by simplifying, is the insight that *we need to know*. We have an extremely low tolerance for ambiguity.

Most of our initial conclusions about things are absorbed into the network of our belief systems by a force that is as strong and relentless as the pull of gravity: it is a kind of mental gravity, drawing in conclusions with one of nature's other great forces, the human need to know.

We are as oblivious to how we construct our reality as we are to the depth of our addiction to certainty. Our physiology is geared to knowing: to figuring things out, to concluding, to deciding. Our brains are always spinning through data to draw conclusions about our experiences, and when we do not understand something that seems relevant to us, our brains do not give up. They try even harder to make sense of things, because when we do not know what is going on, we feel out of control, and if there is one thing that animals, especially humans, do not like, it is the stress of losing control.

The Need to Control

When you jump in your car and start the engine to drive away, how often do you think about getting into a bad accident? Probably rarely, if ever. What about being in the passenger seat of someone else's car? It is more likely that you will be nervous, and the thought of an accident will float through your mind, depending on your opinion of the driver's skill and style. How about boarding an airplane? Most people have at least a fleeting thought about a plane crash. But the odds of dying in a car crash are about seven hundred times higher than dying in a plane crash, and, for the most part, it does not matter who is driving the car. We feel a high degree of control when we are driving, a lower degree when someone else is, and a lower degree still when we are in a plane and cannot see the pilot.

Psychologists have repeatedly demonstrated just how uneasy we feel when we are not in control of things. In examining the leading causes of stress, neuroscientist Robert Sapolsky notes that "a lack of predictability and control are at the top of the list."[58] When we do not have control, we cannot make predictions, and we cannot respond to the environment in ways that protect us. As Sapolsky points out, all it takes to elicit a stress response from a primate is to move it to a new, unfamiliar cage. When we are confronted with something that we deem important but unfamiliar, our sympathetic nervous system jumps into gear, secreting the stress hormones (cortisol and adrenalin) that activate our alertness responses, which in turn put us on edge until we feel that we have regained control and can relax. Whether it is our car skidding on a slippery road or our first day in a new job, a loss of control creates stress, from mild anxiety to outright panic.

The control that we feel we have in our lives — our ability to influence the factors that affect us — is more highly correlated to a sense of happiness than most other factors, including how much money we make. This is not surprising, given the idea of "learned helplessness" in animals, a concept conceived through experiments in the late 1960s. If animals are made to feel that they have no control over their environment, they exhibit many of the symptoms that we associate with human depression. When we feel that we have little control, we are tense, anxious, on edge; we generally feel off balance. Over longer periods of time, lack of control can evolve into feelings of depression. Some evolutionary psychologists have suggested that the foundation of religious beliefs is the human desire to feel in control (through prayer, or karma and rebirth), rather than surrender to the notion of randomness and meaninglessness in the universe.

How does control relate to knowing? A pervasive biological mechanism is at work: we are biological creatures constructed to respond to our environment in ways that maximize the likelihood of our survival. Our biological system is governed by a "homeostatic impulse": when our body or mind detects an imbalance, it takes action to get back into balance. We are designed to feel uncomfortable when something does not make sense to us, because discomfort motivates us to figure things out, at which point the tension of not knowing is replaced with the calm of knowing, putting our mental state back in balance. Just as calm is restored in us when our car tires regain traction in a skid and we resume steering, so it is restored when our minds get traction on aspects of the world that require understanding and prediction. As mathematician David Orrell notes, "The ability to predict is entwined with the ability to control."[59]

The calm of knowing can be superseded by other emotions that kick in to respond to perceived danger, like the terror of figuring out that a rustle in the bush is caused by a dangerous animal. But it is the tension of not knowing that motivates us to figure things out in the first place, whether it be the mysterious rustle, the confusing betrayal of a friend, the promotion that we did not get or any of the other minor and major conundrums that regularly confront us. The tension of not knowing constantly pulls us to draw conclusions about things, and it never stops: psychologists have proposed that even when we sleep, our minds rehearse scenarios in the form of dreams to make sense of things and prepare us for challenges that await us in our waking moments.

The "illusion of control" is a psychological tendency to overestimate how much control we have in various situations by underestimating the role of luck and all the unknown factors that influence the outcome of an event. All of us are vulnerable to this illusion, even the least superstitious among us, because it is one of many ways that we reduce the anxiety of not knowing.

The Feeling of Knowing

Knowing is a feeling state. This is obvious when we say things like, "I just know — I feel it in my gut." But it is less obvious that knowing is always a feeling and is as automatic as any other feeling. We cut our finger, we feel pain; we lose a loved one, we feel sad; we figure something out, we feel "knowing."

Neuroscientist Robert Burton suggests that "the feelings of knowing ... aren't deliberate" but are "mental sensations that *happen* to us."[60] When we "know" something, we experience the feeling of being right, which is why the feeling of knowing is also referred to by some psychologists as "the feeling of rightness." It is a calm feeling that is pleasurable both for its own sake and as the replacement for the anxiety of not knowing.

Knowing is one of the most powerful feelings that we experience. No other feeling is as relentless; we constantly pursue it as we make our way through the myriad mental obstacles that we negotiate every day. Only a handful of other emotions can match its intensity (jealousy, grief and fury, for example). The feeling state of knowing is so powerful that it can easily overwhelm logic, which is why, as reviewed in 4.3, confabulation is such a prominent feature of our cognitive lives. Our entire neuroendocrine system (combined nervous and hormone systems) is, among other things, geared to figuring things out, to focusing our attention and energy on making sense of the world. For us, the ultimate form of cognitive dissonance is being confronted by something that we intuitively feel is important to understand but that we cannot make sense of. Figuring things out relieves the tension of that dissonance.

Insofar as we are hardwired to avoid unpleasant feelings like hunger and pain, we seek remedies to assuage these feelings. Insofar as we are hardwired to avoid the unpleasant feeling of tension when we are confronted by something we do not understand and therefore have no control over, we crave the feeling of knowing.

The Addiction to Certainty

We want answers; we *need* answers. We are biologically programmed to find answers, because answers are the only antidote to the tension of not knowing. We yearn for certainty because our physiology motivates us to eliminate the tension of not understanding. We cannot rest until we have answers (quite literally: we suffer insomnia when we go to bed with important unsolved problems). We experience this yearning so intensely that *feeling right is more important to us than being right.*

This is why Christians will not be convinced that Muhammad was a prophet, why Jews will not be convinced that Jesus is the son of God, why left-leaning politicians will not be persuaded by right-leaning arguments, why career academics are hard-pressed to reject their original hypotheses, why when we think that we have been wronged we cannot be persuaded otherwise, why wars are fought. Once we *feel* that we have figured things out, it is very difficult for us to surrender that calm for the discomfort of not being sure.

The flipside of our biological motivation to seek certainty is that we abhor ambiguity. Because we are programmed to figure things out quickly to ease the discomfort of not knowing, our cognitive operating system does not tolerate ambiguity very well. For example, when we rely on our watch for the time, we tend not to second-guess it: the inherent ambiguity of whether it is accurate or not is not obvious, so it goes unnoticed. But if the time on our watch does not match the time displayed on our car clock, the ambiguity induces some stress. It is very easy to avoid ambiguity if it is not as obvious as two inconsistent clocks, and much of the ambiguity of complexity is not obvious because we are not usually on the lookout for it. But we do much more than just ignore ambiguity. When we confront it, we convert it: we take difficult problems, including those that are poorly understood and ill conceived, and invent simple solutions to satisfy our need for the control and calm provided by certainty.

Ambiguity is uncomfortably open-ended. Our need to "close the loop" by figuring things out is reflected in our inborn instinct to complete tasks, known as the "Zeigarnik effect," which reflects the discomfort that we feel when we begin something and leave it unfinished. (This, incidentally, is why one strategy to cope with procrastination is to start the undesirable task and let the cognitive dissonance that arises from leaving it unfinished provide some motivation to keep working on it.) So resistant are we to ambiguity that we will increase our appetite for risk if it gives us an opportunity to avoid uncertainty. This is the "Ellsberg paradox," a phenomenon exhibited in experiments in which subjects choose a risky but specific option over an unknown one. We are not designed to tolerate ambiguity, because we are designed for action and action requires decisiveness.

6.3 When Satisfying and Sufficient Are Good Enough

And so it is that our inborn cognitive miserliness combines with our inborn need to know, which together push us to draw conclusions as quickly as possible. To interpret quickly, we simplify. To conclude quickly, we satisfice. Simplifying our experience to generate possible interpretations, and then accepting the first one that makes sense, is a powerful decision-making strategy.

Naturally, the whole process is largely subconscious, occurs extremely rapidly and ends when we conceive of an explanation that is satisfactory and sufficient, at which point we lock down this interpretation as our conclusion. Once we are satisfied, the search for explanation stops. Psychologist Arie Kruglanski describes the process as "seize and freeze": we seize any evidence that helps us get closure, and we freeze this evidence to the exclusion of any other information.[61] While simplifying is fundamental to cognition because it enables us to reduce the world into digestible bits, satisficing is just as crucial, because it ends the process of generating possible interpretations by locking down a construction of reality that we have confidence in.

Satisficing entails a trade-off between speed and accuracy: accuracy deteriorates as the speed of concluding increases. The faster we draw a conclusion, the less time there is to test it against alternative interpretations, so a degree of accuracy is sacrificed. Satisficing does not necessarily accept the first explanation that comes to mind, since the first one may not satisfy us. But satisficing also does not "optimize" or "maximize," because finding the *best* possible interpretation does not serve the need for speed: that would require a lot of processing time to validate against all other possibilities. Satisficing goes with the first "good enough" explanation, so that reasonable accuracy can be assured in the shortest time.

Our ancestors' first satisfactory explanations were reliable the vast majority of the time, which is why natural selection preserved satisficing as our primary operating principle. Satisficing was a brilliant strategy for survival in a simple but threatening environment: simple visual cues are reliable enough to facilitate quick, satisficed decisions about the straightforward challenges that have historically confronted us. Even today, satisficing works for most of our daily decisions: choosing what we eat, which clothes to put on and whether we can trust neighbours and colleagues. Even for many of our bigger decisions, which are not based exclusively on visual cues, we rely on satisficing, because we have no choice. We cannot interview every potential mate on the planet before picking one; we cannot try every possible job; we cannot test-drive every possible car; we cannot tour every house that is for sale. We go with the best solution we can find after a limited search, at least in the immediate term (before switching lovers, changing jobs, trading cars and moving houses).

Satisficing, geared as it is to speed, does not suit the growing number of important challenges that are too complex to be reliably solved by the first satisfactory explanations for this reason: our lack of expertise with complex problems is not obvious to us, so in our rush to certainty, we can feel that we have got complexity figured out, even when we do not. Our initial interpretations of complexity are oversimplified, but that does not prevent them from generating a sense of knowing, such that we are not inclined to expend the energy searching for counterexamples and alternative explanations.

What is it, exactly, that triggers the feeling of knowing and encourages us to stop the search for deeper understanding? As revealed in 4.3, coherence is the glue that holds our stories together: coherence is the trigger for feeling that we have things figured out.

6.4 Coherence: When It All Makes Sense

We build an expansive net of beliefs about the world that begins as early as in utero, when we start to make some rudimentary connections, like the consistent and reliable sound of our mother's voice. This intricate network of beliefs extends from the banal — things fall down when we drop them — to the sublime — helping others gives us great satisfaction. To be accepted into our network as "true," a particular interpretation has to cohere with the network, unless a section of the network is reworked to accommodate it (as might happen if an atheist finds God or a theist loses faith). The extent to which an interpretation coheres with the rest of what we know determines how easily and quickly it will be accepted. When it is accepted, the tension that accompanies not knowing is replaced with the calm satisfaction of having sorted things out.

We believe things that fit quickly and easily with what we already know. This is not surprising since coherence is precisely how we learn and expand our understanding of the world. What is surprising is the ferocity with which we both welcome beliefs that fit and reject beliefs that do not. It is not just that we test possible interpretations for the degree to which they cohere with existing knowledge; it is also that we unthinkingly and uncritically accept ideas that cohere. As we rush toward certainty, the first explanation that coheres with our web of previously endorsed beliefs invokes the feeling of knowing, generating cognitive confidence.

The urgency with which we seek certainty has an enormous influence on how we test our ideas for coherence. The test is not purely objective; it is subject to meaningful biases that give an advantage to some interpretations over others. These biases help us support the "satisfactory" in satisficing; they usher certain ideas to the front of the deliberating line, enabling them to penetrate the web of our belief systems and take a position within it. Ideas are advantaged if they:

- are based on information that is immediately available to us ("availability bias"),
- can easily be supported with confirming evidence ("confirmation bias"),
- fit with everything else we already believe ("myside bias").

Each of these biases is extremely helpful, because each facilitates quick conclusions by expediting the test for coherence. But because we employ them so automatically, so unthinkingly, they are vulnerable to overuse. The three biases operate together, virtually simultaneously, in the vast, automatic processing of our subconscious thinking. As we contemplate possible ways of interpreting data, we rely on what we already know to determine coherence, putting more emphasis on evidence that supports our fledgling ideas than evidence that contradicts them, and limiting our search for evidence to easily available information.

Availability Bias

"Found art" is a form of art that repurposes everyday objects to create a new visual or sculptural piece. Availability bias is the found art of thinking. In our effort to make sense of things as quickly as possible, we gravitate to using whatever information is easily at hand, which makes availability bias the backbone of satisficing. This may seem obvious: how else can we draw conclusions if not from information that is available to us? This bias is subtler, though, because there is always more information available than we can possibly synthesize, so we gravitate to the information that is *immediately* available. Cognitive miserliness limits our attention to what is right in front of us, rather than searching beyond this narrow spotlight for information that is not immediately present. The absent information is sometimes crucial.

A popular way of demonstrating availability bias, similar to the question posed in 2.4, is to ask whether there are more words that start with "r" or have "r" as a third letter. Most people incorrectly assume that there are more words that start with the letter "r," since it is much easier to recall words that start with "r" even though, in fact, there are more words with "r" as the third letter (car, far, porous, our, more, etc.). Similarly, a current news item about global warming will have more impact on us than a book on the same topic that we read a year ago, even if the book was more thoroughly researched.

This bias is not restricted to our limited capacity for remembering things. When we sort things out, our assessments are dominated by the facts and beliefs that immediately come to mind. "Presentism" is the name given to our tendency to overweight our immediate thoughts and feelings, even as we reflect on the past. Presentism encourages us, for example, to judge slave owners by modern-day standards instead of by the mores of their time. One source of easily available information is the ideas offered to us by perceived sources of authority, as when we assume that if something is reported in the news, it must be accurate. "I heard on TV that such-and-such a food is linked to cancer." Why be skeptical if it takes less cognitive energy to just accept what is reported? This kind of thinking disregards how fallible all sources of information can be, especially media on tight schedules to turn out interesting stories whose depth is limited to a two-minute narrative or a few columns in a newspaper. Unfortunately, complexity does not easily conform to the evening news.

Availability bias takes the form of confirmation bias when the evidence we examine is limited to the immediately available information that supports our early hypotheses.

Confirmation Bias

This bias was noted and named by psychologist Peter Wason in his simple but revealing "2-4-6 problem."[62] Subjects are asked to guess what underlying rule generates the three-number sequence; they are also asked to propose other three-number sequences that matched the 2-4-6 sequence. In this experiment, people rarely, if ever, propose a sequence that contradicts their hypothesis; they attempt to confirm their hypothesis rather than disprove it. Guessing 6-8-10, for example, confirms the belief that the rule is "increase by two," whereas guessing 1-2-3 would disconfirm that hypothesis. The rule, in his experiment, is nothing more complicated than that the subsequent number must be higher than its predecessor, but people rarely get to this rule because they do not attempt guesses that contradict it, like 2-4-3. By relying on examples that confirm their hypotheses, they miss out on the crucial information revealed by disconfirming examples.

We work much harder at confirming than we do at disconfirming, because confirmation relies on easily available information, whereas disconfirmation often requires a search for information that is not readily apparent. In our rush to get to a conclusion, seeking to confirm our first impressions is much faster than seeking to disconfirm and having to start from scratch. Confirmation bias is based on a "positive test strategy," because it restricts the search to evidence that supports the hypothesis. Like a seasoned prosecutor, we focus attention exclusively on the facts that support our case.

A proclivity for confirming makes perfect evolutionary sense, stemming as it does from the convenience that a straightforward environment provides. After our minds put together a story from readily available information, the only thing left to do before locking down a conclusion is to confirm the initial intuition with whatever evidence we can quickly access. We lose our keys, and our first thought is that our spouse borrowed them; we ask her when she last had them, and if she cannot remember, that is all the evidence we need to be convinced that she misplaced them, even though, as we later discover, we left them in our own coat pocket.

Confirming is easier and faster than disconfirming because of what is referred to as "prior acceptance." We generally accept propositions before we reject them, as a means of understanding: before determining the truth of a proposition, we accept it, try to understand it and only then work to validate or disconfirm it. The idea of something precedes the idea of its negation. This concept was first introduced by philosopher Baruch Spinoza in the seventeenth century.[63] He proposed that understanding something requires us to have a mental vision of it — an initial assumption that it is true — since it is impossible to reject something that we have not first contemplated in its positive form. We accept ideas before we reject them because comprehension requires initial acceptance.

This view of how understanding works has been tested in many ways. For example, children do not develop a strong ability to challenge ideas until later in their cognitive development, and a positive statement is much more intuitive for them than a negative one. If children are instructed to guess a number between one and one hundred by asking questions, and they then ask, "Is the number greater than fifty?" and the response is, yes, they will be excited; if the response is no, they will be disappointed. Even though the answer in both cases is equally informative, there is a bias toward acceptance in their reaction. Similarly, adults evaluate true statements much more quickly than false ones. Statements that are unequivocally true ("The sky is often blue") are processed much faster than statements that are patently false ("The sky is never blue").

Our own emotions can play a big role in confirmation bias because they are readily available cues that can help reinforce our conclusions.

If we conclude that someone has wronged us, our own anger supports this conclusion. Even if we have completely misinterpreted an innocent gesture as a slight, we tend to resist alternative explanations, because the anger itself builds momentum to support the initial reaction. The feelings triggered from our initial interpretations reinforce those interpretations in a feedback loop generated by confirmation bias: we "know" that we have been slighted; otherwise, we would not feel angry!

Confirmation bias works on hypotheses that are "warm," that are established early and are ready for quick validation. In our hurry to conclude, we grab whatever information we can access as quickly as possible to confirm our hunches, and the most readily available information, besides our immediate sensory experience, is our pre-existing beliefs.

Myside Bias

The faster we can match our present experience with what we already know, the faster we can reduce the tension of not knowing and respond. It is not surprising that we choose to believe things that fit with everything else we know: we carry around our beliefs, values and preferences all the time, so they are readily accessible. What is surprising is the extent to which we resist anything that does not fit quickly and easily with our mental belongings.

"Myside bias" was coined by David Perkins to describe how our existing beliefs influence our interpretations.[64] Perkins asked subjects to record their opinions about various social issues and then to record arguments both in favour of their opinion ("myside arguments") and those that challenged their opinion ("otherside arguments"). Unsurprisingly, people recorded far more myside arguments than otherside ones. We favour evidence and arguments that support our initial position, or at least our initial leaning. Myside skews our view of the world by constantly building and reinforcing the belief systems we have already adopted. Notes Daniel Kahneman, "In the competition with the inside view, the outside view doesn't stand a chance."[65]

Myside and confirmation biases are features of what is referred to as "motivated reasoning": the path that our thinking takes is geared in a certain direction, based on what we already believe and our proclivity for confirmation. Some social psychologists go even further, claiming that the central, evolved purpose of human reasoning is to be argumentative — to persuade others of our point of view. They propose that the evolution of reasoning and language was in service of convincing and manipulating others. From this perspective, we are more like lawyers arguing for one side of a case than like judges gathering all of the relevant information and making a dispassionate assessment.[66]

Just as the flipside of certainty addiction is the abhorrence of ambiguity, and that of confirmation bias is the disregard of disconfirming evidence, the flipside of myside bias is our strong proclivity to reject ideas that do not fit our belief systems. Ideas that do not fit what we already know are disadvantaged. Most such ideas, like flying cows or talking trees, are disadvantaged for good reason: they do not conform to our experience. What about ideas that are not as obviously preposterous, that happen not to fit our belief systems? We are unlikely to accept and integrate into our pre-existing network ideas that contradict our existing beliefs. But when ill-fitting ideas are hard to deny, we run into a problem: myside has an interesting corollary, which is our innate desire to eliminate cognitive dissonance.

We suffer tension when we do not understand, which motivates us to figure things out. What if figuring things out entails reexamining beliefs that have already been firmly established in our minds? This can get a little messy: cognitive dissonance describes the uncomfortable tension that arises when we hold competing beliefs, when we are attracted to two ideas that do not fit together and therefore cannot both be true. How do we combine seemingly contradictory beliefs, such as, "My neighbour is completely self-centred, but he offered to take in my mail while I'm on vacation," or "God is both omnipotent and benevolent, but the earthquake killed thousands of innocent people"? We are confronted by dilemmas like these every day, and the vast majority of the time, we resolve cognitive dissonance in favour of our prior beliefs.

Cognitive dissonance is a particular form of the tension of not knowing: we have difficulty putting beliefs together that individually seem true but do not jibe. The lack of fit signals that there is something about the world that we do not understand. Our need to know is so powerful that it will not leave a dilemma, like two inconsistent beliefs, unresolved. Something has to give, and unravelling a whole system of interlinked beliefs is not a quick solution, which is why myside bias favours pre-existing beliefs over new ideas that may be reasonable on their own but do not fit with what we already "know." So we reject the idea whose falsity is less consequential for us. Why blow apart an entire belief structure and invoke the psychic chaos that would ensue, if we can simply figure out a way to reject the idea that does not fit? It is very daunting to second-guess ourselves when a belief system in its entirety (sometimes referred to as a "memeplex") would otherwise have to be deconstructed, introducing a lot of incremental tension. Myside bias preserves the integrity of pre-established beliefs, and, if necessary, it invokes confirmation bias to assist. For example, "The neighbour's offer is probably a scheme of some sort, probably to snoop around my house."

Myside and its counterparts, the availability and confirmation biases, are useful ways to expedite decision making. They are just not useful in all circumstances. The satisficing process ends with the feeling of knowing, which cements our conclusions. This feeling of confidence in our interpretations is based on how they cohere to our belief systems. This cognitive process is fundamental to how we function as human beings.

The difficulty, however, is that coherence and confidence are often the enemies of a reliable understanding of complexity. In our rush to certainty, satisficing often directs our thinking to the wrong place and holds it hostage there.

7

When Satisficing Meets Complexity
Missing Alternatives

Ignorance more frequently begets confidence than does knowledge.
Charles Darwin*

Unfortunately, our underdeveloped expertise with complexity does not deter us from rushing to conclusions: as explored in this chapter, our need to know is so powerful that we are compelled to conclude as soon as we feel we have figured things out.

Consider this saga:

> In the late 1990s, the Gates Foundation began a multiyear program of granting over one billion dollars to expand the number of small schools in the education system, based on the belief that smaller, more intimate learning environments were more conducive to academic achievement. The statistics appeared to support this assumption, because top-ranking schools were disproportionately represented by small schools. Statisticians Howard Wainer and Harris Zwerling analysed the statistics in depth and discovered that, while the best schools were indeed smaller, the worst schools were also disproportionally represented by small schools (a fact that was missed by the Foundation researchers). Small schools, consisting of fewer teachers and students, demonstrated much higher variation in academic results; larger schools were typically more consistent. In fact, in the case of secondary schools, the larger ones tended to score better than

---

* *The Descent of Man*, Penguin, 2004.

most small schools, in part because of greater teacher diversity, more course options and more extracurricular activities.[67] In 2005, the Gates Foundation abandoned its plan to convert large schools to smaller ones.

A quick and shallow interpretation of statistics can support our initial intuitions, which is why "quick and shallow" is not productive for assessing complexity. Our first reasonable interpretations, however good they may feel, are rarely good enough. We over-rely on an early test of coherence based on our preconceived intuitions about how a straightforward world works. And we over-rely on the feeling of knowing based on the confidence we have in these intuitions — a confidence developed for straightforward problems.

7.1 Coherence Taken Too Far

The benefits of the three biases that expedite coherence — availability, confirmation and myside — can be offset by their disadvantages: invoking them for complex problem solving overextends their use, leading us astray.

Availability Bias Taken Too Far

To varying degrees, availability bias permeates absolutely everything we think about; our thinking processes focus on immediately available sensations and information and typically neglect information that requires time and energy to find. The bias has very little downside when it comes to straightforward challenges where the search for hidden data is not important, because everything we need to interpret is right in front of us. As the complexity of our challenges increases, however, immediate cues are useful but nowhere close to sufficient. With complexity, our neglect of hidden but crucial information has severe implications.

One example of hidden information lies in the realm of mutual fund investing. Most investors choose mutual funds on the basis of the funds' track records. Past performance is more readily available than any other information about a fund, so it is intuitively presumed to be a good indicator of future performance. We buy cars, computers and appliances based on reports of their track records; we take vacations in places that have received favourable reviews; we hire contractors because their work has impressed others. Why should investing in a mutual fund be any different? The answer is that the future performance of a fund depends on many more subtle and ambiguous factors than that of a car, vacation or renovator. A vast amount of crucial information is submerged beneath the easily available performance record of a fund, and this hidden information is crucial to a meaningful prediction of future fund performance. We rarely expend the effort to search for the hidden information, not just because of cognitive miserliness but also because we are oblivious to the relevance of the hidden information, having been strongly influenced by the availability of past performance.

Past returns do not reveal two crucial pieces of information: the continuity of the stock-picking approach, and the degree to which luck, versus skill, played a role. Take a fund with impressive returns over a variety of periods and durations. The hidden information is whether or not the same manager using the same style was responsible for the decisions throughout all these periods. If not, the numbers do not represent the skill of the current manager. If it is the same manager, the numbers do not reveal if her skill is applicable to the current market trends and the size of the fund she currently manages, both of which could have changed significantly. Even if there is continuity in manager style, how can we be sure that the fund manager is truly skilled? A large part of any stock picker's performance has nothing to do with skill and everything to do with luck — both good and bad. Separating luck from skill in fund managers is one of the most difficult, if not impossible, challenges for investors. A number of academics have demonstrated mathematically that even a twenty-year period is not long enough to accurately reveal skill, because there is just too much random noise in the past performance returns.

Hidden information may not be deeply buried but is still not top-of-mind when we draw conclusions based on comparisons that are too narrow. Our tendency to limit our comparisons to easily available information makes our comparative base rates insufficiently representative (referred to as "base rate neglect"). The classic example, illustrated in 2.4, is whether the soft-spoken book lover is more likely to be a librarian or salesperson: most people assume librarian, ignoring the broader base rate of hundreds of times more salespeople than librarians.

True story: A company president proudly announced to his board that the expensive new sales training program that he had implemented generated a 5 percent increase in sales. Everyone applauded. But the base rate he used to quantify the increase was the prior year's sales. Looking at a broader base rate, the entire industry increased sales by 7 percent over the same period.

Misleading base rates are not limited to business. Sociologist Becky Pettit undertook a comprehensive analysis of inmates in the American prison system to broaden the base rate that government officials used to assess improvement in the quality of life of African Americans over two decades.[68] Prisoners are not counted in the U.S. Census Bureau and very little information is gathered about them, which makes them, as Pettit puts it, "invisible." Since half of the incarcerated prisoners in the U.S. are African American, she concluded that the living standards of this segment of the population are significantly overstated, especially for young black men. When the incarcerated are added back to the comparative statistics, the unemployment rate among the black population is much higher than reported; likewise, contrary to what is reported, their educational standards are much lower and their quality of life is not trending higher. The Sentencing Reform Act of 1984 made prison a more likely result of criminal activity and extended jail terms. As more low-income and undereducated young males — disproportionately represented by African Americans — were charged with a variety of crimes, the census and unemployment statistics began to underrepresent this segment of the population, because they were not accounted for.

Sometimes the useful base rates are inaccessible. A marketing campaign (or sales training initiative) may seem to have been successful, but it cannot be compared with the ideal base rate, which represents sales results had there been no marketing campaign (or sales training). Similarly, if we hire someone and she does an excellent job, we cannot make a definitive assessment of our hiring decision, since we cannot compare her performance with that of someone we rejected who might have performed even better. Same with a choice of careers, mates, neighbourhoods, and so on: we cannot compare our decisions to the base rate of outcomes generated by alternative decisions, so definitive assessments are impossible.

Nassim Nicholas Taleb emphasizes this challenge with what he calls "silent data."[69] He points out that consumer advocate Ralph Nader is rarely celebrated for the tens of thousands of lives that have been saved by the seat-belt laws that he championed. For Taleb, it "is much easier to sell 'Look what I did for you' than 'Look what I avoided for you.'" The positive consequences remain hidden in this case. Just as startling to contemplate are the negative consequences of decisions that are typically invisible to us. Taleb uses the example of the millions of dollars spent to rebuild New Orleans after hurricane Katrina. He does not question the effectiveness of the city's rebuilding but points out that the same money might have gone to cancer or diabetes research, since more people die every day from these diseases than were killed in total by Katrina. Death by cancer and diabetes cannot compete, however, with the graphic footage of hurricane devastation, which elicits a strong emotional response from voters who expect their politicians to do something about it.

When we employ readily available information in the service of validating our initial hypotheses about complex phenomena, we are at high risk of overextending our bias for confirmation.

Confirmation Bias Taken Too Far

Another example of base rate neglect occurs when we neither seek out nor pay attention to evidence that contradicts our hunches. Psychologists Robin Hogarth and Hillel Einhorn demonstrated the reluctance to disconfirm hypotheses by asking subjects to assess the expertise of a stock market analyst who advertises that he has always predicted stock market rises and points to his published recommendations, all printed before each market rise.[70] Most people are more impressed than they should be, because they neglect the important *disconfirming evidence* of what his recommendations were before stock markets *fell*. If he predicted a market rise every year, and we check his predictions only against the years that the market went up, we will fail to discover that he also predicted the market going up when it actually fell. We fail to consider both up years and down years because he did not offer all of the statistics; our bias to confirm leads us quickly down the path of accepting his claim.

A balanced investigation takes time and energy, which is why, typically, we are confirmers, not falsifiers: most of our straightforward problems do not warrant the effort of disconfirmation. Our thinking is biased toward confirming hunches: it is asymmetrically skewed toward confirmation and away from disconfirmation. But as a shortcut, the bias contravenes the logic of certainty, which requires that the testing of hypotheses should be asymmetrically skewed toward disconfirmation. This contravention is not problematic for straightforward problems that require less testing, but it is very problematic in the face of complexity. For complex problems, *our minds are asymmetrical in the wrong direction.*

The most celebrated example of knowledge asymmetry is the case of the black swan, which was first used by philosopher Karl Popper.[71] If you want to validate the hypothesis that all swans are white, you have your work cut out for you, because there are a lot of swans in the world whose colour has to be checked. You can count all the white swans that you find, but since it is impossible to inspect every single swan on the planet that has ever lived or ever will live, you still will not have proved that *all* swans are white. There is only one definitive conclusion that you can make about *all* swans: if you can find one that is not white, you can claim with certainty that *not all* swans are white. (Popper liked the swan example since Europeans believed that all swans were white until 1697, when an explorer discovered a black swan in Australia.) The aim of science, according to Popper, should be based on the search for disconfirming evidence.

This view of science has been challenged by philosophers and scientists by appealing to the approach that scientists actually take (as will be outlined in greater detail in 15.1), which is responsible for the considerable progress that science has made. Scientists do not rely exclusively on 100 percent certainty to make meaningful claims; they test their hypotheses and raise or lower their confidence in their beliefs based on accumulated evidence. Popper was nonetheless correct in demarcating the line between claims that we can be certain of and claims that we cannot: he extended David Hume's insight that inductive reasoning does not admit certainty by pointing out that the only way to establish the definitive truth of a hypothesis is to prove that its not being true is impossible (referred to as null hypothesis testing). Popper did concede that a falsifiable theory can still be useful even if it has not been disproved, as long as it is corroborated by a lot of evidence: ultimately, theories that resist falsification have a higher likelihood of being correct, which is why Popper believed that science should focus on disconfirmation.

Popper and Induction

Knowledge asymmetry reflects what is known as the problem of induction, which David Hume described two centuries before Popper. Hume pointed out that you can never make a universal claim about something based on evidence from experience, because you can never inspect every single instance of that thing.[72] Popper tried to solve this problem by introducing the notion of falsifiability, which, contra Hume, he contended would make science possible. Although we cannot prove things from experience with certainty, we can still disprove them with certainty, as long as the claims are falsifiable, i.e., they are capable of being disproved.

Popper used falsifiability to criticize Freud's psychoanalytic theory and Marx's interpretation of history. The two thinkers, he insisted, failed to make scientifically meaningful claims, because their claims cannot be disproved. Even though confirmation bias generates the illusion that these theories have scientific credibility, their ideas are essentially just conjecture; the evidence supporting these theories may be interpreted in many different ways.

Although he was not the first to make this observation, Popper's insight — that scientific claims can be proven wrong, but they cannot be proven right — was influential and extremely powerful. The reason for this asymmetry lies in the difference between deduction and induction. Induction cannot achieve certainty since it relies on repeated observation to arrive at conclusions, but there is no number of observations that guarantees that the next one will match all of the previous ones. Deduction can achieve certainty because it does not rely exclusively on empirical observation but on logical structure. In the Popperian model, the only viable approach to science is the logically deductive argument that disproves a hypothesis (called

> *modus tollens*). It follows the general structure of "If my hypothesis is correct, then the evidence would support my claim; but the evidence contradicts my claim; therefore my hypothesis is not correct."

How could we have survived this long if our thinking opposes logic — if we gravitate to confirmation even though certainty is available only through disconfirmation? Because speedy confirmation is good enough and a lot better for fast survival decisions than the time-consuming process of falsification. But when satisficing meets modern-day complexity, we cannot simply invoke the same confirmation bias and expect the same rate of accuracy with the same degree of confidence.

Myside Bias Taken Too Far

No one is immune to myside. The theory of general relativity indicated that the universe was not static, but either expanding or contracting. Initially, Einstein could not accept this implication, because he thought that it was impossible for the universe to be moving. So he recalculated his work to eliminate the implication of a non-static universe that did not fit with his preconceived ideas. The Hubble Space Telescope demonstrated that the universe was indeed expanding, so Einstein eventually conceded his error. They do not come smarter than Einstein, nor do they come without ingrained myside bias.

The most quickly available information we can use to confirm something, second only to our immediate perceptions, is our pre-existing beliefs. They are with us all the time, ready to be used instantaneously — in fact, automatically and subconsciously. Myside is an important shortcut to integrate or reject new ideas efficiently and reliably in a straightforward environment. In a complex environment, however, we are at risk of over-relying on what we already know. Because complexity is configured differently, it requires more creativity than reverting to the storehouse of what we already believe.

One of the disadvantages of myside is that inaccurate or unproductive beliefs get reinforced as we add supporting beliefs to our belief systems. There is "path dependence" in the formation of our beliefs, because new beliefs come down a path that is determined, in part, by prior beliefs that have paved the way. Once a child accepts the idea of

Santa Claus, for example, it is an easy jump to the Easter Bunny and the Tooth Fairy. Or more worrisome, if a child is teased and made to feel that he does not fit in, it is a clear path to the belief that he is deficient in some way as a person. Or if an illness is misdiagnosed by a doctor, it is a quick trip to the idea that naturopathy is a superior form of treatment. The architecture of our belief systems is an intricate web of many interrelated ideas; we test ideas by how they cohere with one another (associative logic), rather than by strict adherence to analytic reasoning (deductive logic). Over time, people construct massive belief systems that, largely because of the path dependence of their creation, can appear wildly hypocritical to others, even if they are perfectly coherent to the individuals, such as devout Christian racists or penny-pinching billionaires.

Path dependence plays an enormous role in our lives: we go to a certain school where we meet certain people who influence us in certain ways that are completely unpredictable at the time, even though a clear path can be established when we look backwards. The path that leads us to marry a certain person or choose a certain job is discernible only in retrospect, but that does not mean that each event is completely random in and of itself. Each event is the product of a series of preceding events that created the path to a specific end result. Path dependence does not imply that all events are predetermined, just that long, complicated chains of causal factors lead to specific effects. Myside can have a highly skewing influence on the adoption or rejection of new beliefs, since we test the coherence of new ideas against existing ones, which is not a neutral standard but a highly path-dependent one.

While some individuals are more inclined to myside than others, some ideas are also more inherently accommodating of myside than others, based on the intricacy of the web of interlocking beliefs that supports them. This is why it is easier to convince people that dark chocolate and red wine are healthful than that their religion makes no sense. Some beliefs become so entrenched and integrated into our larger belief system that we cannot help but "know" that they are right, oblivious to the somewhat contingent process that gave rise to this "knowledge."

7.2 Confidence Taken Too Far

Once the tension of not knowing is replaced with the calm feeling of knowing, we assume that our work is done. Just as the search for coherence is enabled by availability, confirmation and myside biases, the lockdown of knowing is enabled by a bias of sorts that originates from a central operating assumption buried deep in our minds. The feeling of knowing is possible only if our intuitive starting assumption is that we are capable of the kind of unconstrained and reliable rationality that gets us to the truth. If we believed that we were meaningfully limited in our ability to arrive at conclusions that represent the way things really are, then we would not be able to achieve the confidence that stops the search for interpretations. We stop because we have confidence that we know, which is based on the underlying assumption that we *can* know.

The Assumption of Rationality

What gives us the confidence to assume that we have arrived at truthful conclusions? What keeps us from the self-doubt that even a modicum of intellectual modesty would call for? The answer lies in the fundamental assumption at the base of satisficing: *we are competent thinkers, eminently capable of the kind of rational thought that accesses truth and knows when it has found it.* If we did not assume that the answer to a problem was attainable, we would not have started the process of searching for a solution in the first place, let alone arrived at a conclusion that we had confidence in. We operate with an implicit assumption that our thinking is good thinking — good enough, at least, to bring us to conclusions that we do not need to second-guess. We do not view our conclusions as "first satisfactory": we see them as "the truth."

There are four underlying elements to the large-scale assumption that truth is ours for the taking. We assume that:

- we are objective (we assess things with a neutral, bird's-eye perspective);
- we are accurate (we assess things without invoking cognitive errors);
- we can achieve explanatory depth (we assess the whole problem, not just parts of it);
- we can access truth (ultimate answers are available to us).

For straightforward problems, all of these assumptions are generally warranted. For complex problems, not so much; each of these is precarious.

Objective Neutrality

We cannot climb out of our own heads to see the world "as it really is," so we are forever constrained by the perspectives through which we interpret everything. We can never see the world from every possible perspective, certainly not from the perspective of our neighbour, boss or lover. To take a completely neutral, unbiased view of things is impossible. This constraint is especially significant in light of the path-dependent formation of our belief systems, as well as the assumption of naïve realism that we default to. We are not programmed to reflect on the limits of our neutrality, because we are not biologically designed to postpone concluding by asking ourselves: "Have I neglected hidden information? Have I challenged my beliefs with disconfirming evidence? Have I considered alternative interpretations that do not cohere with my prior beliefs?"

We take for granted that our conclusions are based on a fair, objective and neutral stance, which is why we are so confident in them. But we rarely test our conclusions with the rigour and discipline that a detached observer would employ. In a complex world, the illusion of objectivity makes us extremely vulnerable to unwarranted confidence. The best we can do is to *strive* to be *as objective as possible*, and acknowledge the impossibility of perfect neutrality.

Accuracy

Being prone to naïve realism — assuming the world works exactly the way we perceive it — leads us to conflate interpretation and truth: we forget that our view of how things work is not necessarily identical to how things actually work. This assumption is not of great concern in the case of straightforward problems. In this realm, viewing our thinking as riddled with logical flaws, irrational quirks and faulty memories would have us second-guessing every conclusion to the point of useless, even dangerous analysis paralysis. Natural selection has equipped us to be blissfully inattentive to just how prone we are to errors in our thinking. The assumption that our beliefs are accurate is a concern, however, when applied to complexity, where the basic models we use to interpret the world and the biases we rely on to expedite our conclusions are anything but foolproof. Solving complex problems requires a lot more humility.

Explanatory Depth

This assumption is subversive. It lies below our awareness until we are asked to explain something that we thought we had a handle on and find ourselves hard-pressed to describe in detail. Kids catch us out with their simple follow-up questions: after we have explained how babies are made, we may find ourselves stumbling on a question like, "If the sperm carries the man's genes and the egg has the woman's genes, how come all brothers and sisters don't look the same?" We thought we had a pretty good understanding of reproduction, but all of a sudden, we realize that there is no depth to our knowledge.

The mind is a multilayered processing unit that is not entirely transparent to its owner, which is why it is so easy for us to jump quickly to the feeling of knowing without achieving true depth of understanding. The "illusion of explanatory depth" originates from the race to release the tension of not knowing: once we sense that something makes sense, we rarely continue to probe further. The ease with which we can formulate an explanation is a cue to us that it coheres with what we already know and therefore must be valid, even if it is not. The Dunning-Kruger effect, named after the two psychologists who documented the bias, refers to the inverse relationship between confidence and knowledge: the less we know, the more confident we are, since the deeper we attempt to understand something, the more we uncover ambiguities, contradictions and further questions.[73] Lack of explanatory depth feeds our certainty, as Darwin pointed out in the epigraph to this chapter.

Truth Availability

Like all animals, we constantly scan our environment for relevant information. Humans take the environmental scan a step further. We are not satisfied to operate exclusively in the present moment with our immediate perceptions: we ruminate about the past and the future. Much of this "out-of-the-moment thinking" is geared toward looking for answers. Satisficing would not work if we did not assume there were answers to our questions, truths that we could reliably arrive at. And there are answers for us when we are operating in a straightforward world. But when we come up against problems that do not yield to simple, solitary solutions, we are left with the tension of an uncertainty that begs to be resolved.

Our tendency is to force fit single explanations onto these problems. Unlike run-of-the-mill difficult problems, which have many steps but a single-path to solution, such as optimizing the production output of a factory, complex problems defy single-path solutions. In fact, many of our complex challenges have no available answers: the best we can do is to experiment by interacting with the problem and continually tweak our responses based on the feedback we get. Solving versus tweaking is what philosopher Abraham Kaplan had in mind when he differentiated problems from predicaments: the former can be solved; the latter can only be coped with.[74] Predicaments are permanent dilemmas that have no discernible, distinct causes to be isolated and addressed; for example, the challenges we face in our close relationships, our jobs, our child rearing and our quest to create meaning for ourselves. When we treat predicaments as if they were solvable problems, we become vulnerable to frustration. Most of our problems are of the straightforward type that converge to a single solution; however, these are not the problems that present significant challenges for us today.

Hindsight Bias

Confidence taken too far is reflected in the observation by novelist Fyodor Dostoyevsky who wrote that everything seems stupid when it fails,[75] which is a pithy summary of hindsight bias, the "I knew it all along" attitude that we often have when we reflect on past events. We all know that hindsight is 20/20, but what is less obvious is our tendency to believe that the way things unfolded was more predictable after the fact than when we assessed their likelihood before knowing the result. We tend to assume that we knew more in the past than we did, that our previous foresight was greater than it really was. "Should have known better" is an easy criticism to make in retrospect, when more information comes to light. We tend to forget that a lot of what becomes current information was not available or easily accessible in the past. In fact, hindsight is so prevalent that we are just as likely to say to ourselves, "I knew I should have listened to my gut" as we are to say, "I knew I should have thought that through more carefully."

This bias is pervasive because it is very difficult for us to review history without the lens of what we currently know. Indeed, hindsight is a component of how we learn: we reflect on how events could have unfolded differently had we taken different action. Hindsight updates stale knowledge of the past so that it can be used for future reference. But when we revisit past events that are complex, by overlaying a simplistic interpretation retroactively — one that eliminates the true complexity of a situation — we lose the opportunity to explore alternative possibilities that are not as self-evident. Hindsight bias misleads us into thinking that there were two and only two alternative paths that emerged from a point in the past: the way history did unfold, and the way we think it would have unfolded if a different decision had been made. Yet for complex situations, we are far too quick to assume that clairvoyance was available to anyone in the past who could have just "opened their eyes" and "faced facts." In fact, we often use hindsight to unfairly assess others' decisions, judging them for what we think was knowable in the past. We criticize their poor decision making because it was "obvious" what the "right" decision was.

Hindsight surfaces when our need to know inflates the confidence we have in retrospective explanations that relate events in a simplistic way, based on a particular causal path. This narrow look back is at the expense of all the other possible paths that could have evolved from the same set of starting conditions. A telling example is the March 2003 invasion of Iraq, which is almost universally viewed as an unequivocal mistake because it turned out that the country had no weapons of mass destruction and Saddam Hussein was never a true threat to global security. In fact, as an enemy of Iran, he inadvertently provided the West with some counterbalance to Iran's strength in the Middle East.

This assessment may be accurate, but it assumes that all of today's information was available at the time of invasion. Notwithstanding the absence of WMDs, we will never know how Iraq and the Middle East would have evolved had the U.S. not invaded. Might Hussein have acquired enough chemical or nuclear warheads to become a serious threat? A headline in April 2003 could have read, "American troops discover stockpile of biological weapons in Saddam's palace." We know *now* that this could not have happened, and even at the time, intelligence reports suggested it to be unlikely, but that does not mean it was not a possibility. Or, if there had been no invasion, a headline four years later, in April 2007, could have read, "Iraq launches biological attack on Israel; warns it will activate nuclear warheads if U.S. intervenes." After all, Saddam had violated the 1991 Gulf War arms control agreement and many UN resolutions, including the requirement to allow regular, unencumbered inspections. The invasion of Iraq may well have been a mistake, but we cannot compare the outcome of invading to the alternative paths that would have unfolded had there been no invasion, so we are not entitled to confidently and definitively declare it a mistake. It is far too facile, as tempting as it may be, to revisit history with a "should have known better" perspective.

Locking Down Too Soon

Confidence is not only a motivator of action but also an energy-saving mechanism and a way of circumventing the limited capacity of our working memory: once we have concluded, we have freed up space for other problems. The whole mechanism is undermined, however, when closure is invoked prematurely. When satisficing meets complexity, the result is overconfidence.

Psychologists Jerome Bruner and Mary Potter explored our proclivity for premature conclusions in experiments in which subjects were exposed to blurry photographs that only gradually came into focus.[76] The longer the photos took to become clear, the less likely the subjects were to correctly identify the objects. Bruner and Potter concluded that the subjects' thinking was influenced by their initial hypotheses about what the objects were before they came into focus, hypotheses that they were reluctant to abandon. Subjects who initially saw the photo clearly and then were exposed to its blurring were more consistently accurate than subjects who started with a blurry photo that eventually became a clear picture. We are accustomed to locking down our ideas as quickly as possible, and this does not always serve our interests.

We need the feeling of knowing to make decisions and take action. But, as Daniel Kahneman points out, the confidence that we feel in our judgments is not necessarily a reflection of a deep, reasoned evaluation of the probability that our conclusions are correct. Rather, it is a feeling that reflects how well our ideas cohere.[77] Even though it may seem reliable, Robert Burton notes, "felt knowledge" is not the same as knowledge that results from testable observations.[78] Feeling that it is right does not make it right. In a complex world, we are often much more certain than we are entitled to be, because we rarely pause to seek out information that is not readily available, and even more rarely seek out evidence that disconfirms our initial hypotheses. Concluding is premature when the lockdown feature of confidence precludes the consideration of better interpretations, in which case confidence becomes overconfidence. Locking down one interpretation, as if it were the true perspective, does not convert a complex problem into a straightforward one; rather, it robs the decision maker of a deeper understanding. The best explanations for complex problems are usually not immediately obvious or first in line. In our rush to disambiguate reality, we lose meaning and we shortchange the flexibility and creativity that we need for tackling complexity, thereby reducing the efficacy of our responses.

Knowing and Fertilization

The feeling of knowing, which feeds our addiction to certainty, triggers a lockdown of our thinking the way a fertilized egg locks out competing sperm.

Most of the three hundred million sperm that are released at a given moment in search of an egg will perish in the acidic walls that greet them. But a few thousand will continue the struggle to meet the ovum as it floats down the fallopian tube. The surviving sperm will be guided up the tube by the chemical signal that is sent by the egg to attract them. As many as two hundred sperm will eventually surround the egg, but the first sperm that lines itself up just perfectly will bore a hole in the egg's wall by releasing enzymes that break down that wall, allowing the sperm's nuclei to enter and fuse with the nuclei of the egg.

Following the sperm's successful penetration of the egg, a number of chemical reactions are triggered, one of which initiates the release of granules that fill spaces in the egg's wall, making it larger and harder

> so that other sperm are incapable of getting through the wall. Many healthy sperm could have made it into the egg, some of which would likely have been genetically stronger than the victorious sperm, but they were not first, so they are rejected forever.
>
> The mind is like an egg, and sperm are like the myriad possible explanations for any given problem that the mind tries to solve. Just as a few thousand sperm are stronger than the millions that perish along the way, some ideas are fitter than others. Of all the possible ideas that might generate a good explanation for an event, only a small number will be considered. Once our mind is impregnated with the first, reasonable ("good enough") idea, a feeling of knowing rushes in, and we shut out alternatives.
>
> Just as the first sperm that merges with the egg is not necessarily the best (in fact, it can occasionally lead to an unhealthy fetus), the ideas that we accept are not necessarily the best ones: they are the first satisfactory ones, not the best of all possible ideas. Accuracy often loses the contest in the face of the strength and urgency of our need to know.

Overconfidence arises from automaticity, when our mental processes operate without conscious supervision and intervention, so that our concluding is reflexive. Automatic concluding is effortless, so it serves the need of cognitive miserliness, but it does not work in situations in which we need more conscious oversight. Overconfidence is the product of underappreciating our cognitive limits; we think that we have got things figured out, oblivious to how our unsupervised thinking can lead us astray in the realm of complexity. When we overconfidently ignore the distinction between our constructions of reality and reality itself, we conflate truth and interpretation and forego the most important strategy for coping with complexity, which is the examination of alternatives. When we operate instinctively on the basis of naïve realism, we assume that we are capable of objective, neutral, dispassionate thinking: thinking that is unencumbered, clear, logical and sufficiently penetrating to get us to the answers. If only it were so.

7.3 A Problem on Top of a Problem

Like all addictions, the need for certainty is deeply ingrained. Unlike other addictions, we cannot abandon or resolve it cold turkey, which makes it a difficult addiction to manage. We cannot just get ourselves "off" certainty, because we need it throughout our daily lives for all the straightforward decisions we need to make quickly. Satisficing is the best way to make fast decisions with limited brainpower in an environment in which straightforward challenges are characterized by reliable visual cues, and in which alternative interpretations are unnecessary, making a high level of confidence in our initial conclusions warranted.

Satisficing limits our consideration to only what is easily available, easily confirmable and easily fitted to our pre-established beliefs, and while these shortcuts work well for straightforward problems, they result in a meaningful sacrifice in accuracy when we confront the complexity of the modern world. Whether overcoming interpersonal conflict, managing a career, raising children, solving climate problems, deterring terrorism, starting a business or assessing public policy, the first satisfactory conclusion is rarely the best one — or anywhere close to being the best. Today's complex challenges require more information than is immediately available; they are not the kind of look-and-solve problems that we excel at resolving. The speed and definitiveness of satisficing do not accommodate the deliberative, critical thinking necessary for seeking and analysing information that lies below the surface. Oversimplifying in itself would not be as problematic as it is if we did not accept our oversimplified conclusions as final, thereby adding a problem to a problem.

For complex problems, the speed–accuracy balance of satisficing sacrifices too much accuracy for the sake of speed: the premature lockdown shortchanges explanations that might approximate truth better. Because our expertise in handling complexity is underdeveloped, we have no choice but to sacrifice some speed in order to achieve greater accuracy: handling it better requires that we adjust the speed–accuracy balance *away from* rushed conclusions and *toward* more accurate ones. Improved accuracy depends on, first, our becoming more attuned to the nuances of complexity and the traps we are vulnerable to in thinking about it — which has been the focus of this part of the book — and second, invoking more sophisticated models to dissect complexity and draw conclusions about it — the focus of the part that follows.

Part III

The Brain–World Problem
When Intuition Meets Complexity

8

Intuition as Expertise
Ten Years of Quality Practice

*The development of genuine expertise requires struggle,
sacrifice, and honest, often painful self-assessment.
There are no shortcuts.*

K. Anders Ericsson et al.*

According to the theory popularized by writer Malcolm Gladwell, becoming an expert in a particular activity requires ten thousand hours of practice.[79] This insight was first proposed in 1973 by Herbert Simon and William Chase, who called it the ten-year rule.[80] Their study of top-ranked chess players revealed that none of them had fewer than that many years of experience. Twenty years later, K. Anders Ericsson and two other academics, quoted in the epigraph to this chapter, published a seminal research paper that reviewed a number of studies on expertise, including their own analysis of violin and piano students at a German conservatory.[81] Their conclusion supported Simon's work: expertise requires about ten years of practice — typically twenty hours a week for ten years, or ten thousand hours. But the number of hours is not sufficient: the individual must be engaged in a particular kind of practice, what they call "deliberate practice." Only then can reliable intuitions — the kind that facilitate expertise — become deeply embedded.

Developing more reliable intuitions about complexity is precisely the purpose of this book — of parts 4 and 5 in particular, where different models for thinking are examined. The purpose of this chapter is to

---

* "The Making of an Expert," C. Coke, K.A. Ericsson and M. Preitula, *Harvard Business Review*, July-August 2007.

investigate more deeply the notion of expertise as it relates to intuition, because perceived expertise based on unreliable automatic thinking is the source of the gap.

We toss the word "intuition" around a lot, but we do not always mean the same thing by it. It connotes a form of thinking that is automatic and not fully transparent: we do not have to put effort or concentration into intuition, and the source of our intuitions is often the result of unclear or obscured logic. Herbert Simon (the originator of the terms "bounded rationality" and "satisfice") defined intuition as the recognition of patterns that are stored in memory.[82] Simon's definition is generally endorsed by cognitive psychologists, but it needs one clarification. Although intuitions have subconscious origins, some are largely fixed, like an intuitive fear of spiders or sudden, loud noises; these forms of intuition are more usefully categorized as "instinct" — inborn tendencies, sometimes called "fixed-action patterns." The broader form of intuition is more flexible since it is shaped and tweaked as we gather experience, such as an intuitive view of whom we can trust and whom we cannot. These kinds of intuitions are developed over time with learning and are the source of expertise.

8.1 Expertise

Expertise is based on the development of deep and reliable intuitions, the kind that an athlete employs when she automatically positions her body and racquet to return a serve, or a firefighter uses to assess the danger of a burning house, or a pilot depends on to land a plane. Repeated practice allows us to fine-tune our subconscious thinking so we can respond without invoking the kind of slow, conscious, analytical thinking that a beginner or non-expert must invoke.

Psychologist Gary Klein has demonstrated that as we gain experience, we rely increasingly on "recognition-primed" decision making.[83] We do not start from scratch to understand something; we examine the cues a situation generates to recognize patterns that are familiar to us based on previous experience. Experts usually attend to cues that are invisible to the rest of us: the batter scrutinizes every detail of the pitcher's movements as the ball is thrown; the tennis player dissects the stance and motion of her opponent as the ball is served. The more experience we accumulate with a particular situation, the higher our reliance on recognition-primed decision making, and the deeper our intuitions.

Experts hone their intuitions through hours of practice. But it is not just the quantity of practice that matters; it is the *quality* of the practice that generates mastery. A self-taught golfer who undertakes years of playing without any coaching will develop and reinforce bad habits; even though he will have deep intuitions about swinging a club, he will not become a champion golfer no matter how strong his innate proclivity for golf is. A less obvious aspect of high-quality practice is a tolerance for the psychological discomfort of constant failure. Mastery is a function of the time and attention devoted to overcoming weakness. The concentration and perseverance required for this kind of "deliberate practice" is highly demanding, which is why most of us are not Olympians or virtuosos.

Expertise, then, is the consolidation of sufficient quality and quantity of learning into reliable intuitions. Our intuitions are useful when they incorporate appropriate cues, ignore noise and take advantage of relevant feedback. Deliberate practice is of the painstaking kind, but it is fruitful only in situations that actually lend themselves to the development of reliable intuitions. Klein and Daniel Kahneman have worked independently and together to document the two conditions that are necessary for developing high-quality intuitions: first, the situation has to be structured in an orderly way with regularities that provide useful and accessible cues; second, there has to be an opportunity to learn from these cues by getting clear, timely and consistent feedback.[84] Both of these conditions are hard to meet where complexity is concerned.

Our intuitions about driving a car can be very reliable, once we have had sufficient opportunity to practice with the mechanisms and learn from the feedback, whereas our intuitions about playing a slot machine, no matter how many times we play, are useless because the machine is driven by a randomizer: there are no useful cues to assess, and the feedback we get when we put our money in the slots has no meaning.

What if the task is less clear-cut than the random mechanics of a slot machine or the relatively structured problem solving of driving a car? What if the problem is how to respond to a defiant child, a belligerent spouse or a demeaning boss; choosing a marketing strategy to grow a company; championing an unpopular public policy initiative; motivating a dispirited team? Our intuitions are not as dependable in tackling these dilemmas, because a sufficient amount of high-quality practice is usually not available to learn from.

8.2 Intuition and Complexity

Complex problems are tricky to develop reliable intuitions about for two reasons: they are difficult to assess because their cues are obscured, and they are difficult to learn from because their feedback is highly ambiguous.

Obscured Cues

Accurate interpretations depend on picking out relevant data: separating out useful signals from useless noise. Complexity makes it difficult to differentiate signal from noise, because the cues we need are not easy to recognize, so we are vulnerable to missing them, as well as overinterpreting the noise that surrounds them.

Robin Hogarth emphasizes that we learn by what we experience, not by what we do not experience, and this can be problematic for us.[85] Hogarth uses the example of a waitress whose initial intuition about customers is that poorly dressed ones do not tip as generously as well-dressed ones, because they have less money. The waitress devotes more of her time and energy serving the better-dressed customers, and in appreciation, they tip well; the other customers' tips, of course, reflect the markedly less-attentive service they receive. The feedback experienced by the waitress reinforces her intuition.

Her learning is flawed, however, because it is skewed by feedback that excludes important hidden data: how poorly dressed patrons tip when they receive a high level of service. Had her first couple of months on the job been characterized by an equivalent level of service for both groups, she would have discovered that the correlation between dress and tipping was actually quite weak. She would have made more money overall by delivering reasonably good service to everyone, rather than superior service to one group and substandard service to the other.

This example represents how easily we develop and reinforce

intuitions that are inaccurate, based on cues that we ignore because they are hiding. Just as the waitress does not see the tips that she could have earned, real estate agents do not see the referrals they never get from unsatisfied customers; doctors do not see their misdiagnoses of patients who change medical professionals; managers do not see the success of prospective employees they choose not to hire; and companies do not see foregone profits on products that they decided were too risky to launch. Failures and mistakes that are hidden can dramatically skew intuitive learning, because we strengthen connections between things that are obvious to us and miss the opportunity to learn from those that are not readily available.

Missing "the rest of the story" is a source of friction in virtually all interpersonal relationships. Our intuitions about other people are often extremely shallow, because we emphasize what we see and pay little thought to what we do not. True story: A woman was diagnosed with cancer, and during two years of intensive chemotherapy and radiation, her husband cared for her while holding down his job and looking after their children. A couple of weeks after the doctors announced that the cancer had been eliminated and that her future would likely be cancer free, her husband informed her that he wanted a divorce. The woman was outraged, telling her friends that nothing could be more egregious and immoral than requesting a separation from someone who was recovering from an awful disease. They agreed with her and ostracized the husband. What she did not know was that her husband had been planning a divorce two years earlier, before she was sick. Upon the news of her cancer, he selflessly postponed his plan until she had recovered; he chose not to tell her that he had delayed the divorce because he did not want to hurt her further. Her friends, who lacked this information, thought he was evil; his friends, who had access to it, thought he was saintly.

Hogarth proposes that our intuitions are expedited by whatever we can easily and quickly visualize. The corollary of his hypothesis is that whatever is not immediately present or is difficult to imagine requires more effortful, conscious deliberation. Because we are very selective about what we expend conscious energy on, we are disinclined to search for hidden data when we believe that we can construct a coherent story without it.

Another true story: Customer service ratings at a large bank were consistently ranked as among the lowest in the industry. The bank had undertaken countless initiatives over the years to improve its standings: all staff were trained in the finer points of customer satisfaction; tellers were empowered to reimburse customers for errors; greeters were hired to welcome customers as they entered; and a special year-end bonus was offered to branch managers who improved their branch ratings. Still, the bank's industry ranking lagged.

A new president was determined to solve this problem once and for all and mandated a senior executive to "do whatever it takes" to improve service. The executive assigned to "fix the problem" undertook her own investigation of the bank's competitors. She observed that good and bad customer experiences occurred at all of them. Each bank had twenty thousand or more individuals working in over a thousand branches; each trained their staff with the same off-the-shelf initiatives. The executive became convinced that whatever distinguished the top bank from her last-place bank was not a simple matter of more smiles, faster service and fewer mistakes. Something was missing from the equation, and maybe that something was not the service.

The executive decided that the solution was more a matter of improving the industry ratings than it was a matter of improving service — two related but distinct problems that historically had been treated as the same. After interviews with staff and customers from both her bank and competitor banks, she concluded that what differentiated the highest-scoring bank was a relentless and long-term marketing strategy that put customer service front and centre in all of the top bank's communications. The bank's advertising always featured a commitment to top-rated service, and all of its branch posters trumpeted its #1 standing, year after year after year. This bank's strategy of branding themselves as the leader in customer service had more impact on its clients' perception of the bank than the clients' actual branch experience, which was not differentiated in any meaningful way.

In contrast, the executive's bank changed marketing messages every other year. Every time a new head of marketing was named, the new executive inevitably put his or her own stamp on things by inventing a new slogan and reinventing the brand strategy "to freshen up the positioning" and "make our marketing more relevant." The branch posters never mentioned customer service. It was no wonder that customers rated one bank as the leader and the other as the laggard; after all, the leading bank told them every day which was #1, and the executive's bank avoided the topic entirely. The key to higher ratings lay in the cues that lay outside of the service itself, hidden in the communication strategy of the bank to its customers.

Equally problematic to missing important information is imparting meaning to useless information. Because intuition is a form of recognition — identifying familiar signals amidst the noise — expertise depends on adeptly differentiating relevant from irrelevant cues. When it comes to complexity, our expertise in separating noise from signal is meaningfully compromised, because we are inclined to attribute information value to randomness. The subtle and hidden cues that unlock complexity, and the manner in which randomness masquerades as signal, are explored in chapters 13 and 14. But even when we are able to correctly identify the appropriate initial signals, the subsequent signals — in the form of feedback — can be even more daunting.

Ambiguous Feedback

The quality of our intuitions depends on consistent and usable feedback. Timely and unambiguous feedback is what makes experience useful in predicting the outcomes of our actions. Feedback is what we use to assess our interactions with the world and fine-tune our responses. In a straightforward world, we develop expertise because the feedback is "clean": it is delivered quickly and is directly related to our interactions. The feedback generated by outcomes in complex situations, however, is "dirty": it may be far removed in time from the decision, and the outcome may be loosely related or even unrelated to the decision. To the undiscerning eye, feedback can be exceptionally deceiving where complexity reigns: it often appears clear and informative when it is anything but. As Kahneman suggests, "the feedback to which life exposes us is perverse."[86]

Feedback signals from complexity are easy to misinterpret when they are buried in a lot of noise. If we are assessing the success of a company to decide whether to invest in its stock, or assessing an individual to decide whether to hire him, we have to make sense of a lot of data, some of which may appear important but actually be just random and not useful. In complex environments, separating luck from skill is highly problematic, as will be reviewed in 14.6. A mutual fund's performance record illustrates how difficult it is to set time frames for evaluating outcomes: at what point do outcomes warrant definitive conclusions? If a mutual fund manager has an impressive ten-year record but a terrible eleventh year, does that mean she has lost her touch, or is she still one of the best managers in the industry? If a superstar employee screws up, how do we weigh his one-time poor judgment against years of previous success? With complexity, outcome is often a function of an arbitrary endpoint. Think of the "Mission Accomplished" banner erected in the early days of the Iraq invasion.

One of the reasons for the limited value of outcome feedback in complex decisions is that we cannot compare outcomes the way we can with straightforward decisions. If a car is speeding toward us as we cross the street, our experience with speeding cars is sufficient for us to imagine the hypothetical outcome of a decision to stop walking. But if we are choosing a particular career path or mate, hiring one person over another or moving to a different city, how do we run through alternative scenarios in order to compare hypothetical outcomes? Where is the useful feedback that will allow us to assess our decisions and learn from them? That feedback is permanently hidden in the comparisons of alternative paths that we can never take: we cannot go back in time and repeat the experiment with different decisions and then compare outcomes. We cannot run scenarios in which all the variables stay the same except one, in order to compare the different outputs. In a complex environment, there is no reliable base rate, and controlled experiments are virtually impossible.

Yet some of the most important information is contained in the alternative paths that we never see and rarely contemplate. We choose one job instead of another; we buy one house over another; we pursue one lover over another. We will never know how the alternative paths would have unfolded if we had made different decisions: the different people we would have met, the different events that would have influenced us. The feedback that we receive from our decisions is limited, because alternative outcomes are invisible. Our learning is constrained because we cannot factor into our thinking what never happened but could have. Invisible histories unfold from every one of our decisions and are forever lost, taking with them the feedback that would be so useful to us.

Never mind alternative histories: we do not even see many of the actual consequences of our decisions — the unintended consequences of our decisions that materialize but are not obvious. For example, the government legislates the addition of ethanol in gas, because biofuels release less greenhouse gas than fossil fuels, but this inadvertently leads to higher demand and therefore higher prices for corn (corn-based ethanol is the leading gas additive), which, in turn, inadvertently increases the food shortage in less-developed countries that rely on cheap, often imported corn to feed themselves. This albeit oversimplified example demonstrates how well-intentioned public policy can (and often does) explode with unintended consequences that are difficult to predict because they are so far removed from the initial policy interventions.

Psychologist Philip Tetlock has studied the ability of self-professed experts to forecast outcomes accurately in their domain of expertise.[87] His research is extensive, based on his accumulation of many forecasts over many years. His conclusions are unequivocal: predictions made in complex arenas, such as economics and political forecasting, are shockingly poor. His research also reveals that the more training people receive, the more prone to error they are likely to become, because they are fuelled by an exaggerated confidence in what they are able to understand and forecast.

The problem is that the obscurity of complexity's cues and the ambiguity of its feedback do not deter us from plowing ahead confidently; it is not always obvious when the quality of our learning is poor as a result of missing or misinterpreting both cues and feedback.

8.3 Intuition and Overconfidence

Our addiction to certainty pushes us to conclude by seizing whatever data are available and locking down conclusions quickly as if we were experts. The more ambiguous the situation, the more latitude we have to interpret it in a way that conforms to our pre-existing beliefs — that which feels familiar and definitive. Without clear signals that contradict our satisfied interpretations, we are lulled into a false sense of confidence.

Outcome Bias

World #1 trains us to assume that the outcomes of our decisions represent the quality of our decision making, because outcomes are a useful signal with straightforward problems. The feedback from complexity, however, is delayed and ambiguous, but our rush to certainty does not accommodate ambiguous feedback. If, for example, our children grow up, get great jobs and are happy, then people assume that we did a great job raising them; if they drop out of school and get arrested, they assume we blew it. Not quite. We are highly vulnerable to overinterpreting outcomes as signals of the quality of the underlying decision making.

Psychologists demonstrate "outcome bias" by asking subjects to rate the quality of decision making in identical scenarios with different outcomes. In one experiment, a scenario was posited in which a doctor decides to perform surgery on a patient with chest pain, knowing that the surgery has an 8 percent mortality rate. Subjects were asked to evaluate the doctor's decision, after the results of the operation were revealed. The subjects who were told that the operation succeeded rated the quality of decision making higher, on average, than the subjects who were told that the operation failed. The decision, however, should not be evaluated solely by the outcome, since there are too many factors beyond the knowledge and control of the decision maker that culminate in the final outcome, including luck. The decision should be assessed on the logic of the reasoning, without being unduly influenced by the end result. Psychologist Jonathan Baron, who pioneered some of these studies, emphasizes that we have a strong tendency to assume that a good outcome signals a good decision, even though good fortune can dominate results, independent of the quality of decision making.[88] A desirable outcome does not necessarily prove a good decision-making process, nor does an undesirable outcome necessarily equate to poor decision making. We are not adept at evaluating the decision-making process, because it is faster and easier to judge a decision solely on its perceived outcome.

The loose connection between outcomes and decisions makes learning difficult, but even when there is a tight connection, learning in one scenario does not transfer tidily to other complex situations. Understanding one situation does not guarantee reliable predictions about similar situations, unless we can account for every single variable, including the hidden and the random ones. This is why geologists are adept at explaining the causes of earthquakes but cannot predict them with accuracy. (In fact, this was precisely the scientific community's defence of a group of seismologists who were convicted of manslaughter in an Italian court for not alerting the population to an earthquake in 2009 that killed three hundred people.)

Complexity and the Feeling of Knowing

Without the feeling of knowing, there could be no intuition. Intuition is *self-perceived* expertise, based solely on how confident we feel. Herein lies a problem, and it is a huge one: we can have powerful intuitions that feel like they are grounded in expertise but are not. We can "know" intuitively that something must be true, even when it is not. If, in our rush to figure things out, we can craft a story that feels right, we accept it as such. We experience the same feeling of knowing with both valid and invalid intuitions: as long as our subjective test of coherence is met, the accuracy of our intuition feels the same.

It is precisely because its cues are hidden and its feedback is ambiguous that complexity lends itself to interpretations that we can feel confident in, notwithstanding how misguided we may be. In a straightforward world, we know when we are wrong because it is obvious; in a complex world, our errors are not as conspicuous. In World #2, we are more vulnerable to an unwarranted feeling of knowing because it is easier to be wrong and not know it.

Contributing to the illegitimate feeling of knowing is a sense of recognition based on faulty familiarity. One complex scenario is never identical to another, no matter how similar they may appear on the surface. But superficial similarity between complex situations gives us a false sense of knowing. Kahneman coined the term "substitution" to describe the ease with which we invoke simple rules to interpret complex problems.[89] For example, in contemplating how happy we are with our lives, we will substitute the simple question of how we are feeling in the present moment; or in predicting a person's likely career success, we will substitute the question of how attractive the person is. Substitution is just one example of how we treat complexity as something more familiar and accessible to us than it usually is. We think that we are dealing with the kinds of problems that we have expertise in solving, but we are not.

Substitution allows us to ignore the ambiguous feedback from complex problems: we can interpret ambiguity in such a way as to justify our intuitions, even if we have to rationalize discrepancies away. We are oblivious to how easy it is for us to miss important cues, overinterpret noise and misinterpret feedback, so it is easy for us to force fit familiar patterns onto complex phenomena. In so doing, we filter out what is unique in a given situation — often the very distinctions that matter the most in understanding complexity. Most of us have not had sufficient high-quality practice to recognize the hidden regularities of complexity, notwithstanding how confident we may feel about our interpretations. True experts know the bounds of their expertise; amateurs do not know what they do not know.

Cognitive psychologists have determined that it takes until about the age of eight for children to develop sufficient mental flexibility to begin to challenge their own intuitions. Before this age, they stubbornly cling to their visceral assumptions about how things work. Psychologist Bruce Hood devised an experiment that demonstrates how obstinate children's intuitions can be: unable to overcome their all-powerful intuition of gravity, they will continually look for a ball that has been dropped down a tube, directly below the drop point, even when the tube is obviously twisted to eject the ball about a foot away from where they are looking.[90]

We adults may be amused by the rigidity of children's intuitions, but we are just as vulnerable to unyielding intuitions, even in the face of contradictory evidence. Just as young children disregard the twist in the tube because they do not perceive its relevance to the problem they are solving, we adults are oblivious to cues in complex problems because we are not accustomed to paying attention to them.

What are these cues, and why are we oblivious to them? This two-part question is the pivotal concern of this book. The short answer is that complexity's cues are those identified by complexity science (notably systems theory, as will be reviewed in chapters 13 and 14). We are oblivious to these cues because our mental models have not evolved fast enough to keep pace with the complexity of modern-day problems. (The long answer to this question starts on the first page of this book and ends on the last.)

Intuition, when it is reliable, reflects expertise: the sum of our experience with situations whose dynamics we have mastered. We can master only the situations with which we have consistent, high-quality practice. The quality of this practice depends on the opportunity for repeated learning from high-quality and consistent signals, both initial cues and those that arise from feedback. The catch, however, is that complexity rarely offers these conditions, making it difficult for us to develop expertise. This is why our intuitions are not consistently reliable, even if we feel they are — when the mismatch is not obvious. Our challenge, therefore, is to become more familiar with how complexity works; in particular, how differently it works than non-complexity. With the right kind of high-quality practice, we can develop more refined and useful intuitions about complexity: thinking that is more productive in the second of the two worlds we live within.

9

A Tale of Two Worlds
Living in Complexity

---

*The ancestors of all of us alive today were still living
in yesterday's world until 11,000 years ago.*
Jared Diamond*

Gradually at first, then gaining considerable speed, one world became two.

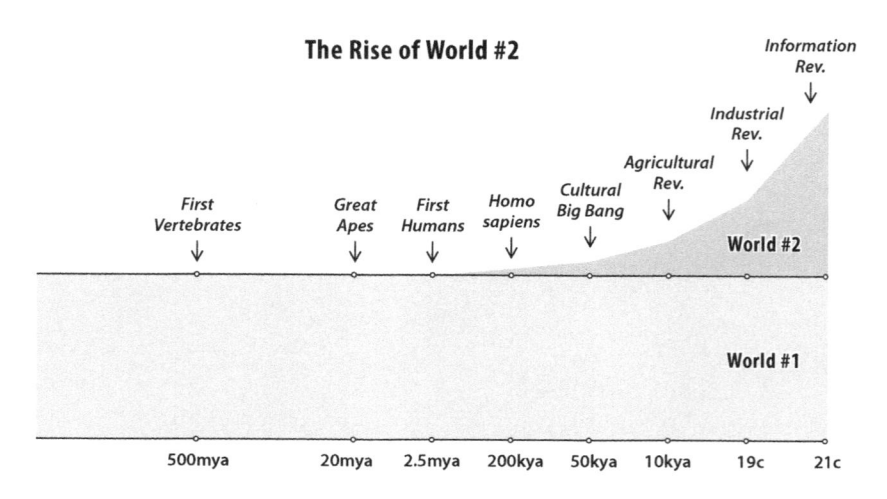

The Rise of World #2

First Vertebrates — 500mya
Great Apes — 20mya
First Humans — 2.5mya
Homo sapiens — 200kya
Cultural Big Bang — 50kya
Agricultural Rev. — 10kya
Industrial Rev. — 19c
Information Rev. — 21c

World #2
World #1

World #2 snuck up on us: as recently as two hundred years ago, most of us were farmers, using clear and timely feedback to learn the best ways to raise crops and animals. At the time of World War I,

---

* *The World Until Yesterday: What Can We Learn From Traditional Societies?* Viking Adult, 2012.

a mere century ago, farmers were still the largest segment of most countries' population, with the major exception being Britain, where the Industrial Revolution had taken hold and blue-collar industrial workers were growing in number. The industrial workforce rose quickly in developed countries, peaking in the 1950s, and then gave way to what is now the largest segment: "knowledge workers," as management guru Peter Drucker labeled them.[91] Knowledge workers cannot rely exclusively on simple cues and timely feedback to make decisions.

As this chapter illustrates, we are now living in a world that is not as friendly to our intuitions as the world lived in for thousands of years by our farming ancestors, or the one lived in for hundreds of thousands of years by our hunter-gatherer ancestors. For 98 percent of the time that humans have been on the planet, we have lived in just one world, and then complexity exploded for the human species. Consider the many ways our challenges are different now:

- **The quantity of the information that we produce and share bears little resemblance to that of the past.** Thirty thousand years ago, written communication was limited to cave paintings, which evolved to simple hieroglyphic symbols by five thousand years ago, eventually giving rise, around 700 BCE, to the Greek and then Roman alphabet. The amount and availability of information exploded with the invention of the printing press in the mid-fifteenth century (before this invention, most people were illiterate). The binary, digital language of computers was invented in the late 1930s by Claude Shannon: computer language enabled unimaginable amounts of information to be coded and stored in minuscule packets that could be transmitted and decoded almost instantly. The growth rate of digital information is accelerating based on the ever-increasing processing power of computer chips and increasingly sophisticated software programs.

- **Wealth is created in very different ways.** Just a century ago and for thousands of years before that, wealth was directly tied to the control of physical resources. The global economy has transitioned from hard assets, like land and crops, to soft assets, like scientific discoveries and service-industry innovations such as financial engineering. Just a handful of generations ago, 90 percent of us earned our living in farming communities, where we all had similar daily routines. In developed countries, this percentage was still 60 percent one hundred and fifty years ago; today it is only 2 percent. No single occupation has replaced farming because the advent of the computer has dramatically expanded the variety of professions.

- **Social complexity has mushroomed.** The research of anthropologist Robin Dunbar suggests that our brains have evolved to maintain about one hundred and fifty social connections: the "ideal" human community consists of roughly that number of people.[92] Even after the Industrial Revolution, the average person met at most twice this number of people in his entire lifetime, since his travel was restricted by the speed of a horse. Today we meet hundreds of thousands of people in our lifetime through a web of interconnections that has completely altered the way we live together. Since the dawn of civilization, the dominant social organization has been hierarchical, in which the power base of different classes was based on property ownership, with most of the power in the hands of a select few and usually one all-powerful individual who orchestrated everybody else's activities. The organizational structure of today's society is increasingly represented by networks: power is distributed among many individuals and groups of people. Whereas a hierarchical system cannot be any more complex than its master controllers at the top (who control the pyramid of people below them), a socio-economic system based on a network formation can be more complex than its parts, because power arises from the multiple interconnections within the network.

- **More of us are around longer.** Our life expectancy averaged about twenty years until the agricultural revolution, when it started to climb gradually; by the time of the Greek and Roman civilizations, we lived on average twenty-five years. As the Industrial Revolution took hold in the eighteenth century, we lived to about thirty-five. These average ages reflect a much higher level of infant mortality than exists today, skewing the averages lower. Nevertheless, safer living conditions and modern medicine have bolstered life spans in developed nations to close to eighty years. More of us live longer, and face far more options regarding how to spend our extra time on the planet.

- **Our goals, desires and values have all taken on a degree of complexity that our great-grandparents never could have imagined.** Just over a century ago, a marriage contract was largely an economic agreement that, for example, underpinned the operation of a farm: the man provided for and protected his family, and the woman raised children and managed the household. Divorce in an agrarian society is highly impractical and therefore was rare. The idea of switching partners because one's romantic expectations were not being met was unheard of. Issues like "poor two-way communication" and "un-nurtured self-esteem" were unintelligible. Modern obstetrics alone has reduced infant mortality to the extent that women do not have to endure multiple pregnancies to create a family, thereby liberating them to pursue other goals. Advances in science are forcing us to confront moral questions that were not relevant to our ancestors, such as stem cell research and euthanasia. As we collectively spend less time and energy worrying about harvesting crops and having babies, our low-order discontent is being replaced with higher-order discontent. The more our basic living needs have been met, the more pronounced have our higher-order needs become; for example, our desire for intellectual stimulation, for influencing the community and for creating a sense of personal meaning. The increasing complexity of our personal objectives translates into the increasing complexity of our relationships with one another, as our personal choices interact with the choices of others.

Our ancestors inhabited the world in which all animals live: a world composed entirely of straightforward survival and reproductive challenges; a world in which getting through another day depends exclusively on immediate sensory experience. This world's signals are clear, because they are easy to differentiate from the noise that surrounds them, and they are easy to interpret. This is World #1, a world in which we spend little time thinking about our decision making, because natural selection has equipped us to navigate it almost without effort. This world continues to feature prominently in our lives today.

But we are also living in another, quite different world. Once our brains reached a tipping point somewhere in the stretch between a hundred thousand and fifty thousand years ago, our earliest forms of conscious awareness enabled language, culture and innovation, and we began to create a new world for ourselves. Eventually, the number of us living on the planet began to increase exponentially. More people generate more innovations.

Having good ideas is one thing, but being able to communicate them so they can be reproduced and built on by others is a hallmark of human development. We are not limited to the gradual, vertical transmission of new habits to our progeny; we have the ability to share good ideas instantly with a large population. This "horizontal transmission" of ideas means that the impact on our lifestyles can be altered quite radically over a short period of time. Cumulative cultural evolution appears to be unique to humans: other animals can learn new foraging techniques from one another, but there is currently no compelling evidence of their ability to accumulate innovations from one generation to the next. We do not have to wait for natural selection to pick out survival-enhancing features over many generations; in fact, we are capable of subverting natural selection, as in the case of modern medicine, which protects weaker members of our species from perishing.

The explosion of human innovation has translated into greater interdependence, both with one another and with the planet; in other words, more complexity. Our geological epoch is referred to formally as the Holocene, which began when the last ice age ended, nearly twelve thousand years ago. But an increasing number of geologists are using a different term to describe all or part of this epoch: the Anthropocene. This designation is intended to reflect a unique geological development: the increasing influence that humans have on the world's ecosystems.

9.1 Two Distinct Worlds

The theme of two worlds is reflected in the writing of various academics in different fields. Although the work of each has a distinct objective and context, there is a commonality to their terminology:

	World #1 ⟷ World #2	
Daniel Kahneman/Gary Klein	High Validity (high regularity and predictability)	Low Validity (low regularity and predictability)
Robin Hogarth	Kind (relevant feedback)	Wicked (irrelevant feedback)
Nassim Nicholas Taleb	Mediocristan (normal distribution)	Extremistan (non-linear distribution)
E. F. Schumacher	Convergent Problems (arising from non-living things)	Divergent Problems (arising from living creatures)
Kenneth Hammond	Physical Perceptions (correspondence is test)	Conceptual Frameworks (coherence is test)
Keith Stanovich	Benign (heurisitic-usable cues)	Hostile (lack of heuristic-usable cues)

Each of these distinctions reveals two categories of problem sets that we face.

For Kahneman and Klein, a high-validity environment is one in which cause and effect are closely linked and the cues available for interpreting are valid and easy to discern.[93] The patterns that facilitate interpretations are consistent, and predictions within this environment are reliable. In contrast, a low-validity environment is one in which the regularities or patterns are difficult to detect, and they are sometimes non-existent. The cues that are available are often misleading, because they are ambiguous and easy to misinterpret. Anesthesiologists work in a high-validity environment in which feedback from the patient in response to the dosage of a drug is clear and nearly instant. Psychotherapists are generally limited to low-validity environments: the effects of their therapy are difficult to distinguish and are separated from their interventions by long intervals. Short-term weather forecasting operates in a high-validity environment; long-term weather forecasting operates in a low-validity one.

Hogarth classifies the two-world distinction as kind versus wicked.[94] A kind environment is easy to understand and navigate, because the feedback we get from it is relevant; mistakes can be easily identified and corrected. A wicked environment is difficult to decipher, however, because the feedback it generates is often irrelevant by virtue of its ambiguity and high level of inherent randomness; mistakes are likely to be significant and difficult to correct. A professional tennis player developing intuitions about how best to play the game is operating in a kind environment. A human resources recruiter, attempting to determine which top-tier job candidates are most likely to succeed, operates in a wicked environment because learning from the candidate selection process is severely limited. The feedback on the candidates' ultimate success is delayed and "dirtied" by many other factors, such as the projects that they are assigned to and the supervisors who oversee them. Rejected candidates are never assessed to determine how they would have done if they had been given the opportunity.

Taleb names his two worlds Mediocristan and Extremistan.[95] Mediocristan is the world we are accustomed to thinking about. It is defined by a normal distribution of variances around the average or typical example; there are exceptions, but they remain within a limited range. If we consider human height, which varies within a tight range from shortest to tallest, the average height is one that many people share. Extremistan is the world in which there are no typical examples because the exceptions can be enormous in size and consequence. It is defined by unanticipated, extreme events; randomness is much less predictable in this world because it is so difficult to statistically model. "Black Swan" events like severe stock market crashes and 9/11-type terrorist attacks arise in Extremistan: they overshadow whatever regularities can be detected.

Economist E.F. Schumacher noted that convergent problems have single solutions. Divergent problems, however, have more than one possible solution, because there is no single way to definitively solve them.[96] Non-living things generate problems that lend themselves to convergent solutions, whereas living things generate problems that diverge to many solutions, with no way to determine which is the ultimate answer or best solution. Transportation on two wheels is a convergent problem, for which the bicycle provided a single solution: while the bicycle has been tweaked over the past century, its basic design has not changed because its structure is the best solution. But the education of children is a divergent problem that lends itself to numerous approaches, many of which continue to be experimented with.

Psychologist Kenneth R. Hammond highlights how we rely on our physical perceptions to achieve empirical accuracy: the correspondence between what we perceive and how the world really is.[97] But where conceptual frameworks are concerned, we cannot ensure perfect correspondence with reality, so we depend on seeking conceptual rationality, defined by the coherence of our ideas both with each other and within a logical structure. These two forms of truth — correspondence and coherence — define two ends of a spectrum of problem types. At one end of the spectrum is clinical medicine, in which experiential learning allows doctors to rely largely on correspondence to make reliable diagnoses. At the other end is theoretical physics, in which the coherence of the overall framework is a determinant of the framework's intelligibility.

In Keith Stanovich's view, benign environments offer cues that are exploitable by our shortcut rules (heuristics), but cues in hostile environments are both obscured and unusable by these same heuristics.[98]

Although each writer uses different terminology and has a different approach, each draws attention to a distinction between two different problem environments: one that is kind to our intuitions because of its regularities, predictability and easy-to-verify accuracy, and one that is hostile to our intuitions because its regularities are much harder to detect, the feedback it offers can be very misleading and arriving at singular, definitive conclusions is unlikely.

World #1 is static and represents the set of challenges that resembles those of early *Homo sapiens*. World #2 originated only as our species began to develop the higher-order thinking that enabled more complex social interactions and technological innovations; it continues to evolve into greater degrees of complexity. Today, we occupy both worlds simultaneously. Each world contains different problem sets that require different types of thinking. Our intuitions are well suited to World #1 because we have developed solid expertise from our experience in this straightforward world. But these same intuitions fail us in World #2, because they are not as useful and we have not yet developed sufficient expertise with complexity. When we fail to distinguish these two worlds, we end up treating World #2 as if it were World #1. Because our automatic intuitions mislead us in the second world, mistaking it for World #1 is the root of so many of our individual and collective challenges.

It is not that World #1 is easy. There is nothing uncomplicated about a hungry person hunting for food, or a commuter figuring out the fastest route to her office. These tasks involve multiple steps based on the different forms of experience needed to execute them. As difficult as these challenges can be, they are nonetheless straightforward: the tasks are well defined, the desired outcomes are clearly envisioned, the cause-effect cues that define the problem are tightly linked, the steps required can be laid out and the relevant feedback along the way indicates progress toward the goal. World #1 is not an easy world to live in: it is not benign or kind in the sense that everything is taken care of for us. But the challenges it tosses our way are not difficult to interpret, so responding to them is not guesswork.

World #2 is different. Consider the challenge of negotiating an armistice with an erratic and nuclear-equipped despot. Much of the important information that we need to make reliable predictions is hidden, such as the ultimate motive of the despot: he likely has many motives, some of which conflict with one another, and his priorities will change depending on the situation he finds himself in. In reaction to our interventions, he may or may not respond the way that other despots have in the past, or the way that he himself has responded in the past.

In a straightforward scenario:

- The cues we need to interpret the situation are immediately available.
- The patterns that these cues form represent regularities that we can recognize, because they match those that we have stored from previous experience.
- The predictions that we make are narrow and reliable.
- The feedback we get is instantaneous and direct.

But in a complex scenario, such as the negotiation:

- Many of the cues are hidden: we cannot be sure if the despot's threats are real or a bluff.
- The patterns formed, by whatever cues are available, cannot be easily matched with "how typical despots respond," because there is no typical despot in a typical situation.
- We can make many predictions about how he will respond, but we cannot be confident in any of them (game theory, for example, can take us only so far in understanding a madman's motives).
- Whatever we do, the feedback that we receive does not necessarily increase our learning about the situation — just because he backs down this time does not mean that he will do the same next time.

In World #2, we cannot rely as heavily on visual cues to make sense of the numerous, intricate interrelationships that define a complex problem. A three-leafed ground plant presents sensory data that, when mixed with our past experience of and learning about dangerous plants, generate conclusions closely linked to the visual cues that we detect. But making sense of a rebellious child, a mean-spirited neighbour, a failing company or an economic collapse requires more construction on our part, because the immediate visual cues are only loosely and indirectly linked to the proximate and ultimate causal explanations. Making sense of World #2 requires our unique human powers of hypothetical thinking (a.k.a. imagination) and critical deliberation. More active cognitive assembly is required to extract relevant cues and model interpretations. The more we have to construct in order to interpret, the greater the possibility of interpretive error, since there are fewer "hard sensory facts" to justify our conclusions. This is why our confidence in World #2 is often unwarranted.

Complexity is a gradient: the two worlds overlap at the point where problems are neither straightforward nor wildly divergent. But beyond this overlap, World #2 is not just "the same as #1, just more complicated." World #2 is meaningfully different in the way that a worm's physiology is meaningfully different from a human's, even though intermediary species of varying complexity lie between them.

Which brings us to the structure of World #2 and the nature of its complexity. What is it that gives rise to cues that are ambiguous, feedback that is deceiving and learning that is so difficult?

9.2 World #2 Complexity

The everyday use of the word "complex" captures much of what the more scientific version includes: something is complex if it is composed of many interconnected parts arranged in a way that makes understanding the thing difficult. The challenge of complexity stems from those many interconnected parts. It is not the *number* of parts that generates complexity; the number is a contributing, but not an exclusive, defining factor. The key to complexity is the *intricacy of the interactions* among the parts.

A car engine consists of hundreds of distinct parts, but the interactions among the parts are mechanistic: nothing too mysterious happens under the hood of a car, at least not to a well-trained mechanic. The straightforwardness of the interactions renders the engine understandable, predictable and therefore not complex.

A human being is also composed of a limited number of parts, some of which interact in complicated but straightforward ways. But many of the interactions are sufficiently intricate that describing them as complicated does not do justice to how they operate. We could spend an entire lifetime taking every possible course on the biological and psychological workings of a human being, but all that learning would not give us anything close to the ability to make highly accurate predictions about how a person will respond in a large variety of situations. Unlike the complicated workings of a car, which can be captured in a thick book or taught in a course, we cannot condense the information needed to fully understand human beings in order to make consistently reliable predictions about how they will behave in all cases.

Complexity scientists typically measure complexity by the amount of information required to describe something: the longer the explanation or formula needed to capture its behaviour, the more complex it is. Whereas it does not take a lot of information for us to describe and predict the response of a pigeon reacting to seed being thrown on the ground, it does take a graduate degree in international relations and years of working in the diplomatic service to make reasonable predictions about a despot's behaviour under different circumstances. The challenge of understanding complexity is to ignore enough extraneous data to make the phenomena intelligible, but to not ignore so much that we miss key information. This challenge is easy with straightforward situations, because signal and noise are readily differentiated: simple systems have patterns that are easy to identify because their cues are obvious, clear and familiar. Purely random events have no patterns and therefore no cues that can be used to interpret or predict their short-term behaviour (although, as reviewed in chapter 14.6, randomness can reveal some long-term patterns).

Between simple systems and randomness lie complex systems. They have patterns because, unlike purely random noise, useful information is encoded in them. Their regularities are difficult to detect, though, because the cues that signal these patterns are obscured by a lot of noise, and the patterns themselves are often unstable, shifting and changing, forming and dissolving.

Complexity had its start as a science in the analysis of the interactions between things: it was formalized in the field of cybernetics in the late 1940s by the American mathematician Norbert Wiener.[99] One of Wiener's main interests was randomness and how to separate out noise from signal in electronics. The same year, 1948, that Wiener published his book *Cybernetics*, Claude Shannon, another American mathematician, published a pivotal paper on information theory, which demonstrated how the combination of simple elements can create complexity through their interactions.[100] *Complexity arises from the interconnections of parts, the number and intricacy of which preclude a model from fully capturing all the properties of the system.* The last word of the previous sentence is pivotal: in the language of complexity science, "complexity" is shorthand for "complex system."

Systems

Everything can be viewed as a system — as the interaction of parts. But some systems are more complex than others.

The system that defines a rock falling to the ground is about as "least complex" as you can get: there are two parts and one force that make up the system — rock, ground and gravity. The causal relationship is straightforward, because it is a one-way direction (the rock has less mass than the earth, so it moves toward the earth) and because it is linear (the speed of the rock's descent accelerates at a constant rate). But put that single rock in a large collection of rocks, lying on the edge of a steep mountain, and we have a system that is more challenging to understand and predict. The collection of rocks is more complex because its interactions are more intricate.

In the rock pile, the earth's gravitational pull is still the primary driving force, but many other factors are in play. Each rock in the pile has a direct influence on the rock that it is touching and an indirect influence on all the other rocks. Throw in a bit of wind, some precipitation and some additional rocks that may tumble down the mountain and land in the pile, and we have the makings of a complex system that eventually culminates in an avalanche. We can examine the pile of rocks very carefully, as well as all the rocks on the mountain that may tumble into the pile, and we can make very detailed assessments of likely weather patterns, but we will still be hard-pressed to predict accurately when the pile will start to crumble, how it will fall apart and which rocks will be left in the pile after it collapses. All the computer modelling in the world still limits us to best guesses, because the system that contains the making of an avalanche cannot be reduced to a finite amount of information. There are a lot of variables to account for, and they interact with one another through feedback loops that can dramatically alter the behaviour of the system as a whole. For example, a small pebble tumbling down the mountainside and landing on the pile might be all that is needed for the entire pile to collapse and an avalanche to ensue.

It is these features that make the interactions of a complex system more intricate than those of simple systems: multiple causal factors that make up the interactions (multivariate causation); and the back-and-forth influences that these causal factors exert on one another, which are often disproportionate to their size (nonlinear feedback loops). The dominant model of basic, linear cause-effect that works so well for straightforward things does not capture the behaviour of complex systems, where effects become causes of other effects, in a recursive and constantly changing "dynamical" cycle.

So numerous and intricate can the interactions between parts be, that complex systems give rise to behaviours that are not found in the parts themselves — characteristics of the systems that *emerge* from the parts but are not present in them. Emergent behaviour makes predictions about complex systems difficult because combining the parts changes the parts themselves, so we cannot just add up the behaviour of the various pieces to arrive at their combined behaviour. We cannot model or predict a traffic jam by examining individual cars, since the interaction between them is not contained in each separate car. We cannot model or predict an ant colony by dissecting individual ants. We cannot model or predict a stock market by examining each company. Nor can we model or predict human consciousness by examining the neurons and synapses that connect them, out of which emerges what we call "mind." None of the component parts of a brain are conscious, but the incredibly dense web of interactions among all the parts gives rise to consciousness. The macroscopic properties of the whole (e.g., consciousness) are not shared by the microscopic components (e.g., neurons) from which the macroscopic properties emerge.

Emergent behaviour makes modelling and prediction especially difficult when you get to the myriad systems that spring out of human interaction, such as those contained in a marriage, a business, a political movement or an economy.

9.3 The Nature of Complex Problems

World #2 problems are qualitatively different from World #1 problems. Schumacher's distinction between convergent and divergent problems — those that converge to a single solution versus those that diverge into many possible solutions — represents the difference between the two worlds. World #1 problems are of the former variety: they are tidy. Many of our World #2 challenges, however, diverge into many possible solutions, none of which we can be sure is optimal. Unlike what mathematicians refer to as "intractable problems," which are theoretically solvable but there is not enough computational time or power to solve them, divergent problems have more than one solution. The choice of what career path to pursue, for example, diverges into many possibilities, since most of us have more than a single talent and single interest.

The more extreme divergent problems are "wicked" because they are difficult to pin down: they are constantly shifting; they do not sit still long enough for us to be able to understand and solve them. Solutions are rarely final, because as soon as we interact with these problems, we change their dynamic. Consider, in contrast to the career-path choice, a newly married couple who argue frequently about household expenses. Their problem is divergent because there are many ways to solve the friction in their marriage; different marital therapists will recommend different approaches. But it is an especially difficult divergent problem, because the underlying causal factors of their tense interactions are legion and can change depending on circumstances or their moods on a given day. Contributing to the problem are the habits that they absorbed as children in distinct households, their individual values about money, their hard-to-define values about saving and living for the future rather than for the present, their individual tolerance for risk versus security, their need for the status that material consumption signals and so on.

There may be many possible strategies that would help them reconcile their conflicting spending habits, but none can be guaranteed as a long-term solution. The couple can experiment with different strategies, but even if they have success, the problem might resurface with a vengeance if one of them loses a job or has a sibling who asks for a loan or one of their kids crashes the family car.

There is no definitive line that, when crossed, earmarks a divergent problem as a wicked one, but the list of wicked problems is endless.

- Other people are "wicked" insofar as their goals and operating styles are rarely transparent, and they are frequently difficult to predict and interact with. Despite myriad books touting facile solutions, developing satisfying romantic relationships is among the most difficult challenges that human beings confront. Raising children is no less wicked a challenge, and again, there is no shortage of books on the subject, many of which contradict one another. It would be naïve to expect that there is a single parenting solution for every child in every situation. Some situations and moods are amenable to one approach, others to a different approach. Interacting with others is so wicked partly because they anticipate your moves and respond to your attempts to influence them in often-unexpected ways, thereby changing the dynamic between you. (This example demonstrates why Schumacher proposed that divergent problems arise from living things.)

- Company strategy diverges into many possible opportunity sets. There rarely is one best way forward for a company that operates in a highly competitive industry; rather, there is a series of options, all of which make sense but none of which can be proven optimal even after one has been chosen.

- International terrorism is a classic wicked problem, because determining its causes is so difficult: Is it poverty that makes people prone to terrorism, or indoctrination from an early age, or an imbalance in the male–female ratio in a country that makes young males feel hopeless, or religious fundamentalism or (most likely) some combination of these and other factors? If you cannot clearly define the root problem, how can you implement optimal solutions? How do you determine if it is even productive to negotiate?

- Most large-scale public policy initiatives are wicked: dealing with the rising cost of health care, narrowing the income disparity between rich and poor, arresting urban decay while simultaneously reducing budget deficits, improving access to high-quality education and so on. The unintended side effects of policy initiatives usually come back to haunt, despite best efforts to plan and predict.

It is tempting to assume that there is always a single best choice, because that is exactly how our minds function: we want optimal solutions. It is even more tempting to review past decisions with the hindsight bias that makes choices look straightforward in retrospect. Hindsight, however, does not convert divergent problems to convergent ones, because we can never compare the chosen path with the outcomes of alternative paths. As we become increasingly interconnected and our systems become increasingly complex, convergent problems morph into divergent ones, of which an increasing number are the wicked variety, based on multiple interrelationships between multiple causal factors.

Problems That Cannot Be Solved

Early Eastern philosophers were acutely aware of the confusing nature of unsolvable problems. Hinduism and Buddhism both propose that contradictions arise when we contemplate the world, because reality does not fit our tidy, conceptual frameworks. We misunderstand the holistic nature of reality by trying to compartmentalize it; however, we cannot "freeze" bits of reality with our mental models and expect that these bits will all fit together in a coherent way.

Kant demonstrated the problem of conceptual contradictions when he documented his antinomies, which purport to demonstrate how two mutually exclusive and contradictory claims can be both logical and intelligible.[101] His first antinomy notes that time and space cannot be both limited by having starting points and unlimited by having none; yet both propositions make sense on their own. They are both intelligible but contradictory, Kant argued, because we use reason illicitly when we assume space and time are real structures of things-in-themselves instead of mental constructs.

One of the interesting paradoxes of being human is that we are capable of asking questions that seem intelligible and important but that we may not be able to answer. What is our collective and individual purpose? How does consciousness arise from neuronal complexity? How can a physical being have free will? Cognitive scientist Stephen Pinker notes that just because we can imagine the existence of an infinite number of "thinkable thoughts" does not mean there are cogent explanations for every problem we conceive.[102] "Cognitive closure" is a property of the mind that philosopher Colin McGinn sums up as the insufficiency of "our organ of reason" to attain truth for every conceivable philosophical problem.[103] Some Western philosophers argue that playing with ideas like freewill and goodness is equivalent to a cat playing with yarn — fun but pointless. This conclusion seems plausible when we survey the historical back-and-forth thinking on these seemingly unsolvable problems, but most contemporary philosophers would still prefer to say that we have a more coherent and intelligible sense of abstract ideas today than did earlier thinkers.

Unsolvable problems may represent the outer boundary of our cognitive limits, where the problems themselves may be poorly defined by language attempting to go beyond what it can usefully describe. To assume that there is an answer for every conceivable question we pose is to misunderstand the nature of our minds and the limits that define them. Sixty years after Kant, the philosopher Søren Kierkegaard wrote, "The supreme paradox of all thought is the attempt to discover something that thought cannot think."[104] A century later, philosopher Albert Camus defined absurdity as the result of reason ignoring its own limits.[105]

It is not that complexity was absent throughout our entire history: it has always been with us. But complexity's various forms were not as relevant to our distant ancestors because they did not need to understand them in any depth in order to survive. And they had little control over large-scale forms of complexity, like prolonged droughts. It is not just that the world is throwing more complexity at us; we ourselves aspire to grander, more complex ambitions. Not only do we face monumentally complex problems as a species (exponential population growth in underdeveloped countries, environmental degradation, and nuclear/biochemical technology proliferation, to name just a few), but we also, as individuals, face an increasing number of personal challenges unique to our modern world.

As we maneuver through our twenty-first-century world, complexity matters more and more, because our challenges demand a deeper perspective to address them: an understanding unconstrained by the interpretations and conclusions that our automatic intuitions limit us to. Architects do not need to worry about the curvature of space when designing buildings; they get by with Euclidean geometry, despite its limitations. But astrophysicists do have to contend with curvature, so they need Riemannian geometry to accomplish their objectives. Similarly, our basic intuitions work just fine in World #1 — hence our dependence on them. But an increasing number of human challenges are located in World #2, which operates outside of the familiarity of World #1. In this new world, we are vulnerable to the gap between our default way of thinking and how World #2 really works.

10

The Intuition–Complexity Gap
The Brain That Did Not Change Itself Fast Enough

*Our sophistication continuously puts us ahead of ourselves,
creating things we are less and less capable of understanding.*

Nassim Nicholas Taleb*

The human brain is a vast system of systems. But despite the growth in its size and intricacy over thousands of years, we still live with the mental constraints of a very limited conscious working memory and a very slow conscious processing speed. We have astoundingly deep expertise in managing the myriad straightforward problems that confront us, but when we apply our intuitions to problems where our proficiency is underdeveloped, the results are usually disappointing, and sometimes even calamitous. The following tragic story not only illustrates this gap vividly but also serves as an analogy for how we misapply our intuitions in contexts for which they are ill suited.

> When John F. Kennedy Junior stopped at a gas station on his way to the Essex County airport, on a warm Friday evening in July 1999, he had a lot on his mind in addition to the flight he was about to pilot. His plan was to take his wife, Carolyn, and her sister, Lauren, first to Martha's Vineyard, where Lauren would be dropped off, and then proceed to Hyannis Port for his cousin's wedding, scheduled for the following day. They were supposed to depart at 6:00 p.m., but that afternoon Lauren had requested a delay to their departure, and by the time she had joined Kennedy at his office in New York City for the

---

* *Antifragile: Things That Gain From Disorder*, Random House, 2012.

ride to the airport, traffic had swelled with everyone leaving the city for the summer weekend.

Kennedy was likely distracted as he picked out a banana and a bottle of water. It had been a busy day of meetings, and he was in pain: the day before, he had a plastic cast removed from his left ankle, which he broke six weeks earlier while paragliding. His marriage was purportedly on shaky grounds, and the magazine that he founded — *George* — was in serious financial trouble.

After arriving at the airport around 8:00 p.m., he did the preflight checks on the Piper plane he had purchased a few months earlier. The Piper was notably faster than the small Cessna he had previously flown. At 8:26 p.m. the sun set, and at 8:38 p.m. Kennedy taxied to the runway with his two passengers and took off.

The plane never arrived at its first destination. No one heard from its pilot or passengers. The Kennedy family became worried and contacted the Coast Guard. It took five days for the plane to be located in the Atlantic Ocean. The three bodies were recovered, still strapped in their seats. The National Transportation Safety Board identified the probable cause of the accident: "The pilot's failure to maintain control of the airplane during a descent over water at night, which was a result of spatial disorientation."

What happened? Kennedy was a new pilot. He had flown just over three hundred hours; fifty-five hours of that had been at night, including nine in his new plane. He had received his pilot licence the year before, but had not completed his instrument training, which solidifies a pilot's ability to rely exclusively on the cockpit's instrument panel, rather than the visual cues available through the windows. In fact, had the visibility been a little worse before his departure, Kennedy would not have been allowed to make the flight without a certified instructor accompanying him.

When you are flying during a clear day, you have the benefit of seeing everything around you: you have a reliable, visual sense of where you are heading, which is confirmed by the instrument panel indicators. If you are flying toward a target, you can see it; if you are ascending, you can see the sky above you; if you are descending, you can see the earth approaching. The visual cues are easy to detect and unambiguous.

But at night, it is completely different: you are flying blind. When the plane starts to careen, you cannot decipher your direction by looking out the window, so you have to focus on the instrument dials. But you also have to resist whatever signals your own body is sending. Your inner ear has mechanisms designed to detect the pull of gravity and give you a sense of balance, but the inner ear is designed to work in conjunction with visual cues, and, as we all know from spinning around quickly, our sense of balance and orientation can easily be thrown out of whack. In a plane, your sense of balance and orientation will be compromised if your brain does not have the benefit of visual cues; a plane can descend quite rapidly even though you feel that you have not changed altitude at all. If your internal signals conflict with the instrument readings, you have to decide whether to believe the technology or your own intuitions, and the pull of your intuitions is extremely strong.

An experienced pilot has developed the required discipline to override his intuitions and prioritize the panel readings; inexperienced pilots tend to default to their instincts, especially when they get anxious. At night, your intuitions can get you into a lot of trouble very quickly. When your brain is telling you that you are level and the instrument panel is telling you that you are heading down, it is very hard, in a moment of panic with terrified passengers screaming at you, to assess the situation speedily and accurately and take corrective action. Flying into Martha's Vineyard is difficult at night if you approach it from above water, as Kennedy did. It is difficult to differentiate the water, the sky, and the land in the horizon: everything just looks dark. To make matters worse, a thick haze hovered near his destination that night. A review of the radar history of his flight before the crash revealed an erratic flight pattern: the plane had ascended and descended repeatedly and abruptly, culminating in a rapid descent. A downward spiral gathers a lot of speed very quickly if you do not know how to quickly level the wings and gently pull up the plane's nose. It took less than a minute for Kennedy's plane to make its final rapid descent, hitting the ocean with tremendous force.

You do not have to be a pilot to appreciate the challenge of poorly calibrated intuitions. When the car you are driving abruptly starts to skid, your brain will tell you to steer aggressively in the direction opposite to where your car is sliding. A trained and experienced driver knows that to get out of a skid, you have to point the wheel in the direction of the skid. Skidding requires a mental override, because our default intuitions are unreliable in this situation. Most of us do not have sufficient experience with skidding to have developed useful intuitions about it, just as new pilots' night-flying experience is limited, making them vulnerable to over-relying on their daylight flying intuitions. The driving and piloting skills that serve us so well on dry roads and in clear skies are insufficient for more complex situations, which require more sophisticated responses than our default intuitions offer.

The three deaths were tragic. They were also instructive, not just in flying, but in life. We have been extremely well trained by our experiences, and those of countless generations before us, to cope with straightforward "daylight flying" challenges that have confronted us for millennia. But we are not well trained for the complex "night flying" challenges that have exploded onto the human scene, and continue to multiply exponentially. Unlike night flying, however, where we sense that the dark environment is different, we are not adept at distinguishing the straightforward from the complex, so we are even more prone to overconfidently rely on our default intuitions, tail-spinning without even realizing it. Unlike inexperienced night pilots, who are prone to panic, we often have a false sense of being in control when navigating complexity, until our expectations of how things will unfold are not met. As reviewed in 2.3, the Flynn effect suggests we are getting smarter. But we are not getting smarter fast enough. World #2 has exploded onto the scene like a second big bang, and as quickly as we are learning to think more abstractly, we are not keeping pace.

One of the greatest obstacles to faster conceptual development has been the lack of pressure to think differently: we are still here because we have developed highly reliable intuitions about how to interpret and handle World #1 problems. The few and fairly inconsequential errors our mental shortcuts generate are a small price to pay for the fast decision-making expertise that serves our survival in the face of straightforward challenges. As reviewed in 2.5, this is the paradox that traps us: we are only as smart as we are *most* of the time precisely because we are not that smart *some* of the time. We could not be as quick as we need to be to respond to our World #1 environment if we were constantly ruminating about the accuracy of our intuitions or the potential of alternate interpretations for the situations we face.

But the *some-of-the-time* occasions, where automatic thinking does not work, are increasing: the shortcuts are generating more and larger errors. The "most-some" trade-off that has historically made our brains powerful is no longer balanced because our problems are increasingly of the World #2 variety. Being unaware of this brain–world gap makes the problem more severe and results in responses that are often unproductive and occasionally quite destructive.

Somehow, we have to retain the speed and effectiveness of our default automatic thinking for World #1 tasks, while applying a different approach to World #2 problems. So, the relevant question is not, *"How smart are we overall?"* but, *"In which circumstances are we smart and in which ones is there a gap?"*

10.1 Errors of the Gap

The brain–world gap in World #2 arises for two reasons: our tendency to oversimplify (by overextending the simple heuristics that work in World #1), and our tendency to be overconfident in our conclusions (by overextending the satisficing that works in World #1). These are the two fronts in the battle to improve our thinking. Oversimplifying and overconfidence encourage us to see things differently than they are in World #2.

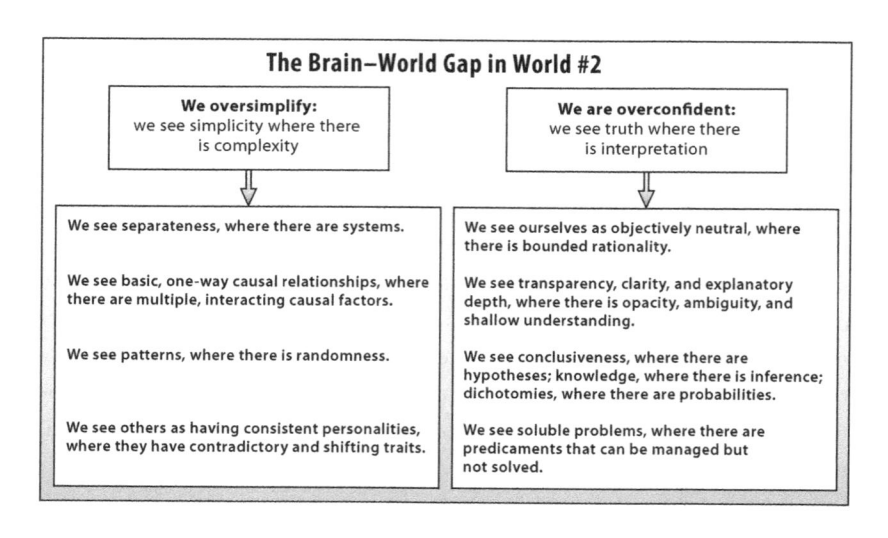

The Brain–World Gap in World #2

We oversimplify: we see simplicity where there is complexity	We are overconfident: we see truth where there is interpretation
We see separateness, where there are systems.	We see ourselves as objectively neutral, where there is bounded rationality.
We see basic, one-way causal relationships, where there are multiple, interacting causal factors.	We see transparency, clarity, and explanatory depth, where there is opacity, ambiguity, and shallow understanding.
We see patterns, where there is randomness.	We see conclusiveness, where there are hypotheses; knowledge, where there is inference; dichotomies, where there are probabilities.
We see others as having consistent personalities, where they have contradictory and shifting traits.	We see soluble problems, where there are predicaments that can be managed but not solved.

Our tendency to oversimplify (sometimes referred to as the "heuristic bias" or "bias for reductionism") generates our misunderstanding of complexity. Our tendency to be overconfident (sometimes referred to as the "analytic bias" or "functional fixedness") means we miss alternative perspectives that are more productive.

The Oversimplifying and Overconfident Brain

Our brains are information compressors; we take in a lot of data and categorize it with just enough detail to be useful for us. We do not have the time or cognitive energy to select, model and storytell in excessive detail, but we cannot afford to miss important cues, either. It may be irrelevant to note the number of veins on a patch of poison ivy, but if we perceive it as just greenery, we will fail to note the trifoliate structure of the plant that signals danger. World #2 requires a deeper analysis of detail than we are accustomed to, because the cues that reveal how it operates are not readily apparent to the untrained mind. World #2 demands that we compress data into chunks that are less coarse and more refined than World #1 — otherwise we will miss the signals that are crucial to understanding and responding to complexity.

We have good reason to be confident in our judgments about World #1. The majority of data we process is visual and, except for the occasional optical illusion, our eyes are (more specifically, our visual cortex is) quite reliable. Considering the hundreds of micro judgments we make every minute, the accuracy rate of our visual processing is very high — so high that we run our day-to-day lives based on the assumption that our view of reality is perfectly matched to the way things are and so there is rarely any need to second-guess ourselves. We select data that are relevant, model the data with heuristics that are suitable, and generate interpretations that are feasible, all the while assuming that the first reasonable story we come up with is a good bet.

The speed and decisiveness of satisficing does not accommodate the deliberation required for understanding World #2 because the early conclusions we draw about complexity are less reliable than those we draw about straightforward phenomena. Our constructions in World #2 are necessarily more elaborate because the cues we need to understand complexity are not exclusively visual: we have to make a greater number of inferences to interpret, and the further we get from relying on visual data, the less reliable is our information because our input into the construction process rises exponentially.

Because our inferences in World #2 are more elaborately constructed, and are not immediately verifiable with our instantaneous perceptions, most of us, most of the time, do not challenge the certainty of our judgments. Yet second-guessing and exploring alternatives are precisely the cognitive strategies that constitute high-quality thinking in World #2, where our expertise is underdeveloped.

As a result of our modern-day abundance of simple-carbohydrate foods, there is a gap between the environment our bodies were designed to thrive in and the environment our bodies currently occupy: a gap we address by planning balanced meals, dieting, and going to the gym. It is not that our automatic craving for the sweet and the salty is no longer useful: it is still important that we crave sweet fruit and salty protein. But these default desires are too restrictive and too inflexible to be useful all the time now. In the same way that we understand the origins of our automatic food cravings and correct for them, we need to understand the origins of our automatic thinking and correct for it when necessary.

10.2 Closing the Gap

Other people's problems are usually much easier to solve than our own. It is not uncommon to find ourselves asking, "Why doesn't she just leave him already?" or, "Why doesn't he just get another job?" or, "Why haven't they bombed those nuclear plants by now?" The answer to these questions is the same: people do not behave the way we think they should because we do not have all the information they have. We see other people's problems more clearly than our own because we do not have to integrate into our analysis all the bits and pieces that are important but invisible to us as outsiders.

The friend who "should just leave her husband" may portray her spouse as inconsiderate, but that is because his negatives are top-of-mind, more interesting to talk about with others and more likely to elicit sympathy. She knows that he also has many qualities that offset his negative qualities, so the struggle between the pluses and minuses is difficult for her. Deep down, she would rather be with someone who is a mix of good and bad than be alone. Similarly, someone may have good reasons for not quitting even if he is complaining about his job: maybe he figures that the risks of changing jobs far outweigh his short-term boredom or conflict with his boss. And, of course, we are never privy to all the behind-the-scenes complexities of international relations, despite how many newspapers we read. We may think we are privy to all the information, but that is only our certainty-addicted brain assuming that if we feel right, we are right, and if we are right, then we must have all the relevant information.

Key to understanding complexity is the hunt for hidden information. *Closing the brain–world gap depends on ferreting out the hidden cues that unlock less obvious and less familiar patterns: the patterns that we have not developed expertise in identifying.* Complexity requires us to expand the spotlight of our attention beyond the narrow focus we rely on for World #1; it requires us to move our attention spotlight around so we can get a better look at the important but fuzzy information in our periphery, just as we move our eyes in saccades to overcome the limitation of our foveal vision.

So our challenge in confronting complexity is twofold. First, we need to learn what hidden cues to look for: this is our "scientific" challenge — the challenge of avoiding oversimplification. Second, we need to resist the powerful temptation to lock down on early conclusions, by proactively searching for alternative interpretations: this is our "psychological" challenge — the challenge of avoiding overconfidence. What are the antidotes to each of the two cognitive challenges? We need the sophisticated scientific models of complexity science to avoid the errors that interpretive shortcuts generate in World #2. And we need the psychological strategies of cognitive science to avoid the errors that decision-making shortcuts generate. Marrying complexity and cognitive sciences dramatically improves the effectiveness of our interaction with World #2, making us better natural scientists (understanding how World #2 works), and better cognitive scientists (managing our thinking processes).

Complexity Science

The foundation of complexity science, and the solution to oversimplification, is systems theory. The key to understanding and responding to complexity is appreciating how complex systems work and how they can be influenced. These are precisely the objectives of complexity science, a discipline that cuts across many fields, including physics, economics and psychology, to examine the intricate ways that things interact. Because the field is fairly new, originating alongside the development of computer science, it can take different forms, but the unifying principle is the study of systems that are situated between simple, predictable ones at one end and random disorder on the other. Whereas standard science relies on reductionism — breaking something down into its core components to analyse its parts — complexity science focuses on how component parts add up to bigger wholes — how behaviours emerge from interactions. The concepts of systems theory are the focus of chapters 13 and 14.

Cognitive Science

Cognitive science is a broad field. It includes psychology, neuroscience, sociology and comparative studies between human and animal thinking and between adult and child thinking. An important component of the science is its examination of how we construct our version of reality. As the principles of systems theory are key to generating more sophisticated interpretations of World #2, the key to more sophisticated concluding and decision making lies in a deep understanding of constructive realism. Cognitive science reveals that naïve realism is not an innocuous operating assumption in World #2, the way it is in World #1. It also teaches us that disregarding the bounds of our rationality, which does not hinder us too much in World #1, can be disastrous in World #2. As systems theory emphasizes interconnections, constructive realism emphasizes provisional truth.

The conclusions we draw about complex systems are always open to reinterpretation, because they are not the same kind of conclusions we draw in World #1. A conception of truth that is too definitive or final entices us to lock down conclusions that need to be updated and refined as we interact with and learn about complex phenomena in World #2.

Does it make sense to have two different versions of truth? The short answer is yes, because the two worlds operate so differently. The long answer is not really, but this equivocation is reconciled by understanding that both forms of truth are just two different points on a spectrum, with 100 percent likelihood of being correct on one end, and zero percent likelihood of being correct on the other. The implications of provisional truth are the focus of chapters 15 and 16.

10.3 Complex Thinking for Complex Problems

A paradox: Twenty-first-century communication, especially social media, has evolved to the point where we now find ourselves treading water in a sea of information at the same time that social, economic, political and industrial complexity is exploding, making us more vulnerable than ever to the illusion of explanatory depth. A prediction: The illusion of explanatory depth will become more pronounced as we continue to organize information into simple cause-effect bundles that belie the complicated causal chains that define wicked problems. Tweets and sound bites give us simple explanations and exhort us to adopt simple solutions to increasingly difficult problems. We have been conditioned to treat our intuitions as reliable in all situations, rather than in just a subset of them.

The only way to close the gap is to tackle the increasing environmental complexity of World #2 with greater mental complexity. Complex systems require less "blinking" (to use Malcolm Gladwell's expression for automatic thinking[106]) and more effortful thinking; less intuition and more deliberation; less automatic response and more human reflection; less ancestral thinking and more complex thinking.

Returning to the thinkers in the last chapter, who each had a different description of the two kinds of problem sets we confront, the following table illustrates how each articulates the kind of thinking that works in each world.

	World #1 ⟷ World #2	
Daniel Kahneman	*"Fast Thinking"*: Intuition based on experience and practice	*"Slow Thinking"*: Deliberation based on effortful analysis
Robin Hogarth	Outcome feedback is usually sufficient	Process feedback is necessary
Nassim Nicholas Taleb	Normal distribution probabilities: Binary decisions	Fractal distribution probabilities: Empirical skepticism
E. F. Schumacher	Analysis	Integration of irreconcilable solutions
Kenneth Hammond	Correspondence-based intuition: Goal is empirical accuracy	Coherence-based analysis: Goal is conceptual rationality
Keith Stanovich	Type 1 processing	Type 2 processing

World #1 can be accommodated with the quick, intuitive Type 1 thinking that suits straightforward challenges, whereas World #2 demands slower, reflective, Type 2 thinking. In World #2, it is much more difficult to validate correspondence with the real world, so we are forced to rigorously assess the coherence of our beliefs as we integrate them. A heavy dose of skepticism keeps us honest in World #2, a world where we do not have the luxury of binary or dichotomized options, and where the outcomes of decisions are not always reliable forms of feedback on the accuracy of our interpretations.

Our cognitive operating system and the various programs it runs come up short in the face of complexity. A key flaw in the operating system is our reluctance to second-guess our initial conclusions. And the application programs that run on this operating system were developed for games that were fashionable for millions of years, not for some of today's more challenging ones. As it happens, modern humans are uniquely endowed with powerful utility software for making fixes to some of our weaker programs. These fixes are the unique human form of metacognition that facilitates self-assessment and self-correction.

We cannot afford to rewrite all of our software, because it works extremely well for the myriad straightforward scenarios we face each day. We do, however, have to add some functionality to it so we can handle both the old games and the new ones: functionality that is explored in the next part of this book.

Increasing the complexity in our heads to match the complexity outside them means we cannot be exclusively captive to our automatic intuitions. We need to engage more deliberative, aspirational thinking — the kind that allows us to see, understand and close the gap.

Part IV

The Brain–World Solution
Complex Thinking

11

Two Types of Thinking
Automatic and Effortful

> *The human individual ... possesses powers of various sorts that he habitually refuses to use.*
> William James*

The human brain may be the most sophisticated "algorithmic compressor," or "Bayesian processor," on the planet, simplifying what it takes in and making predictions about its environment, but it still expends most of its energy performing a more elaborate version of what all animals do. Our version, however, is intricate enough to be categorized into two kinds: thinking that is automatic, like the kind most animals are limited to; and thinking that is not entirely automatic, but voluntary, deliberative and effortful. Paradoxically, while the second type of thinking helped create modern-world complexity, it also expanded the gap between our brain and that complexity by not finetuning its own functioning to master this new world.

The solution to closing the gap is to improve our effortful thinking, the starting point of which is the subject of this chapter: understanding the difference between effortful and automatic thinking.

11.1 Two Types of Thinking

The idea of two thinking processes dates back to early Eastern philosophy, with its emphasis on the distinction between quotidian thinking

---

* *The Energies of Men*, Kessenger Publishing, 1998.

and contemplative meditation. Sigmund Freud is often credited for inventing the distinction between unconscious and conscious thinking, but many thinkers got there before him. Two hundred years before Freud, the philosopher Gottfried Leibniz wrote that, in addition to animalistic reasoning, humans have a special, reflective consciousness that he estimated accounts for one quarter of their thinking.[107] He noted that we have perceptions and memories that lie below our awareness but influence our behaviour. Philosopher Arthur Schopenhauer, writing eighty years before Freud, believed that we repress painful ideas such that they become inaccessible to us.[108] In the late 1860s, a group of British physicians, including William B. Carpenter, argued that the greater part of our cognitive operation is unconscious rather than conscious.[109] This theme was picked up by Francis Galton, two decades before Freud's work, when he suggested that consciousness was essentially the "helpless spectator" of all the work performed by the subconscious.[110]

Freud's description of the unconscious, at the end of the nineteenth century, is quite different from how scientists think of the unconscious today. He conceived of it as the location of repressed thoughts and feelings, as something to do battle with, as something to be fixed.[111] Freud distinguished two kinds of thinking: one that focuses on seeking gratification and avoiding pain, and another that focuses on inhibiting our gratification-seeking behaviour, when it is not socially acceptable, by repressing those desires. The modern view of the unconscious is more aligned with the pre-Freudian perspective: the unconscious is an important processor of information that enables us to navigate the world, in order to survive and reproduce.

To his credit, Freud dramatically expanded the scientific and philosophical significance of non-conscious mental processing and the notion of two kinds of thinking, what is now referred to as the "dual-process" model of human cognition. This model of thinking has been an important feature of cognitive psychology since taking hold in the late 1960s; by the late 1970s, psychologists were actively researching and describing the dual-process model.

The distinction between the two types of thinking is obvious in many cases: eating a meal as opposed to cooking one; reading a novel as opposed to writing one; watching tennis as opposed to playing it. There is automaticity to the first set of activities: we just do them without thinking about them. The second set requires concentration: these activities do not just happen without a concerted effort to initiate and sustain them, and because they are more effortful, we are more aware of what we are doing when we undertake them.

Automaticity describes the thinking that we invoke when we meet someone for the first time. We instantly size them up by corralling myriad micro assessments, including the scrutiny of their hand movements, their posture, their eye contact, and the volume, tone and cadence of their voice: we do this without any effort or awareness of the process that generates our impressions. Effort and deliberation describe the thinking that we invoke if later we are asked to explain exactly why we liked or did not like the person: we focus our attention on the relation between the cues we observed and the conclusions we drew; we become aware of the thought process that underlies our judgments.

Cognitive scientists characterize the two types of thinking in different ways. The left column in the following table represents the kind of thinking used for activities that are automatic, like eating. This thinking is ancient in the sense that we have always relied on it, as have all animals. Our version of automatic thinking may be more elaborate because it is based more on learning and less on innate instinct, but it is structurally similar. The right column represents the kind of thinking used for activities that require a degree of effort, like cooking. This thinking is new in the evolutionary sense, and finds its fullest expression in humans.

There are five hallmark differences between the two types of thinking: automaticity, speed/capacity, awareness, flexibility and dependence.

Ancient, Automatic Thinking (e.g., for eating)	New, Effortful Thinking (e.g., for cooking)
Involuntary: Thinking that happens without having to invoke or initiate it; it is effortless in the sense that it proceeds on its own automatically. You do not have to consciously deliberate on how to eat a fried egg.	**Deliberate:** Thinking that requires varying degrees of mental exertion, from a little (frying an egg or looking for a parking spot) to a lot (preparing a three-course meal or parallel parking on a busy street with cars honking at you).
Fast and high capacity: It works through parallel thought processes that proceed simultaneously. It relies on intuitive shortcuts based on developed expertise.	**Slow and low capacity:** Its dependence on working memory limits capacity and speed. Working memory is what cognitive scientists describe as the mental space or "blackboard" where we hold ideas temporarily in our mind as we contemplate them; it can only hold a small amount of information and can only manage a single train of thought at a time, which forces effortful thinking to proceed serially or sequentially.
Below awareness: We are generally oblivious to this thinking process, because it takes place below the level of our conscious awareness. We are hard-pressed to explain in detail all the steps involved in walking, even though we mindlessly perform this task all the time.	**Within awareness:** We are typically conscious of the fact that we are thinking, especially when we are concentrating very intensely. We are aware of what we are doing when we calculate a restaurant tip, and we can easily explain the steps we take to figure it out.
Rigid: We have limited control over its activity, influencing it only when we successfully raise aspects of it into our conscious awareness. Ordinarily lacking self-awareness and therefore self-doubt, this thinking typically concludes with "I know."	**Flexible:** Working memory allows us to decouple our perceptions from reality, so we can escape the confines of present-moment sensory experience and engage the mental flexibility of hypothetical thinking. It enables self-scrutiny, which permits counter-examples and self-doubt, thereby allowing us to conclude with the more open-ended "I believe."
Stand-alone: This thinking is the dominant kind in the animal kingdom: it operates independently. In humans it is a massive network of processing that churns below conscious thinking.	**Dependent:** This thinking is underpinned by automatic thinking, without which it could not function. Working memory, which effortful thinking relies on, gets much of its content from tapping into a whole infrastructure of subconscious memories, beliefs, models, preferences and values.

While cognitive scientists generally agree on these five points of contrast, the terms employed to describe each type of thinking are nowhere close to being universal, as this sample of the different descriptors indicates:

Automatic Thinking	Effortful Thinking
Intuitive	*Analytic*
Instinctual	*Controlled*
Emotional	*Rational*
Want brain	*Should brain*
Caveman brain	*Human brain*
Basic Mind	*Super Mind*
Sub- or unconscious	*Conscious*
Reflexive	*Reflective*
Implicit knowledge	*Explicit knowledge*
Heuristic	*Algorithmic*
Associative	*Propositional*
Experience-based	*Theory-based*
Type 1 process	*Type 2 process*
System 1	*System 2*

If "reflexive" and "reflective" did not sound and look so similar, they would be ideal terms to highlight the contrast. The most common terminology in cognitive psychology literature is System 1 and System 2 thinking (created by Keith Stanovich and popularized by Daniel Kahneman).[112] These terms suggest that within our brains are separate systems that support each type of thinking. Matthew Lieberman, among other neuroscientists, has demonstrated that when we transition from automatic thinking to effortful thinking, activity in one circuit of the brain appears to switch over to a different circuit.[113] Neuroimaging provides strong support for the dual-process theory; however, the same research also reinforces how complex and overlapping the two cognitive processes are. There do not appear to be completely separate systems for each kind of thinking. To avoid the implication that there are actually two separate brain systems, some psychologists, including Stanovich himself, have adopted the terminology of Type 1 and Type 2 processing. What "Type" lacks in poetic flair it makes up for in practicality: the word represents a cluster of characteristics without implying two completely separate systems. Reasonable substitutes for each type are "automatic" and "effortful," which will be used interchangeably with Type 1 and Type 2 in this and subsequent chapters.

Each type of thinking has advantages and disadvantages. We have always relied, and will always rely, on Type 1 to survive. The benefits and drawbacks of Type 2 are the inverse of Type 1, and that is no coincidence: many evolutionary biologists believe that higher-order human consciousness arose from the environmental pressure to overcome some of the vulnerabilities of automatic thinking. An animal that could supplement speedy, danger-sensitive thinking with a slower, more creative approach would be well equipped to handle increasingly complex survival challenges, like figuring out how to grow food and how to get across an ocean that is not frozen. But the flexibility of effortful thinking comes at the cost of its narrow scope.

Gary Klein draws an analogy between human vision and the two types of human thinking to demonstrate how crucial Type 1 thinking is to us.[114] Our vision is made up of two distinct parts. Foveal vision is the tiny area right in the middle of our visual field: it is precise and clear. Peripheral vision is the rest of our field of vision: it is rough and imprecise. If we damage our fovea, we are unable to focus with clarity on anything. If we damage our peripheral vision, we have only a precise but very small circle of vision, about the size of a quarter, right in front of us: we would have to move our eyes and head all over the place to get any sense of where we are, to see the big picture.

Klein points out that a small bit of precise vision is less useful to us than a whole lot of imprecise vision. So too is the quick and gist-like peripheral nature of Type 1 more important to us, because it incorporates such a vast amount of information compared with the deliberative and clear foveal Type 2, which can be focused only on a single problem at a time. And just as our peripheral vision alerts us to focus on something if it picks up the gist of an important or unexpected signal, Type 1 will engage the deliberation of Type 2 when it senses something that needs deeper thought.

The analogy suffers, though, from one important limitation. The fovea and periphery can operate independently of one another, but Type 2 cannot function without the underpinning of Type 1. While effortful thinking can override automatic, it is always dependent upon Type 1's incredible processing power, without which it could not function.

The Strength of Type 1: Processing Power

The conscious awareness that we have of ourselves belies the vast amount of processing that occurs below our awareness, keeping us alive without any help from our consciousness. Automatic thinking has enormous capacity; it operates at an incredible speed, processing an unimaginable amount of data — literally unimaginable since our conscious awareness is incapable of keeping track of all this automatic processing.

Brain circuitry is typically described by cognitive scientists as modular, meaning that neurons are organized into circuits that extend through different parts of the brain; each circuit has a particular function, such as face recognition, motion detection and language acquisition. These circuits run simultaneously (in parallel), so a lot of information can be processed at the same time. Type 1 processing happens all the time; it begins in utero and continues until we expire, including the constant processing that continues while we are sleeping.

Brain studies have confirmed that we do not need to be conscious for Type 1 to function. Patients with frontal lobe damage who cannot recognize a screwdriver are often still able to handle it with proficiency. "Blindsight" describes patients who have no awareness of visual stimuli but are nonetheless adept at "guessing" the objects that are put in front of them: they see without being aware that they can see. Stroke patients who are unable to identify faces still register recognition responses, picked up by skin-conductance equipment, when confronted by familiar faces. Similarly, cognitive scientists have also pointed out that the speed of the pitch in professional baseball is far too fast for a batter to perceive it and react to it consciously. Without even being aware of it, batters react to all the cues that Type 1 can accommodate in the short period it has: the pitcher's body position and arm motion, the ball position upon leaving the pitcher's hand and so on. The batter is relying on his Type 1 best guess as to the direction and speed of the ball, because the ball is too fast for Type 2 to be helpful at the moment of the pitch. The subtlety of cues that Type 1 picks up is astonishing, as is the speed with which it does this.

The Weakness of Type 1: Inflexibility

Processing speed comes at a cost, and that cost is flexibility. Type 1 does not engage extensively in hypotheticals; it does not have time for elaborate what-if scenarios; it has no patience for alternative hypotheses or counterfactual contemplations. It simply matches the data it receives with the information and patterns that it has on record. Its job is to reach a conclusion as quickly as possible, and second-guessing itself does not serve that end. It force fits the world into the mental models that it has already developed. Like a starving man who grabs at anything edible with minimal discrimination, Type 1 grabs on to any cues it can, even if they only vaguely seem to fit its models.

Stanovich uses the following basic syllogism to demonstrate how Type 1 "contextualizes" all the problems it faces by using any cues available to it: *All living things need water; roses need water; therefore, roses are living things.*[115] The question is whether the conclusion follows from the two premises that precede it. Almost 70 percent of respondents reply that the conclusion does follow, which is not that surprising, because the argument feels right: of course roses are living things. Here is where it gets interesting: only 22 percent of the respondents think that the following argument is valid, even though it follows the same structure as the rose argument: *All animals of the hudon class are ferocious; wampets are ferocious; therefore, wampets are animals of the hudon class*. Neither the rose nor the wampet argument leads to a conclusion that follows from the premises, but most people think that the first argument holds because they cannot help but be influenced by their prior knowledge of roses. Type 1 creates context out of whatever cues are available, whether relevant or not. Type 2, when not unduly influenced by the pull of Type 1, can decontextualize information to analyse it more carefully.

It is not just its speed that makes Type 1 inflexible, but also its opacity, which makes it more difficult to analyse and influence. While René Descartes is traditionally blamed for cementing the Western tradition of separating mind and body into two distinct ontological categories, his equal if not more conspicuous error was his all-encompassing assumption that he could generate "clear and distinct ideas" about himself and the world, based on the full transparency of his mental operations.[116] The tradition championed by Freud overturned this Cartesian assumption by advocating that our access to much of our cognitive activity is constrained, because it is submerged below the level of our conscious awareness.

To demonstrate the opacity of Type 1, Jonathan Haidt concocted hypothetical but controversial scenarios to quiz people about.[117] When asked to read a short story about two siblings having protected sex, or a family eating its dog after it died, his subjects all reported a level of disgust and a strong conviction that the protagonists were wrong to do what they did. When asked to give a convincing and coherent explanation for their views, they were stumped, unable to properly access the thinking that led to their conclusions. They just "felt" that the siblings and pet owners were wrong.

Fortunately, automatic thinking is not completely opaque and therefore not entirely rigid. Aspects of it are more translucent, since we can investigate and talk intelligibly about our subconscious thinking if we expend some conscious effort. Indeed, Haidt's research is geared toward understanding the origin and functioning of our submerged ethical preferences. But there are constraints on our ability to influence automatic thinking because it is not fully transparent.

In addition to its opacity and rush to certainty, a final reason for Type

1's inflexibility is how it develops, especially in our earliest years. We are born with a brain that quickly forms an increasing number of connections between neurons (i.e., synapses) and then, starting at about age three, it eliminates the unused connections. We overproduce connections for maximum learning, and then prune back the connections that are not used, to preserve mental energy. This Hebbian process (named after the psychologist Donald Hebb) gives us the opportunity to learn a vast amount of things, but it helps us preserve energy by limiting the period in which our brains are maximally "plastic." This is why it is so much easier for young children to learn a new language than it is for adults. The preserved connections are those reinforced repeatedly with experience, but once these connections become strongly established, our automatic thinking patterns and behaviours become cemented and difficult to unlearn. This inflexibility in our automatic thinking arises from what psychiatrist Norman Doidge calls "competitive plasticity": once connections are formed and reinforced through repetition, they take up real estate in our brains and are difficult to dislodge.[118]

The process of human brain development is paradoxical: we are as smart as we are because of the breadth of learning that we establish from experience in the first years of our life, but the rigidity of these connections makes us as dumb as we are when we stubbornly rely on those fixed thought patterns in situations where they are mismatched to the world.

The Strength of Type 2: Flexibility

The flexibility of Type 2 is as awe-inspiring as the processing speed of Type 1. The what-if creativity of effortful thinking is much more flexible than the "this matches what I already know" approach of automatic thinking. It is the contemplation of possibilities that are not represented in our immediate experience that makes Type 2 special. Working memory enables us to separate our in-the-moment perceptions from a larger view of reality; it allows us to run thought experiments based on our ability to decouple our representations of reality from the world itself. By decoupling our perceptions from our ideas, by separating the world from our ideas about the world, we have evolved the ability to transcend the automatic workings of our mind and reflect on them:

to think about thinking; to assess how we think and how we can think better; to determine if our default mental shortcuts suit a particular problem; to gauge our vulnerability to our intuitive biases; to ask ourselves if we are entitled to be as confident in our conclusions as we feel we are.

This form of mental flexibility — thinking about thinking — is called metacognition, and it enables humans to exploit the plasticity of their brains by deliberately choosing to reinforce new habits and ways of looking at the world. Shy people can practice social interactions; anxious people can practice relaxation exercises; quick concluders can practice suspending judgment. Although cognitive habits are difficult to alter because Type 1 is stubborn and powerful, they can nonetheless be influenced with the application of our unique form of metacognition.

The Weakness of Type 2: Subordination

The power of Type 1 to interfere with Type 2 is reflected in the Stroop effect, named after John Ridley Stroop, who first wrote about it.[119] The pace at which we read a list of colours (e.g., blue, green, yellow) can be significantly slowed if the words are printed in colours that do not match the words themselves. When the word "blue" is written in green ink, we stumble because Type 1 interprets the colour green much faster than Type 2 can process the word blue: Type 1 gets ahead of, and hinders, Type 2.

Type 1 has been in development for over five hundred million years. Type 2 is "brand new," its prototype originating in great apes sometime in the last fifteen million years, and its fuller, metacognitive expression in humans sometime in the last thirty thousand years. In fact, the human version of metacognition is, in some fundamental respects, still at the prototype phase: we are still in the early stages of learning how to think about how we think, and how to manage how we think.

The best we have got (so far) is a somewhat clumsy ability to check our automatic thinking, the effort and discipline of which does not come easily to us. Despite Type 2's impressive accomplishments (as in Shakespearean plays and Einsteinian physics), it is slow and inconsistent, easily distracted and readily influenced by the slightest of factors, such as how much sugar is in our bloodstream at a given moment or how much sleep we had the previous night. At its best, the limited capacity of working memory means that we cannot handle what-if scenarios that are too intricate, and we cannot work on multiple things at one time. To top it off, most people can focus on a single, high-concentration task for a maximum of ninety minutes before their productivity declines precipitously; a forty-five-minute stretch is more typical. As Herbert Simon noted, attention is one of the scarcest of human resources.

Whereas Type 1 is always chugging along as consistently as our hearts beat and our lungs expand and contract, Type 2 works at its highest potential only periodically. It is no surprise, therefore, that Type 1 thinking, without direct intervention, dominates. There are many analogies that have been invented to capture the power of Type 1 over Type 2.

David Hume: Reason is the slave of the passions; it can only serve and obey them.[120]

Jonathan Haidt: Automatic thinking is the powerful elephant on whose back conscious deliberation rides as an advisor, attempting to use its broader perspective to cajole the elephant.[121]

Michael Gazzaniga: Split-brain experiments reveal that we have an "interpreter" module in our left hemisphere, the primary job of which is to interpret the subconscious conclusions that make it into consciousness, after the fact.[122]

Steven Pinker: The conscious mind is a spin doctor whose job is to create a story that explains the actions that the unconscious mind initiates.[123]

Owen Flanagan: Consciousness is a figurehead president. It has status and putative authority, but it is limited to explaining the real work and output of the hard-working, smart policy advisors working in the background of the subconscious.[124]

Daniel Wegner: Most of what we consider conscious decision making is nothing but illusion: we may feel that we are freely choosing, but subconscious processing is doing the choosing for us and just "letting us know" what it has decided.[125]

Contributing to Type 1's power over Type 2 is the order in which they occur. Subconscious thinking precedes conscious thinking. The best-known experiments on the delay between subconscious and conscious thinking were conducted by psychologist Benjamin Libet.[126] He clocked subjects' awareness of their decision to act at around three hundred and fifty milliseconds after a brain scan had registered that a decision had been made (the electrical activity, or "readiness potential," showed up before the subject reported his decision). Libet argued that this tiny buffer between subconscious decision and conscious awareness of the decision is the source of human free will: it gives our consciousness the opportunity to curtail the subconscious decision to act in certain ways.

Other psychologists have clocked even longer lapses (up to ten seconds) between a subject's report of experiencing an insight and the brain activity that precedes awareness of the insight, as might happen, for example, in solving a word puzzle. Supporting these experiments is the famous gambling task, pioneered by a group of neuroscientists, including Antonio Damasio.[127] They demonstrated that subjects break out in a stress-induced sweat before they are consciously aware of the reason for their anxiety. When playing a card game where the odds are stacked against them, they develop a subconscious intuition about the odds before they are conscious of the likelihood of winning or losing.

The Speed of Type 1 Versus Type 2

There are two Type 1 speeds, both of which are much faster than Type 2. Neuroscientist Joseph LeDoux relates the two speeds to two distinct routes that Type 1 takes: the low road and the high road.[128] The longer, high road proceeds as follows: When sensory data is taken in, it passes through the thalamus, on through the cortex where it is processed and a representation of the stimulus is created. From this point (about one second after the initial sensory intake), it may or may not be elevated to consciousness, depending on whether this Type 1 processing has determined that Type 2 awareness is necessary for additional processing. At the same time as this process is unfolding, the same sensory data also moves along a shorter, low road, passing through the thalamus as well, but tapped into by two almond-shaped structures: each amygdala sits beneath the thalamus and compares the rough, barely processed data with past experiences. If there is any resemblance between the rough data and past danger, the amygdala will initiate a fight or flight response milliseconds before any conscious awareness is involved.

Both the high-road (cortico-amygdala) and low-road (thalamo-amygdala) pathways start at the thalamus and lead to the amygdala, triggering an emotional reaction, but the low-road path is more direct and therefore faster. As survival-enhancing as the fast low road is, it is not as fine-tuned in its interpretations as the slow high road, which is why we can fly off the handle or jump at the sound of a loud bang before our conscious awareness can catch up and intervene. Humans have a lot more cortico-amygdala connections than other primates, giving us some control over our emotional responses, but we are still a long way from our conscious, deliberate thinking being able to consistently control our automatic, emotional responses. By virtue of taking a longer route, Type 2 is late to the party, arriving only after Type 1 has issued an invitation. In addition to its tardiness, Type 2 has a difficult time finding the host: the subconscious thoughts and motivations that give rise to our conscious thinking are usually hidden from our awareness.

Subconscious thinking not only preceded conscious thinking in human evolutionary development (phylogeny), but it continues to do so in infants' cognitive development (ontogeny). It takes about twelve months for new-born babies to develop the brain structures necessary to begin to consciously and intentionally remember things (invoking "explicit" or "declarative" memories), rather than relying exclusively on subconscious, automatic memories ("implicit" or "procedural" memories). While the two amygdalae are almost fully formed at birth, it takes until the third year of life before linguistic processing begins. The first memories we have are stored as rough, hazy images with no accompanying words to describe them, which is why we cannot remember much of anything before age three. Notwithstanding how inaccessible these non-verbal, cognitive impressions are, they can have a significant impact on us: our earliest experiences influence how we view the world, other people and ourselves. These subconscious thought patterns can pull us in directions that often confuse us, and our higher-level reasoning is hard-pressed to untangle and decipher them.

While the scope of our awareness is much broader than that of other animals, it is still severely constrained, which is why Type 1 dominates most of our thinking. We can and do override Type 1, as when we pass up a second helping of dessert, quit smoking or reconsider our beliefs. More often than not, though, we overeat, indulge in bad habits and go with our gut intuitions. Type 2 is frequently used to simply legitimize behaviours that Type 1 has initiated. Type 2 uses its analytic reasoning to rationalize a decision after the fact, confabulating an explanation for Type 1's decision, which is why Type 2 is sometimes considered our inner lawyer, arguing a case on behalf of Type 1. This is a version of "sour grapes" rationalization: when two beliefs conflict, such as I want those grapes but cannot reach them, or I believe in conserving energy but I like the air conditioning on all day, Type 2 contorts itself to accommodate Type 1's needs.

Although some psychologists argue that conscious thinking is nothing but the pawn of the powerful cognitive operations that reside below our awareness, the vast majority are convinced that conscious, effortful thinking can have at least some influence over the subconscious operations that underlie it. Otherwise, contrary to our everyday experience, any notion of free will would be nonsensical, and we would have no power to change our habits or redirect our goals. But influencing Type 1 does not come easily: Type 1's weapon of choice is the confident feeling of knowing, which instantly disarms Type 2. Type 1 can invoke this feeling extremely quickly, either by excluding Type 2 from the party or disinviting it as soon as Type 1 is satisfied that it can handle things on its own. The very effectiveness of automatic thinking is precisely what gives it the strength and power to dominate, and why we have to be proactive and disciplined in championing Type 2.

12

Climbing the Cognitive Hierarchy
Dual Thinking for Dual Worlds

*All our dignity consists in thought.
Let us endeavor then to think well; this is the principle of morality.*
Blaise Pascal*

The marriage of Type 1 and Type 2 is not an easy one: like any partnership, the strengths and weaknesses of each have to complement one another in the right way in any given situation. Consider the thinking strategy of a grandmaster chess player:

> It is his move: the grandmaster surveys the chessboard, his eyes darting about frenetically. He instantly and unthinkingly sees the few possible moves he could make and, almost as quickly, which one is his best bet. But he hesitates, running through the likely scenarios that could unfold from each possible move, before finally committing to one. After deliberating, he will go with his initial, intuitive choice around 75 percent of the time. But 25 percent of the time, he will go with one of the other options.[129]

Grandmasters achieve their ranking not just by mastering chess, but also by mastering their thinking: by combining Type 1 and Type 2 in a very productive way — precisely the cognitive strategy required in World #2, and the focus of this chapter.

---

* *Pensées*, Penguin, 2003.

12.1 Type 1 and Type 2 in Combination

To depict the relationship between the two types of thinking, the following figure conveys that:

- effortful Type 2 is underpinned by automatic Type 1,
- automatic Type 1 is not entirely transparent to us — it has a mysterious texture to it, and
- effortful Type 2 comes in degrees, depending on how much concentration we invoke.

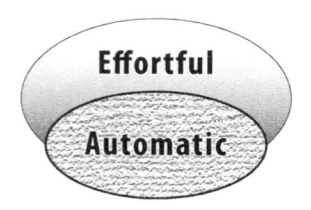

Mental effort is a continuum, from mindless walking at one end to concentrating on mental math at the other, with countless activities between that are neither purely automatic nor intensely demanding. When cutting grass or making scrambled eggs, we are largely on autopilot but still paying attention; organizing a surprise party takes more concentration. This is why the oft-used analogy of consciousness as the tip of the iceberg (with subconsciousness as the larger part beneath sea level) is too limiting: the picture it paints represents only one point on a continuum of conscious–subconscious combinations, ranging from little mental effort to a lot. Type 1 is always functioning; Type 2 is invoked in varying degrees to supplement Type 1. So Type 2 is not simply on or off (unless we are sleeping or unconscious) but can vary in its intensity, depending on our motivation for concentration.

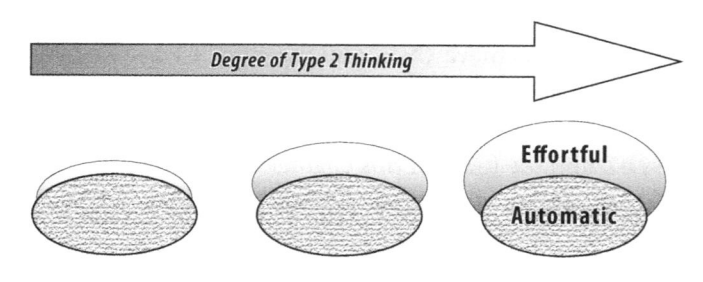

At increasing levels of Type 2 thinking, we broaden the scope of awareness that we bring to a task. The illustration above depicts a constant amount of Type 1 thinking across the spectrum of all degrees of Type 2: the actual amount of automatic, subconscious processing does not decline, since Type 2 is always dependent on Type 1 processing as its base. But at higher levels of Type 2 thinking, we are less captive to Type 1 conclusions. Antonio Damasio uses the variables of intensity and scope to describe the degrees of consciousness that we are capable of: consciousness is dull and has minimal scope when we are falling asleep; it is sharp and broad when we are debating with someone about the merits of, say, gun control.[130] Much of our conscious thinking lies somewhere between falling asleep and debating.

Consciousness and Awareness

A common response to the question, "What is special about human thinking?" is consciousness. But consciousness is not a useful differentiator, whether between humans and other animals or between Type 1 and Type 2. Cognitive psychologist Jonathan Evans has pointed out that Type 1, which is largely below awareness, has an element of consciousness to it, insofar as we are aware of the feelings that arise from this processing. And Type 2, which is largely within conscious awareness, relies on processing and memories that we are not aware of in the moment that we are deliberating.[131]

The other problem with "consciousness" is that the word has no universal meaning; in fact, it is probably one of the most confusing words in cognitive science. Sometimes it means awareness in general; other times it refers to a particular form of awareness, like self-awareness or "awareness of awareness." Still other times, it conveys the ability to voluntarily attend to something. Sometimes it is used to describe the unique form of human cognition, but it is also frequently used to describe the cognition of both humans and other mammals. Cognitive scientists and philosophers use terms such as "consciousness" with "awareness," "intentionality," "sentience," "metacognition" and "theory of mind" inconsistently and sometimes interchangeably. The various forms of the term "conscious" do, however, share something in common, which is the notion of mental awareness.

Many animals have awareness in its most basic sense, as any dog or cat owner will attest. A dog has a degree of awareness of itself and others when it greets its owner with a wagging tail. This brute awareness is typically associated with the ability to have feelings, no matter how rudimentary, based on forming mental representations of the world. It is usually described as subjective experience (referred to by philosophers as "qualia," which is a key feature of sentience), and it is something a worm presumably lacks because it does not form mental representations. To the best degree that scientists can judge, a worm does not experience any form of awareness or thought or emotion when it is slithering through the earth; it responds to its environment in a purely reflexive way. For brute survival, consciousness is a nice-to-have, not a need-to-have. Hundreds of thousands of animal species have no conscious awareness of themselves (the number of insect species alone has been estimated at over three million). They far outnumber the species that do have consciousness, and many of them have much longer histories on this planet than we do.

The vast majority of cognitive scientists agree that, when consciousness is defined as awareness in the sense of subjective experience, all mammals and many non-mammalians are conscious: their behaviours suggest that they do not react to the world in a purely stimulus–response manner but operate with some form of thinking and feeling that indicates a level of awareness.

Where Is Consciousness?

The search for the brain activity that correlates with consciousness (i.e., the neural correlates of consciousness) has revealed that the higher form of human self-awareness is linked to the frontal cortex, which has greater connectivity with other parts of the brain because of very long neuronal axons that reach into these other areas, enabling the integration of information. There are layers and layers of unimaginably intricate neural networks, and from this awesome neural activity somehow emerges consciousness. Scientists have not figured out exactly how it arises: philosopher David Chalmers labels it the "hard problem" of consciousness, because it cannot be solved by dissecting physical processes.[132]

> Some scientists and philosophers, occasionally referred to as "mysterians," believe that we will never be able to understand how consciousness happens. Steven Pinker suggests that our brains are complex enough to give rise to consciousness but perhaps insufficiently developed to facilitate an understanding of how it works: "Our thoroughgoing perplexity about the enigmas of consciousness, self, will and knowledge may come from a mismatch between the very nature of these problems and the computational apparatus that natural selection has fitted us with."[133]

The subjective experience of being aware is first-order awareness, and humans share it with many other species. Where it gets interesting is the human ability to go beyond first-order, subjective experience, to have thoughts about the mental representations themselves. We can reflect on our emotions, desires and thoughts. This higher level of awareness represents higher-order cognition: we can contemplate our contemplations.

12.2 The Cognitive Hierarchy

A first-order mental state is the awareness of experience, such as when you think to yourself, "I really hate my job," or "I don't trust that guy." Worms and insects have zero-order mental states: they do not think about their jobs or other members of their species; they simply respond to their environments.

A second-order mental state has a first-order mental state as its subject: "I should stop ruminating (second order) about how much I hate my job (first order)," or, "I don't know why (second order) I always assume people are untrustworthy (first order)." We can easily master a third order as well: "I'm pretty sure (third order) that my boss knows (second order) that I hate my job (first order)." Second-order mental states can be based on an awareness of others' mental states; third-order mental states can exemplify our awareness that others have mental states about our mental states.

Thinking about what other people are thinking that we are thinking leads right into fourth-order mental states: the kind that are needed for us to converge on Schelling points, named after the economist Thomas Schelling, who described the kind of fourth-order thinking that allows two people to meet in a big city without ever communicating with one another.[134] If I am told to meet someone in Paris but the location is not specified, I will think (fourth order) about what the other person expects (third order) me to think (second order) would be a likely place for her to choose to go to (first order). We will both end up at the Eiffel Tower, because of our ability to think about what each other thinks we are thinking.

Type 1 and Type 2 in Conflict

The unique height humans can achieve on the cognitive hierarchy is most obvious in the case of preference-based mental states. When I think about my penchant for sugary desserts, I may realize that I am drawn to sugary foods because, historically, sweet-tasting things were almost always nutritious and very rarely poisonous. I can contemplate the disadvantages of this preference in a modern world, how simple sugars are detrimental to my weight, health and general wellbeing. Now I have a conflict between the outcomes of my automatic Type 1 and my effortful Type 2 thinking: I wish (second order) that I did not like sugary treats as much as I do (first order).

This kind of second-order thinking is uniquely human. Some animals exhibit hesitation in the face of conflicting motives, such as wanting food but fearing attack if the desired food is in the possession of a competitor, but this hesitation is generated by conflicting first-order preferences (food versus being attacked), and not by conflicting first- and second-order preferences. We humans are riddled with internal conflicts between lower- and higher-order preferences; the stress we suffer from these internal conflicts would seem very peculiar to any other animal because our internal conflicts are a function of the higher-order preferences that are unique to humans.

In fact, there is no stopping us: after ruminating that I wish I did not desire sweets, I may observe the anxiety that I am suffering because of the conflict between my first-order desire and my second-order reflection on the desire, and wish (third order) that I did not care (second order) about my desire for sweets (first order). We rarely have four layers of preferences stacked on one another, because there is negligible utility in this much preferential layering. Nevertheless, our ability to analyse our preferences gives us a degree of freedom that other animals are not capable of.

Whereas there is some disagreement on other animals' capacity for limited forms of second-order thinking, there is virtual unanimity that most forms of second-order thinking, and all forms of third-order thinking, are limited to humans. Some psychologists and philosophers contrast first-order and higher-order awareness as two types of consciousness: first-order is prereflective, phenomenal or core consciousness, in contrast to reflective, higher-order or extended consciousness. But a more productive model of consciousness is represented by a continuum of different levels of awareness, starting from the low end of awareness attributable to many animals (but not worms), moving through to the medium end of awareness represented by the rudimentary metacognition of higher primates and young humans, through to the high end represented by the full, self-reflecting metacognition of mature humans. Conscious awareness arises from a brain state, and the more complex the brain is, the more complex the form that consciousness can take.

In the continuum of awareness, the dividing lines between closely related species is difficult, if not impossible, to ascertain. Somewhere between the very first archaic humans and modern-day humans, full metacognitive thinking arose in *Homo sapiens* and has differentiated us from other animals ever since.

Scope of Awareness (or Levels of Consciousness)

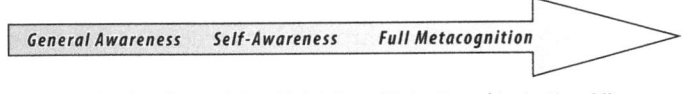

General Awareness   *Self-Awareness*   *Full Metacognition* →

Worm | Turtle | Rat | Lion | Monkey | Chimp | Archaic Human | Modern Human | Aspirational Human

A chimp's cranial capacity is about 50 percent smaller than that of the first humans, which is why archaic humans are to the right of chimps. The first human beings were probably capable of rudimentary forms of Type 2 thinking; as the human brain evolved to become more complex, this type of effortful thinking increased in power and flexibility, accommodated by a greater working memory and an ability to think hypothetically. Human awareness eventually expanded to include recognition of its own thinking processes, the point at which metacognition arose.

Human Versus Animal Cognition

The hierarchy of cognitive states is extremely useful in differentiating human and non-human consciousness: in addition to the first-order awareness that we share with many other animals, humans also have higher levels of consciousness. It is not that other animals have no consciousness; it is that they have lower forms of it. The scope of our consciousness is broader than that of other animals, because we are capable of the higher level of awareness that enables us to rise above our preferences and ideas in order to evaluate them. This insight led philosopher Harry Frankfurt to note that all other animals are what he called "wantons," whereas humans can reflect on and rise above their first-order desires.[135]

An ant operates solely on reflexes. A chimp can engage in a rudimentary form of Type 2, as well as what appears to be a rudimentary form of self-awareness. Only humans can engage in full Type 2 metacognition: the kind of mindfulness that enables us to evaluate and alter our own thinking, the kind of self-awareness that allows us to project ourselves into the future at the same time that we are reflecting on our past (referred to as "autonoetic consciousness"). The flexibility to prioritize our desires and examine our beliefs is the flip side of (and consolation prize for) being plagued by conflicting motives between Type 1 and Type 2 thinking: other animals do not diet to to improve their figure or second-guess the quality of their parenting.

Human children transition through higher levels of consciousness as they age. After the age of three, for example, they develop the ability to create narratives about themselves based on their past experiences, as well as an ability to differentiate their thoughts and experiences from those of other people. It is not until after age five that children begin to acknowledge their cultural identity as part of a group, and until they are older still that they begin to reflect on their own thinking processes.

Just as children broaden the scope of their awareness as they age, humans have broadened their cognitive abilities over thousands of years, eventually becoming a very distinctive species. But just how distinct?

Charles Darwin not only documented a coherent and compelling explanation of how the human species originated from earlier primates, but he also offered an unequivocal opinion on the question of how fundamental the difference is between humans and other animals: he wrote that "the difference in mind between man and the higher animals, great as it is, certainly is one of degree and not of kind."[136] And so it was that the "degree versus kind" debate was launched: whether humans are distinct from all other animals in a more fundamental way than, say, fish are distinct from birds.

Scientific research in the field of comparative psychology over the past century indicates clearly that many animals possess rudimentary versions of human cognition. We have more developed versions (albeit significantly more developed), but most of the behavioural traits that were once considered unique to humans are, we have discovered, shared with some other animals, such as tool-making, communication, punishment of uncooperative behaviour and vicious intergroup aggression. Virtually every form of non-reproductive sexual activity that humans engage in is also present (including, according to Dr. Barbara Natterson-Horowitz, masturbating spiders[137]). The more deeply animal behaviour is examined, the more surprising some of their cognitive skills appear to be. Baboons deceive one another. Dolphins and rhesus macaque monkeys display an ability to monitor and control their thinking by choosing easier tasks with low rewards over harder tasks with high rewards. Crows will bend a stick with their beaks to pick food out of a narrow tube. Rhesus macaques have been observed transferring learning from one task to a different type of task.

Most comparative psychologists agree that human cognitive skills have precursors in other animals, while acknowledging that the gulf between our cognitive talents and those of other animals is enormous. But is it enormous enough to warrant lumping all non-human animals together on one side and humans on another? The degree–kind argument is more a semantic game than an ontological dichotomy: the issue collapses into how we define "kind" and whether there is any meaningful disparity between the notions of "difference in kind" and "huge difference in degree." We are "extra-ordinary" animals; whether this extraordinariness represents a gap or degree is, outside of a religious context, based on arbitrary definitions. Either way, we are the most complex genus of the primate order in the mammalian class of animals.

Humans: Different in Kind or Degree?

The "kind" side argues that there is no compelling evidence to demonstrate that other animals have anything close to the higher-order abstract thinking that humans are capable of, including the cumulative learning that underpins human culture. The self-awareness, flexibility and creativity of the human mind puts humans in a league of their own.

The "degree" side argues that current evidence does demonstrate that some animal behaviour cannot be explained in purely associative stimulus-response terms. All human cognitive activity can be witnessed in very rudimentary form in certain animals: the complexity of our brains translates into more complex versions of animal thinking.

Aside from how the research on animal behaviour is interpreted, there are two conceptual problems with the notion that the difference is one of kind. First, just as there is an enormous cognitive gap between chimps and humans, so too is there an enormous gap between ants and chimps: if the former is a gap in kind, why would the latter not be as well, which raises the question of how many discreet gaps in kind there are between ants and humans. Put another way, on the tree of great apes, are we really that much farther out on our own branch than, say, clever dolphins are on the tree of marine mammals or tool-using crows are on the tree of birds? The degree side does not have to contend with this problem: all animals, including humans, have cognitive powers that lie on a spectrum of varying degrees of complexity. Nor does the degree side have to contend with the second conceptual problem: If the difference is one of kind, when exactly did ancient, apelike humans jump the gap to become modern humans who were different in kind than their immediate ancestors? In the gradual process of ape becoming human, it stretches the limits of intelligibility to envision a jump from non-human to human, except

> over a protracted period of time, in which degrees of cognitive complexity accumulated.

Appendix 2 delves more deeply into the degree–kind argument, with a scripted dialogue between the two sides. We often think that we are arguing about substance when, in fact, we are actually arguing about semantics and just do not realize it until we probe our beliefs more deeply. The argument of this book is based on the optimistic view that we can assess our cognitive vulnerabilities and close the gap between them and the complex world we have created, a feat that is many degrees richer than the limited metacognitive capacities of any other animal.

Two Kinds of Type 2 — 2.1 and 2.2

The dual-processing model is potentially misleading because, in its simplest form, it depicts both types of thinking as monolithic, when, in fact, they are umbrella terms that group many different kinds of thinking. The labels Type 1 and Type 2 apply to a variety of thinking forms that can be loosely categorized as either automatic or effortful. Descriptions like intuitive, subconscious and automatic are useful, but they do not come close to the complex character of Type 1, which can take the form of instinct, learned habits and skills we perform mindlessly, internal body regulation, emotional reactions and moral preferences. Type 2 also expresses itself in different forms: the kind of Type 2 that you employ to solve a math problem is not the same as the kind for negotiating a raise; both are different from the kind you use to resist overeating; and all three are different from the kind you use to create art.

While Type 1 and Type 2 represent a useful catch-all distinction, tackling complexity is not a matter of just *adding more* Type 2; it is about *adding more of the right forms* of Type 2. When Type 2 is engaged, it is not always characterized by our best hypothetical thinking; in its lowest or "laziest" form, Type 2 does not generate alternative hypotheses to compare, but simply confines itself to a single scenario. Cognitive miserliness discourages not only the quantity of Type 2 but also its quality. We default to the shallow form of Type 2 unless we are motivated to engage in a deeper form of it. We start with Type 1 and, if required to engage Type 2, as in an unfamiliar situation where pattern recognition is limited, we will engage the shallow form of Type 2 (which Keith Stanovich calls "serially associative cognition," because it limits mental modelling to associations of what a person already believes and has previously modelled). Only if we are still not making progress — the feeling of knowing is not triggered — will we undertake hypothetical reasoning with the simulation of alternative models (which Stanovich calls "fully decoupled cognitive simulation"). More rarely still will we climb the cognitive hierarchy even higher by engaging the kind of metacognition that allows us to reflect on how we are analysing a problem (which Stanovich calls "reflective-level" Type 2 processing).[138]

So: effortful Type 2 is itself a hierarchy, ranging from shallow analysis to hypothetical reasoning and ultimately to metacognition. We employ the first kind regularly, to calculate restaurant tips, separate laundry and prepare dinner. We use a more intense version of this Type 2 to explore different hypothetical scenarios, like shopping for a new car or picking a restaurant. All of these examples of Type 2 thinking are object-focused — call it Type 2.1. It contrasts with Type 2 thinking that is mind-focused, when we are analyzing thinking itself. We are contemplating thinking itself — ours and others' — when we engage this kind of Type 2, such as considering whether our guests will be pleased or offended if we serve the wine they gave as a gift, or whether our odds of promotion are improved or diminished by challenging our boss's thinking. Call this kind of thinking Type 2.2.

Type 2.2 is the metacognitive thinking underlying questions that we ask ourselves like, "Why did that person upset me so much?" or, "I wonder what she really thinks of me?" and when we tell ourselves, "I really need to figure out how to control my anger" and, "He is so closed-minded." Metacognition is a higher-order mental state, because it broadens awareness to include consciousness itself. Whereas Type 1 operates only at the object level, Type 2 can operate at both this level as well as the higher metalevel, by thinking about thinking, which is foundational to closing the brain–world gap.

Theory of Mind

While second-order cognition can be focused on our own thinking, we typically direct it toward the thinking of others. "I wonder what she is thinking about?" reflects awareness of the cognition of others, and is referred to as "theory of mind" (ToM, for short). The standard definition of ToM includes the attribution of thinking and distinct ideas to others, or mind reading, for short. Most animals and human infants have some awareness of themselves in the same way that they have awareness of external objects, but they are not necessarily aware that others have independent thoughts.

ToM in Animals

ToM is sometimes demonstrated by the Mirror Test, in which different animals are assessed on their ability to identify themselves in a mirror after a red dot is painted on their foreheads. The degree to which the animal can recognize itself in the mirror, rather than respond to the image as if it were another animal, indicates the presence of ToM. A chimp, for example, will reach to touch the dot on its forehead and attempt to rub it off, rather than touching the mirror. This rudimentary version of self-awareness is an important feature of ToM; it indicates the ability to differentiate oneself from others. It is not a true test of ToM, though, because it just reveals a self–other distinction, not necessarily the ability to contemplate the cognition of others.

Attributing ToM to animals is not uncontroversial. Do their behaviours really signal ToM, or are their actions explicable by a basic stimulus–response model? Some of the behaviours of great apes are difficult to explain without assuming some form of ToM, such as the empathetic gestures they make when they witness the pain of other members of their species. Empathy requires an ability to infer the mental states of others, as does the ability to engage in deception, which some great apes are quite proficient at.

While most comparative psychologists believe that some animals do have a rudimentary ToM, this ability does not necessarily entail an ability to introspect. An animal could be aware of itself and even able to "mind read" others, but still be incapable of reflecting on its own thinking.

Humans do not have a developed and functioning ToM until about age four. Before this age, if you show a child a box of Smarties filled with pencils, then close up the box and ask her what she thinks someone coming into the room will think is inside the closed Smarties box, she will say pencils. She has not developed an understanding that others see the world differently from the way she does. It is only between ages four and five that her response will mature to "Smarties," and it is only at this point that a child's ToM is sufficiently developed to engage in lying. Autistic people are considered to have weak or limited ToMs; they are not sensitive to the cues we offer, like smiling or frowning, to interpret the thoughts, feelings and intentions of others; if they ever attempt to lie, they do so poorly.

To differentiate ToM, which is thinking about others' thinking, from thinking about one's own internal thoughts, "metacognition" is sometimes used to refer to the latter. Just as often, metacognition is defined more broadly to include any kind of thinking about thinking; this more encompassing definition is tidier and more useful: ToM is, in this sense, a subset of metacognition.

Mindfulness

The distinction between 2.1 and 2.2 is crucial in the context of coping with complexity: we need more sophisticated models for interpreting complexity (2.1 thinking) and we need metacognitive intervention (2.2 thinking) so we do not over-rely on the automatic intuitions that do not jibe with complexity. World #2 is not decipherable with Type 1 thinking alone, but the solution is more nuanced than just applying more Type 2. World #2 demands that we be more cognizant of *how* we are thinking: both more aware of the simplifying models that we invoke to interpret the world, and more aware of the decision-making shortcuts that we use to draw conclusions. Metacognitive 2.2 is operational when we actively seek evidence that contradicts our hypotheses, when we seek out alternative points of view, when we consider how our biases may be influencing our thinking, when we postpone judgment until we are satisfied that we have conducted sufficient due diligence. These thinking strategies define what it means to be mindful: to broaden our scope of awareness by including awareness of how we are thinking.

Mindfulness is not an either/or mental state; we can be a little mindful or a lot mindful. Mindfulness varies with the degree of effort we apply to thinking about thinking, and it also varies between individuals based on how developed a person's metacognitive skill is. IQ tests do not reflect a person's facility for reflective thinking such as tolerating ambiguity, searching for disconfirming evidence, generating alternative hypotheses or reflecting on their belief systems as a whole. In fact, Stanovich points to research that reveals that there is only a weak correlation between IQ and a tendency for rational, flexible thinking.[139]

The Cognitive Continuum Revisited

The following depiction of our cognitive continuum incorporates the all-important distinction between 2.1 and 2.2.

Continuum of Thinking

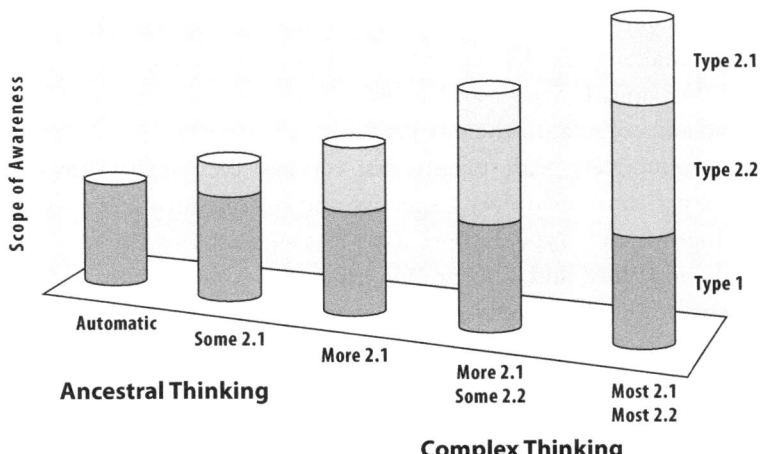

Scope of Awareness

Automatic — **Ancestral Thinking**
Some 2.1
More 2.1
More 2.1 / Some 2.2
Most 2.1 / Most 2.2 — **Complex Thinking**

Type 2.1
Type 2.2
Type 1

The scope of our awareness expands as we move to the right: we go from a perspective of purely automatic thinking to a perspective that builds on automatic thinking by adding hypothetical possibilities, and then on to a perspective that builds on both automatic and hypothetical thinking by adding metacognition. Our traditional or ancestral way of thinking is dominated by automatic Type 1. Complex thinking consists of the same automatic base, without which Type 2 cannot function, and without which it would be impossible to have any sense of confidence in our judgments, since it is Type 1 that triggers this feeling. Complex thinking is not fixed: it has increasing degrees of 2.1 and 2.2.

For the basic tasks of daily life, we do not need much Type 2. We need virtually no Type 2 to chew and swallow an apple, but we need a little bit to cut it up in pieces. We need some Type 2 to pick a restaurant, more to negotiate our preference if our dining companion wants to go somewhere that we do not, and even more to decide what price to offer the owner if we want to buy her restaurant. In addition to the 2.1 thinking that we need to run the financials on her business, examine the competition and industry trends and perhaps create a list of pros and cons of owning the restaurant, we also want to invoke some 2.2 thinking to reflect on how we are thinking about the opportunity. Are we being too optimistic about the prospects of the business because we have always wanted to own a restaurant? Have we been sufficiently critical of the revenue forecasts that she provided? Have we disregarded some risks? These questions are indirectly related to the business venture and directly related to how we are thinking about the opportunity.

Our challenge is that complex thinking is not natural to us: we default to as much Type 1 and as little Type 2 as possible, since Type 1 is cheap and abundant, whereas Type 2 is expensive and scarce. To cope more productively with twenty-first-century complexity, for which we have not developed deep expertise and reliable intuitions, we need more high-quality Type 2: more hypothesis-generating 2.1 for sophisticated modelling and interpreting, and more metacognitive 2.2 for sophisticated decision making. We have to overcome our default way of thinking — the inertia of ancestral thinking.

12.3 The Inertia of Ancestral Thinking

The key difference between ancestral and complex thinking is the degree to which Type 2 is invoked. Whereas ancestral thinking relies on a largely involuntary invocation of Type 2, complex thinking relies more on what Evans refers to as volitional cognition. To see the difference, it is important to understand how ancestral thinking automatically calls up varying degrees of Type 2 and how the automaticity is not always reliable.

Fluency and Invoking Type 2

We are miserly with Type 2: we need sufficient motivation to engage our expensive working memory resources. When we are under time constraints, Type 2 will get short shrift. When our working memory is fully loaded, like driving through a thunderstorm, Type 2 will limit itself to one urgent task. When strong emotional reactions like anger are triggered by Type 1, Type 2 loses most of its influence. The single biggest determinant of how much Type 2 we invoke is the feeling of knowing: when we feel that we are right, we lock down our conclusions and Type 2 is disengaged. As psychologist Valerie Thompson observes, the need for Type 2 intervention is triggered by a metacognitive acknowledgement that we are experiencing little or no feeling of rightness.[140] For Stanovich and Jonathan Evans, the reflective form of Type 2, which responds to the feeling of knowing, will engage the algorithmic form of Type 2 when this feeling is not present.[141] The more quickly Type 1 generates the feeling, the sooner Type 2 is disengaged.

One of the main determinants of how quickly we invoke the confident feeling of knowing is fluency. Fluency is a measure of how easily we process something: if recalling a memory is easy, the recall is fluent; if figuring out a problem comes easily, the calculation is fluent. Fluency is referred to by Kahneman as "cognitive ease."[142] As a subjective assessment of the difficulty of a problem, fluency is a marker of how easily Type 1 can arrive at a satisfactory interpretation. The easier something fits with our stored knowledge, the more fluent the processing will be and the quicker our subjective feeling of knowing is triggered. Fluency can be measured by how long it takes for us to feel confident in our conclusions, including the number of times we change our mind before finalizing. Fluency is so powerful that people will gauge the assertiveness of their own personalities based on the fluency of examples that come to mind, as psychologist Norbert Schwarz has demonstrated. After asking people to list six examples of times when they were assertive, they will report being more assertive, in general, compared with when they are asked to list twelve examples. The fluency, or ease, with which they recall six examples gives them the impression that they are more assertive than when they struggle to recall twelve examples.[143]

We use fluency as a cue for familiarity and fit: if something is easy to process, we are cued to assume we have a handle on it and do not need to expend further cognitive effort in deliberating on it; confidence arises quickly and forcefully. When we confront a novel situation that cannot be fluently processed, it is because there are no easy matches between it and our stored memories, so the feeling of knowing is not triggered and Type 2 is induced to help make sense of the situation.

Confidence in our beliefs has been demonstrated to be directly related to the fluency of processing.[144] A subset of this research indicates the presence of positive, micro facial movements when we are presented with fluently processed problems, versus negative facial movements when we are presented with problems that are not easily processed.[145] Fluency signals familiarity, which breeds confidence, which has a positive effect: it makes us feel good.[146] The calm feeling of knowing eliminates the need for Type 2, which is triggered only by the uncomfortable tension of not knowing.

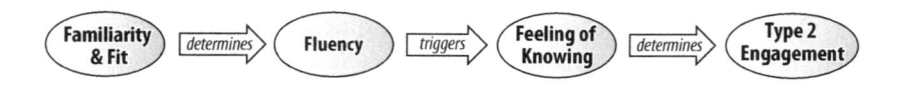

Familiarity & Fit — *determines* → Fluency — *triggers* → Feeling of Knowing — *determines* → Type 2 Engagement

When do we typically invoke Type 2 to supplement Type 1? When Type 1 requests it. How much cognitive effort do we expend in these cases? As much as Type 1 requests. When Type 1 determines that something important needs to be figured out, but cannot quickly process it, Type 2 is brought online, because Type 1 has not alleviated the uncomfortable tension of not knowing.

Fluency and Overconfidence

Fluency is a good indicator of familiarity, and therefore a reliable shortcut to concluding about things we have knowledge of. But fluency can be very deceptive in situations in which the rush to conclude encourages us to assume familiarity based on force fitting complexity into familiar models. As reviewed in 8.3, our lack of expertise can be masked by substituting the familiar for the unfamiliar, generating a feeling of knowing that is unwarranted. This is precisely the problem with an overreliance on "blinking": our intuitions come to us with a high degree of fluency, and this fluency lulls us into a false sense of confidence in scenarios that our intuitions are ill suited to. We think we see familiar patterns that satisfy our need to know, but that familiarity is based on surface-level comparisons, especially if we are taking the simplifying shortcut of attribute substitution — substituting an easy-to-evaluate trait for an unfamiliar one. If something seems familiar and seems to fit, there is insufficient motivation to tap into cognitively expensive working memory for further exploration. If something can be processed fluently, the initial interpretation is welcomed to take up residence in our belief systems by the warm reception of confidence, which closes the door behind it, shutting out other possible interpretations.

Ancestral thinking relies on the confidence that Type 1 creates, as the standard by which more or less Type 2 is invoked. This process is for the most part automatic: our thinking oscillates between Type 1 and Type 2 and back again, until Type 1 is satisfied that a good enough conclusion has been drawn. In complex situations, confidence is not a good determinant of the job being done, because Type 1 will do everything in its power to get to knowing as quickly as possible; it has an agenda to conclude ASAP. It can, and will, contort facts and analyse scenarios in a very selective, narrow way, and it judges the accuracy of its conclusions against its own standards of "good enough" and "makes sense to me." As Robert Burton notes, "'I am sure' is a mental sensation, not a testable conclusion."[147] Unfortunately, that does not stop us from treating confidence as a definitive indicator that we have arrived at a truth. We are just as susceptible to high levels of confidence (strong feelings of knowing) for beliefs that are wrong, as Thompson has demonstrated.[148]

Type 1 works well for the limiting case of problems we have expertise in handling. Beyond this boundary condition, automatic thinking is insufficient. Intuition is only as useful as the accumulated learning that underlies it, and most of our learning in World #2 is constrained by buried, ambiguous cues and misleading feedback. Aristotelian physics is very intuitive, but has limited applications: you cannot use it to fly a spaceship to the moon. Newtonian physics can get you to the moon, but is not sophisticated enough to develop microchip technology, for which we need quantum mechanics. Quantum mechanics is about as unintuitive as it gets. Physicists have to be conversant in relativity, quantum mechanics and a host of other models that are useful in describing reality. Similarly, the aspiring human cannot rely exclusively on intuitive models for interpreting complexity, or on an intuitive sense of when more Type 2 is needed.

While we cannot afford to reject all our initial conclusions, since they are built from a vast number of associations that our conscious awareness cannot fathom, we do not have to be captive to them, either. Burton recommends putting "hunches, gut feelings, and intuitions into the suggestion box" where they can be assessed.[149] This is precisely the strategy employed by professional chess players, who rely on intuition, developed after years of intensive practice, to come up with the move that they most likely will make, but invoke Type 2 hypothetical reasoning to mentally simulate alternative moves before making a final decision. The chess masters employ complex thinking: they use their intuitions as a starting point but do not rely on them exclusively; they leverage their expertise but are mindful enough to perform a mental check on it.

12.4 What Does Complex Thinking Look Like?

Complex thinking:

- differentiates complex problems from straightforward ones, so that
- more high-quality Type 2 can be invoked, specifically:
    - more 2.1 in applying mental models that are suited to complexity, and
    - more 2.2 in applying the mindfulness needed to avoid the decision-making traps of Type 1.

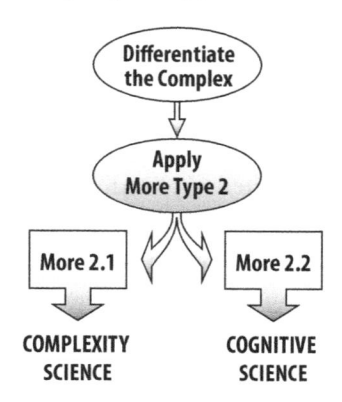

Differentiating the Complex

By failing to differentiate the straightforward from the complex, we default to our automatic ways of thinking, which misrepresent complexity and undermine our effectiveness. Unfortunately, there is no clear dividing line that separates problems into tidy categories, no foolproof algorithm for knowing when to expend energy to override Type 1. But as we become more aware of and conversant with both our mental shortcuts and our proclivity for overconfidence, we can more readily catch ourselves rushing to conclusions with oversimplified interpretations. We can more readily ascertain when a problem demands a different approach, one that is less exclusively reliant on Type 1. As we become more adept at understanding and identifying complexity, our intuitions will slowly be refined, and we will become increasingly more proficient at distinguishing the kinds of problems we face.

The key to recognizing complexity, therefore, is to learn how it works, which is why chapter 13 delves more deeply into complexity science. When we begin to understand the mechanics of systems theory, the complexity of the world becomes more readily apparent, so the cues that we need to make more sophisticated interpretations become more obvious. Differentiating and managing complexity requires high-quality Type 2.

More and Better Type 2

High-quality Type 2 stems from more sophisticated 2.1 modelling and higher levels of 2.2 metacognition. Type 1 is our default and it will not invoke Type 2 unless it has been trained to be sensitive to World #2 cues. Complex thinking requires being on the lookout for these cues, because they trigger the need to manage and supplement our automatic thinking so we are not overly reliant on it.

Psychologists have shown that, given the right triggers, people will interrupt their automatic processes rather than mindlessly default to them: we will eat fewer chocolates and cookies if they are individually wrapped; we will extend the use of gift cards over longer time periods if we receive five $10 cards instead of one $50 card. In a similar fashion, we need to familiarize ourselves with the cues that indicate complexity, so that we are triggered to slow our thinking down.

Type 2.2 mindful thinking brings us as thinking agents into the process of thinking, rather than leaving us on the outside. It exposes the weaknesses in our thinking by bringing our shortcuts to the forefront of our awareness, making explicit the implicit thinking that otherwise goes unnoticed. It appreciates how strongly motivated we are to keep the tension of uncertainty at bay: it increases our sensitivity to those situations in which our subjective feeling of certainty is likely to be misplaced. It grounds us in humility because it acknowledges that the obstacle to understanding and tackling complexity is unwarranted confidence. It avoids the Dunning-Kruger effect, reviewed in 7.2, of inflated confidence based on ignorance, by highlighting that the world appears straightforward and predictable only to someone whose grasp of it is superficial. Mindfulness allows us to pick up at the point where ancestral thinking is insufficiently sophisticated to handle the challenge. It means tipping the balance away from efficiency to achieve greater effectiveness, because the speed–accuracy balance that works for straightforward problems is not suitable for the complex ones, for which "blinking" or gut feel lead us down the wrong path.

Because complex thinking keeps effortful thinking engaged regardless of how strong the feeling of knowing may be, it requires a greater deal of cognitive self-regulation than we typically exercise. We self-regulate to avoid overeating; we self-regulate to exercise; we self-regulate to budget our expenses; we self-regulate every time we are in conflict with others and resist the temptation to lash out at them. Epistemic self-regulation is even harder for us than these other forms of self-regulation, because the need for it is often not obvious to us: we think we have got the answers we need so there is little incentive to use Type 2 thinking to its fullest advantage.

Alcoholics can avoid bars, compulsive gamblers can avoid casinos and dieters can avoid bakeries. But our addiction to certainty is not so easily regulated because we rely on it in World #1. The key to complex thinking is to identify the situations in which we know we are entering the equivalent of a bar, casino or bakery. Only then can we be on guard to do battle with the strong pull of Type 1 to conclude quickly with shortcuts.

Modelling Better and Concluding Better

The challenge of mental complexity, therefore, is twofold: we need to be better scientists by using more sophisticated models for interpreting the world (the what), and we need to be better psychologists by using more sophisticated decision-making techniques (the how). The next four chapters focus on these two challenges.

The trap to avoid when interpreting complexity is oversimplification. Chapter 13 focuses on the newer models we have for interpreting complexity, courtesy of systems theory, a tool of complexity science, which offers us better ways to understand causality and randomness (declarative knowledge). Chapter 14 describes strategies for implementing complexity science's newer models of causality. We have much to learn from this new science. Its focus is ontological (the way we describe the world) with the aim of improving the likelihood of our beliefs corresponding to the world (achieving greater epistemic rationality). Its tools help us move beyond the shortcut model of cause-effect that we habitually apply to interpret the relationships between things: a basic, linear "A causes B" formula that drastically oversimplifies the interactions within complexity. By replacing our everyday interpretative shortcuts, complexity science generates more accurate representations of World #2 and, therefore, facilitates better predictions about it. This science corrects the illusion of simplicity by invoking more rigorous cognitive software application programs for us to run. Rethinking causality is the best 2.1 thinking that we are capable of.

The trap to avoid when responding to complexity is overconfidence. Chapter 15 focuses on how we can be better decision makers when confronted by complexity. It looks at better mental procedures (procedural knowledge) that improve the odds that we will achieve our goals (achieving greater instrumental rationality), by ensuring that our belief systems are coherent. Chapter 16 describes the specific strategies that are derived from rethinking our notion of truth. With a focus that is epistemological (the way we describe how we think), the tools of cognitive science and its siblings, social science and decision theory, broaden our metacognitive capacity, enabling us to resist the pull to premature conclusions by replacing our everyday decision-making shortcuts. Cognitive science corrects the illusion of certainty by upgrading our cognitive operating system. Rethinking truth is the best 2.2 thinking that we are capable of.

In summary:

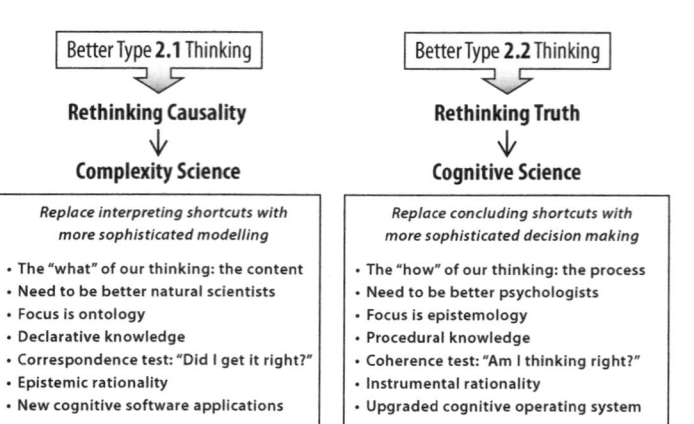

Better Type **2.1** Thinking	Better Type **2.2** Thinking
↓	↓
Rethinking Causality	**Rethinking Truth**
↓	↓
Complexity Science	**Cognitive Science**
Replace interpreting shortcuts with more sophisticated modelling	*Replace concluding shortcuts with more sophisticated decision making*
• The "what" of our thinking: the content	• The "how" of our thinking: the process
• Need to be better natural scientists	• Need to be better psychologists
• Focus is ontology	• Focus is epistemology
• Declarative knowledge	• Procedural knowledge
• Correspondence test: "Did I get it right?"	• Coherence test: "Am I thinking right?"
• Epistemic rationality	• Instrumental rationality
• New cognitive software applications	• Upgraded cognitive operating system

Physicist David Bohm is purported to have said, "Normally, our thoughts have us rather than we having them."[150] Complex thinking requires a reversal of this normality: it requires us to exercise greater control of our own thinking. Only then are we in a strong enough position to think more productively about causality and truth in World #2.

13

Rethinking Causality
Systems Theory Reveals "It's Not So Simple"

Systems thinking is a discipline for seeing wholes.
It is a framework for seeing interrelationships rather than things,
for seeing patterns of change rather than static "snapshots."
Peter Senge*

Any attempt to understand complexity must start with the premise that *there is always missing information*. There is always more to the story.

Sometimes the rest of the story does not matter (as is usually the case in World #1); sometimes it matters a lot (as is usually the case in World #2).

Systems theory is the best we have come up with so far to ferret out more of the story. It offers a model of causality that is more intricate and robust than the basic one that serves us in World #1. Systems' view of causality uncovers much of the hidden information we need, including a deeper (i.e., unintuitive) model of randomness, because causality is only valuable when random noise is properly separated from useful signal. This chapter explores the nature of complexity through the lens of systems theory, leading to the following chapter's exploration of strategies that derive from systems theory.

What is reality really like? There is no answer because the question is unintelligible: we have no direct, unfiltered access to reality; there is no perspective-free reality; no "view from nowhere." What is

---

* *The Fifth Discipline: The Art and Practice of the Learning Organization*, Crown Business, 2006.

intelligible, though, is the insight that our intuitive perspective of reality is a simplified one: a limiting case of a broader, more sophisticated view. In response to the question of what reality is "really like," all we can say is, "More complex than how we see it." But this constraint does not preclude us from developing more useful ways of analysing reality — ways that indicate we are closer to representing the way things work. This is where Type 2 thinking earns its stripes, by overriding automatic, Type 1 intuition with more sophisticated interpretative models.

There are pockets of order within complexity that contain universal properties, allowing us to make some predictions of likely paths. Wars, traffic jams, company life cycles, pandemics, weather, stock markets, romantic relationships, career paths, child development: all of these complex systems defy perfectly accurate forecasting, but all are amenable to a degree of description and therefore a range of possible predictions. The patterns of complexity are only available to us if we become adept at picking out the right signals from all the noise that surrounds them. The challenge to rethinking World #2 causality, therefore, is both to identify complexity's signals and to ignore the accompanying, irrelevant noise.

13.1 Signal and Noise in World #2

The causal factors that contribute to a ball falling to the ground, or a dog chasing a squirrel, are easy to differentiate: signal rises above noise quite explicitly, so it is easy to identify and segregate randomness. The causal factors that contribute to the outcome of a coin toss, a career path, a business's results or a marriage are all difficult to differentiate; they are multiple and interacting. Complexity makes it much more challenging to separate the random causes that are not useful for prediction (called "common causes," since they are not unique to the particular scenario) from the non-random causes that are critical for predicting and influencing behaviour (called "special causes," because they are unique signals). In world #2, we are vulnerable to overinterpreting random noise, by treating it like signal, because we are not adept at recognizing randomness and because our comfort with and tolerance for it is very low. We illicitly convert noise to signal by imposing patterns that we are more familiar and comfortable with, like coincidence ("What a miraculous event!"), various superstitions ("Karma caught up with her") and even skill ("Look at his success — he must be brilliant"). When we incorporate irrelevant information into our interpretations, we severely undermine our ability to predict, because signal and noise get confused.

To separate signal from noise in World #2, the challenge is twofold:

- to avoid *under*interpreting relevant signal by learning the unintuitive patterns of complexity that are revealed by systems theory; and
- to avoid *over*interpreting irrelevant noise by learning the unintuitive patterns of randomness, in particular, streaking and reversion.

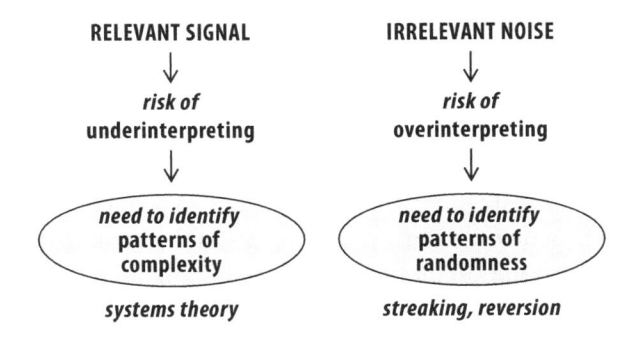

```
        RELEVANT SIGNAL              IRRELEVANT NOISE
              ↓                             ↓
           risk of                       risk of
      underinterpreting             overinterpreting
              ↓                             ↓
       ( need to identify )        ( need to identify )
       (   patterns of    )        (   patterns of    )
       (    complexity    )        (    randomness    )

         systems theory            streaking, reversion
```

Complexity's signals are revealed in the patterns described by systems theory. We have to dig for them both because we are not adept at picking them out and because noise obscures them. We overinterpret randomness (in the parlance of statistics, we "over-fit" the data) by applying our intuitive models to it and then picking out signals that are just noise. So in addition to understanding the nuances of systems theory, we also have to recognize the nature of randomness: the patterns that characterize it, as opposed to the ones we impose on it, specifically, its "streaky" and "reverting" character.

Systems theory and the patterns of randomness underpin a more sophisticated model of causality.

13.2 Rethinking Causality

"Happy is he who knows the causes of things," wrote the Roman poet Virgil.[151] Indeed, causality is the cognitive glue that binds things together for us; cause-effect is our premier model of sense making. But our intuitive model of causality is rudimentary.

Any discussion of causality has to begin with David Hume, who was emphatic that what we think of as causality is, and only can be, a learned habit and not something that is necessarily fundamental to the structure of the natural world.[152] He argued that the constant conjunction of things inculcates in us an expectation of one thing following another, but a true causal connection cannot be logically validated. We can never observe enough instances of conjunction between events to know for certain that one causes the other: there are only two things in sequence; there is no third thing between them to serve as a causal link that guarantees that the two things will always be conjoined in the future.

As philosophically interesting as Hume's point may be, even if we cannot guarantee the necessity of causal relationships, we are nonetheless dependent on the assumption of causality to make our way in the world — a point that Hume himself conceded. What is more interesting is that our efficiency-maximizing minds are in the habit (to use Hume's term) of establishing simple relationships: we look for the single causes that operate uni-directionally to explain the effects we experience.

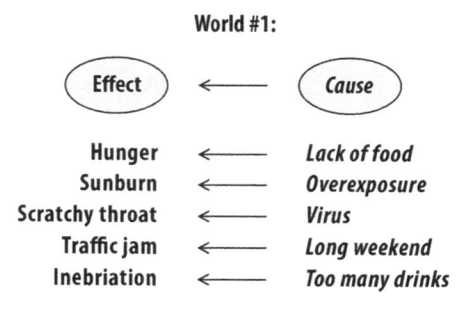

World #1:

Effect		Cause
Effect	←	Cause
Hunger	←	Lack of food
Sunburn	←	Overexposure
Scratchy throat	←	Virus
Traffic jam	←	Long weekend
Inebriation	←	Too many drinks

For every scenario in World #1, there is a simple, one-way causal relationship that our brains search for. There is always missing information in our explanations, but the missing information in World #1 does not matter because our satisficed explanations are good enough for coping with World #1 problems.

In a world that is increasingly defined by multicausal, multidirectional relationships, the same basic notion of causality excludes information that is crucial to understanding and managing complexity. World #2 is not adequately explicable with the simple model that works in a straightforward world, notwithstanding our habit of applying it as though it were:

Volatile relationship	←	*Self-absorbed partner*
Frustrating job	←	*Unqualified boss*
Emergency room wait	←	*Government incompetence*
Economic collapse	←	*Greedy bankers*
Terrorism	←	*Religious extremism*

In the examples above, important missing information lies in many hidden causal factors and the interactions between them. To be more useful in World #2, we need a more sophisticated cause-effect model:

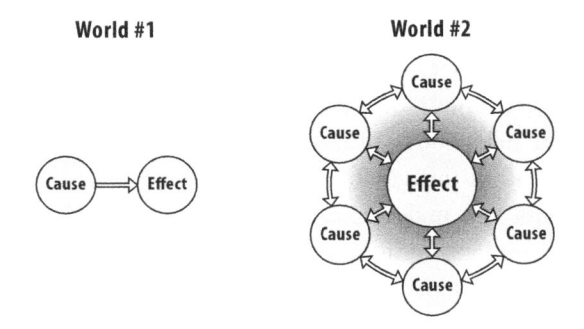

The model in World #2 incorporates multiple causes, and these causes interact with one another ("reciprocal causality"). All the factors, including what we consider to be the effects, influence one another in a series of constant feedback loops that are the defining characteristic of systems theory. Systems theory is a human construct, just as all models are, including special relativity, quantum mechanics and evolution, and it is a particularly useful one for dissecting and tackling complexity.

13.3 Systems in Action

A system is a set of things (inanimate objects or living organisms) that are interconnected in some way. The defining feature of any system lies in the interactions of its parts, because the system's behaviour emerges from these interactions — behaviours that are different from the behaviour of the parts in isolation.

A car engine is a system, as is a dialogue between two people, as is a storm. A system is defined by the boundaries that are chosen for it, so anything can be defined as a system: a single cell is a system of interacting, organic bits, as is the activity of crossing the street, since the pedestrian interacts with traffic signals as well as traffic. The bounds of any system are contingent on what is helpful to include and what is helpful to exclude. Two people who live on separate continents and have never met could be considered a system, in a global village kind of way, but not in a useful way, because their interactions are negligible and highly indirect. The concept of a system is extremely useful, however, for thinking about two people in a relationship: their interactions can be assessed as a system, as can the interactions that define the bigger system that includes their children, or the interactions of the even bigger system that includes their extended families.

While the behaviour of a system's parts often follows straightforward rules, the behaviour of the system itself can be very complicated. The workings of an individual neuron are explicable, but the collection of neurons that make up a human brain is much more difficult to model and predict. Because a system has its own character, distinct from its parts, complexity is aptly described as a state where the whole is bigger than its parts. Understanding the whole is the aim of systems theory, because our default intuitions lead us to be too narrow in what we include in our analysis, which causes us to miss or misinterpret the interactions. Our intuitive way of interpreting often fragments the world into discrete, unrelated parts: we see separateness where there is wholeness, disregarding the connections that reveal the patterns that we need to understand. Systems theory avoids fragmentation by identifying and exploring the connections, enabling us to have more productive interactions with complexity.

Interactions

Carbon atoms are the single component of both graphite and diamond. The exact same element gives rise to completely different materials: one that is soft and dark and one that is hard and clear. The difference is solely attributable to how the carbon atoms in each substance are interconnected. Similarly, it is the nature of the interactions between parts that defines a system's behaviour. Compared with basic systems, complex systems have two-way, rather than one-way, interactions. The street-crossing scenario is a basic system because we have no effect on the traffic signals; we merely respond to them so the feedback is unidirectional. If we walk on a red light into heavy traffic, the system becomes more complex, because now there is two-way interaction between us and the cars, each affecting the other.

How do parts interact in a complex system? Through feedback loops. Just as the key to understanding complexity is systems, so the key to understanding systems is feedback loops: the structures that define the two-way interactions. Interaction of some kind, no matter how basic, is required for something to be called a system; and feedback loops — two-way interactions — are required for something to be called a *complex* system. Feedback loops are the signal we require to decipher complex systems. The better we understand how the parts interact, the better we understand the system, and the better we can manage and influence it. Whereas our intuitive sense-making models are dominated by simple, one-way interactions, systems models focus on the tricky character of feedback loops.

If I toss a stone off a cliff and it falls onto the beach below without disturbing anything other than a bit of sand, there are no meaningful feedback loops between me and the stone: no information is passed on to me by the stone hitting the sand that causes me to change my behaviour, leading to a further impact on the stone. But if I then go home and am criticized by my spouse for being late for dinner, to which I respond by slamming the door, to which she responds by yelling at me for losing my temper, to which I shout at her for serving dinner too early, then we have ourselves an interesting system, a system defined by the informational cues that we pass to each other and respond to, in the feedback loop of communication between us.

Feedback loops, as the defining properties of complex systems, generate behaviours that have some common features: they are nonlinear and dynamical, and they generate emergent properties.

Nonlinearity

The straw that breaks a camel's back: A nonlinear interaction is one in which cause does not produce proportional effect. It is the grain of sand that causes a sand pile to collapse, the gust of wind that precipitates an avalanche, the trivial comment that triggers an explosive argument, the viral video that launches a rock star, the rumor that torpedoes a stock, the assassination that launches a war, the slow driver who causes a traffic jam.

Linearity is more straightforward: a simple system can be defined in linear terms with basic linear equations, because the cause-effect relationships are straightforward. Newton's invention of differential calculus was suited to this purpose — relating small changes to small effects and large changes to the sum of the small changes. In a linear system, you can simply add up all the small changes to see what the end result will be, like tracking the arc of a ball shot from a canon. A linear relationship means that the whole is the sum of its parts, because interactions are proportional.

A nonlinear relationship cannot be described by linear equations, because the parts do not interact in proportional, straight-line ways. It was not until the invention of the computer that non-linear math could be developed in a way that captures the nature of complex systems, which is why weather forecasting has improved over the past few decades. Forecasting weather with precision is still difficult, because non-linear math, as sophisticated as it is, is insufficient for fully capturing the non-linear behaviour of a complex system that changes over time.

Science's yet-to-be-solved mysteries, from defining consciousness to curing cancer, are all non-linear in form. Our world is increasingly characterized by non-linearities, fuelled by exponential growth in technology and population: more people who are more tightly interlinked create more non-linear impacts on one another and on the planet. The feedback loops between all the moving parts amplify change, since positive feedback that is non-linear generates huge impact from initially small changes, such as a few Twitter messages snowballing into a large protest. Non-linearity means that more extreme, unusual events become more common.

Dynamics

Feedback loops cause a complex system to fluctuate rather than stay stable. In the discourse of complexity science, a dynamical system is one that changes over time, and the way it changes also changes: the system is "non-stationary."

The notion of the ever-changing dynamics of reality is as ancient as the first recorded Eastern and Greek philosophies. More recently, G.W.F. Hegel, writing in the early nineteenth century, proposed that every complex situation contains within itself the seeds of destabilization, and that history is the eternal unfolding of perpetual change propelled by internal, conflicting elements (through a "dialectical process").[153] It was not until one hundred years later, writing in the early twentieth century, that the mathematician Henri Poincaré would demonstrate in concrete terms just how unstable a system can be.

Poincaré showed that Newtonian linear equations were insufficient to describe complex systems: Newton's laws of motion were only approximations of the behaviours of real-life objects, because modelling and predicting with precision was impossible. Poincaré pointed out that many real-life scenarios are much more complex than linear math and simple scientific laws account for. He showed that the tiniest, imperceptible interference makes precise modelling and prediction much more difficult than Newton envisioned. This instability explains why Newton's modelling of the solar system was slightly inaccurate: the earth's path is affected by the gravitational fields of all celestial bodies, and incorporating all of their effects on the earth's trajectory is not possible. Newtonian physics achieves precision only in the limiting case of a simple system with no external influences (i.e., a closed system). Even then, Poincaré argued, this closed system would have to be limited to only two bodies. A third body introduces complicating influences as the system unfolds, making precise predictions impossible. Poincaré discovered that the smallest of errors in measuring the starting conditions of a system can accumulate to large deviations from a projected path, which is to say that the system unfolds in a non-linear fashion.

Order, Chaos and Disorder

It was not until the advent of computer processing that Poincaré's ideas were picked up and expanded, notably by meteorologist Edward Lorenz.

In 1961, Lorenz was modelling weather patterns on a computer and discovered by accident that the exact same model created completely different results if the starting point of the calculations was 0.506 instead of 0.506127. By ignoring the last three decimal places in the starting position of a weather system, the same formulae generated meaningfully different outcomes. Lorenz was intrigued by this form of non-linearity and proposed that, in certain systems, the slightest miscalculation in defining the initial state of the system can throw predictions about it way off, even if the model that describes the system is itself perfect. Just as Poincaré had proposed, the equations for setting the starting condition would have to be impossibly precise for predictions to be valid: an infinite number of decimal places would be required to fix the starting position of every single particle in the system.

In a lecture on non-linearity, Lorenz used the analogy of a butterfly flapping its wings in Brazil, ultimately setting off a tornado in Texas. These seemingly unstable systems were later dubbed "chaotic" when chaos theory arose from a branch of mathematics that focused on examining the changes of a system whose behaviour is sensitive to small differences in starting conditions. Prior to Lorenz's discovery and the development of chaos theory, it was assumed that the unpredictability of a system was attributable to excessive randomness that could not be modelled. But chaotic systems are still very difficult to predict even when there are no random elements to them, because their initial states cannot be measured with perfect precision: imperceptible measurement errors produce significant errors in predicted future states. However, because chaotic systems do not contain a problematic level of randomness, patterns *do* reveal themselves, patterns that can be used to make probability forecasts despite the lack of precision — hence, weather forecasting.

The concept of chaos is not just interesting to scientists and weather forecasters; it lends insight to many complex aspects of our lives. First is the realization that the world can operate in pure, causally determined ways, with little or no randomness, and still not be predictable with any precision. What appears to us as random is often just a reflection of our own cognitive constraints rather than a true lack of order or a lack of causal connectivity. Second is the insight that the paths that complexity takes, including the complexity of our individual life paths, can veer off in shockingly unexpected ways, based on non-linear responses to seemingly insignificant events, including events that are imperceptible to us.

Between the extremes of perfectly predictable order and complete disorder (which has no predictable properties) lies complexity, where some order is evident. Some of these complex systems are chaotic. The term "chaos" is unfortunate because it connotes randomness and disorder when, in fact, chaotic systems are not random. Chaos theory is based on mathematical techniques for analysing the hidden order of these systems, which share neither the predictability of orderly systems nor the randomness of disorderly ones. Chaos is the property of a system that follows precise rules that would make the system predictable if it were not for an inability to specify the initial state exactly. Chaotic systems are interesting because they are deterministic (not "stochastic," or random) yet their behaviour still defies precise predictions.

An orderly, non-chaotic system is much more manageable because any errors in measuring its starting position accumulate in a linear way, which means that the forecast error of its path is proportional to the starting measurement error. In a chaotic system, an imperceptible starting error can balloon into a mammoth forecasting error. There is important information in the millionth decimal place of a chaotic system: information that reveals itself only over time as the system's non-linear behaviours unfold.

Consider a weather forecasting model that begins with a starting temperature of ten degrees. Is the exact temperature 10.1, or 9.9, or 10.00203? The difference matters, because even a perfect model will generate different results depending on which of the three variables are used. If you cannot measure the temperature to an infinite number of decimals, your starting state is imprecise. This dilemma is not unusual considering the impossibility of measuring a table precisely: the best we can do in describing it, for example, is to call it four feet plus or minus one-eighth of an inch. No device can measure its length from one edge to the other with perfect precision, down to the last atom.

The challenge of chaos does not preclude us from making predictions: chaos theory works to generate a range of predictions that are possible for chaotic systems. The further out the predictions are, the higher the error rate, since the nature of chaos amplifies the starting error, which grows exponentially. This is why a weather forecast for three hours from now is more reliable than a forecast for tomorrow, which is more reliable than a forecast for next week. It is also why it is so difficult to quantify the long-term impact of higher carbon emissions in our atmosphere, notwithstanding near unanimity among scientists that it is negative.

Complex systems tend to move between perfect order and complete disorder, back and forth, making their behaviours difficult to forecast. Think of an argument between two lovers: it typically follows a period of stability in the relationship, until some event triggers a conflict, which ignites a heated discussion that can move from terse exchanges to shouting, accusations, perhaps even physical violence in the worst-case scenario, followed by relative calm, perhaps some more terse words, and a reconciliation that remains vulnerable to future instability. The dynamical nature of a complex system means that it is constantly shifting as the parts influence one another in different ways; these moves are usually between greater and lesser degrees of order. Complexity does not settle into permanent stability, nor does it break down into complete random disorder. It oscillates between the two extremes, reflecting various forms of orderliness, one of which is chaos. Another example is the stock market, which can gyrate wildly for days in a row, then settle into a narrow trading range before resuming volatility. The space between order and disorder is where we spend our lives, shuttling between various phases. Much of life as we know it occupies the middle ground between stability and complete disarray, constantly moving between the two extremes.

One of the more dramatic ways that a system shifts from one state to another is when it reaches a tipping point (by entering a "phase transition"), such as when water begins to turn to ice at zero degrees Celsius. The system moves from one state of self-organized stability to a different state of self-organized stability, but only after reaching a threshold that is on the verge of change ("self-organized criticality"). This kind of self-organization and self-generated change is characteristic of many complex systems in nature, such as flocks of birds migrating, swarms of bees searching for new nesting locations and colonies of ants foraging for food. These systems are often adaptive: they preserve themselves through their changing behaviour without any formal leadership or influences that are external to the system. Because the changes that complex systems undergo are typically non-linear — disproportionately large changes in response to small inputs — they are usually described by power laws.

Power Laws

The mathematics underlying power laws is based on relationships between variables that increase exponentially (i.e., by powers greater than one). The single grain of sand that triggers the sand pile to level off is an example of a tiny input (the grain) that creates massive change (the pile collapsing), a process that is described mathematically by a power law. The stock market often moves in ways that are represented by power laws, because the ups and downs can be disproportionate to whatever news comes out. So are traffic jams, because traffic can move from a smooth flow to hundreds of cars in gridlock in an instant, if just one car brakes for just one moment. (Traffic jams arise from the asymmetry in braking time compared with accelerating time: it takes much longer to accelerate back to normal speeds than it does to brake to a slower speed.)

The 80/20 rule (known as the Pareto principle) reflects a power law: the majority of the output (80 percent) is derived from a minority of the input (20 percent). Most often, the majority of a company's sales are driven by a minority of its customers; most of a person's career success is determined by a small portion of her work; most of the rain that falls in a given year accumulates over a small number of days; most of a film production company's revenue comes from a small number of all the movies it makes; most Internet traffic is concentrated on a disproportionately low number of sites; most of the wealth in a country accumulates to a small number of people. (The 80/20 phenomenon has many forms, going by many names, such as the Matthew effect, preferential attachment, Zipf's law and Gibrat's law; all of these formulations reflect the principle that a small number of points in a sample represent a disproportionally large share of the outcomes.)

The relevant point about power laws is that extreme events are more relevant than we typically think: power laws explain why stock market crashes, massive earthquakes, multibillionaires, worldwide fame and the like are both more frequent and more extreme than the statistical properties of linear interactions suggests. It is the non-linear feedback loops that generate the power laws that describe World #2 complexity. And much of this non-linearity derives from fractal patterns.

The term "fractal" was coined by mathematician Benoit Mandelbrot in 1975 to describe the irregular patterns that show up in many things: the broken or fractional geometric patterns in nature that are neither smooth nor disorderly but somewhere in the middle.[154] It is not just their irregularity that is interesting: fractals are a mathematical property of objects and systems that can best be described as "self-similarity" — a shape that is composed of copies of itself. The patterns that show up in many ways in both the natural and human-made world reflect repetitiveness. Mandelbrot detected these self-similar patterns on different time scales when he studied the price fluctuation of cotton. He applied this insight to the British coastline, which has the same jagged edge from a distance (from a plane or boat) as it does closer up on a smaller scale (from standing on the shore), as it does even closer up on an even smaller scale (bending down examining a tiny portion of the coast edge). When you look at something with fractal property, you will see the same kind of pattern repeating itself at varying degrees of magnification, no matter what scale you are using.

The self-repeating patterns of fractals are interesting because they can generate complexity from simple rules that are repeated over and over. Many patterns in nature are generated by multiple repetitions, enabling large-scale complexity to emerge from simple starting points. Out of a very simple branching mechanism, pattern replication creates the branching system of circulatory systems, lungs and brains. In fact, fractals appear to be a key element of how DNA works: to contain the entire genome, the tiny DNA molecule can store all the information required to build a complex being, based on a simple set of instructions to repeat certain patterns a specific number of times. Fractal geometry — the repetition of simple patterns — produces the intricacy that emerges from complex systems.

Emergence

The whole is bigger than its parts: Emergence is the concept that explains the gap between the parts and the whole. It is the property of a system that can be observed from looking at a higher level of organization:

an ant colony is the higher-level social system that emerges from the lower-level interactions of ants. Macroscopic properties emerge from the interactions of the microscopic parts of the system. The swarm behaviour of bees emerges from individual bees; Newtonian-like properties emerge from a collection of randomly moving quantum particles; consciousness emerges from non-conscious neurons; innovation emerges from individuals brainstorming together; living beings emerge from non-living compounds.

As physicist Murray Gell-Mann puts it, "You don't need something more to explain something more."[155] If you understand the principle of emergence, you can appreciate how interesting things arise in complex systems without invoking anything mysterious or external to the system. The mystery lies within the system itself — embedded in the interactions between the parts, which are often hard to uncover, yet crucial for understanding the system. You cannot predict the properties of a system by examining the parts in isolation. Just as the behaviour of an ant colony emerges from the ants, to get a full sense of what it means to be human, you have to observe how people interact, since behaviours emerge from interaction that are not present when observing an individual in isolation.

The emerging properties that arise from non-linear interactions force us to think about causality in a different way than we usually do. The Type 2.1 modelling of causality that is suited to complexity is less intuitive in many ways, not the least of which is how randomness works in World #2.

13.4 Rethinking Randomness

"No victor believes in chance," wrote Friedrich Nietzsche.[156] Our nature is to explain, and explanations aim to reduce the role of chance. We are always scanning our environment, looking for important signals by separating them from what we perceive is irrelevant noise. Randomness is a large part of noise; a random event is one unconnected to the event that preceded it. Whether a coin toss turns up heads or tails is not a function of what it turned up in the toss before. Each state of a random system is unrelated to the previous state, so the past contains no information about the future state of a random event: the past and future are unconnected.

Random does not necessarily mean *uncaused*: a multitude of causes can result in a head or a tail — angle and force of the toss, balance and symmetry of the coin, air friction — but all of these interacting causes cannot be measured and integrated to predict the coin's path. Randomness is computationally uncompressible: the causal factors that produce a random event cannot be summarized, which is why random events cannot be predicted.

Since the causal histories of random events are inaccessible and cannot be used to make predictions, the question arises whether randomness is a natural property of the world (i.e., ontological) or a property of our thinking (i.e., epistemological) that merely reflects our inability to describe a causal history of certain events. The answer is both.

Ontological randomness comes in the form of quantum mechanics, where subatomic particles can only be described by probabilities. For example, the past and future of an electron are not connected: if the starting position of an electron were known with theoretically perfect precision, its path would still be indeterminable and unpredictable because no causal conditions link its starting position with its ending position. In fact, two electrons with perfectly identical starting positions and forces on them would still take separate paths for no specific causal reason. The best we can do is describe the behaviour of subatomic particles with wavelike probability distributions, not because we lack more advanced measuring equipment, but because random uncertainty is fundamental to the nature of these particles.

Where Does Subatomic Randomness Go?

Where does ontological randomness go as we move from subatomic particles to our macroscopic, middle world? It is still with us; it does not vanish. But on a larger scale, randomness largely washes out; the odd behaviours of subatomic particles offset one another, and we are left with a more stable and predictable world, like the kind that Newton modelled.

This is why coffee cups do not spontaneously bounce off the

> table. They could: there is nothing physically or logically impossible about a sufficient number of subatomic particles moving in the same direction, causing the cup to move on its own. But the odds are so extraordinarily low that it has never happened. This possibility is easier to imagine when we consider molecular randomness that we can actually see, in the form of Brownian motion, which is the movement of a particle of dust in a cup of fluid. Its motion is the result of being buffeted by water molecules jumping around on all of its sides; it moves wherever the majority of jumping water molecules happens to exert the most force.
>
> In our middle world, the uncertainty generated by ontological randomness is so minuscule, we can assume it does not exist even though it is always lurking in the background.

The "within nature" lack of causality is different from the epistemological randomness that stems from our bounded brains struggling to understand the causal history of certain events that are difficult to decipher. Epistemological randomness is something that we live with every day: a random car accident, stubbing our toe, losing our wallet, meeting our future spouse in an elevator. These examples all have causal histories that link the past to the present, even though the events are unpredictable by us. The challenge of understanding randomness in this case is not related to anything fundamental in the world, but lies in our inability to perceive the causal chains that lead to certain events.

Ultimately, the distinction between the two kinds of randomness is more academically interesting than practically useful. Nassim Nicholas Taleb concludes that the difference is unimportant since, in either case, randomness reflects incomplete information.[157] What does matter, and matters a great deal, is our ability to recognize the randomness that confronts us every day, so we avoid the temptation to interpret it as signal instead of noise.

Overinterpreting Randomness

World #2 has created more noise than we are accustomed to having to filter, so the risk is higher that we will treat it as a signal instead of the noise that it really is. We overinterpret randomness by discerning patterns where there are none, because we are pattern hungry: we have a very low tolerance for patternless data because we cannot understand and control them. We are so pattern hungry that we are incapable of manually creating true randomness: the human mind is not able to formulate a sequence of numbers or letters that is mathematically random. If we put pen to paper and write down a series of twenty Xs and Os that we think are random, such as xoxooxxo ..., we will inevitably create something that a statistician or computer program will find some pattern in. More specifically, we switch our Xs and Os too frequently: random series have an unintuitive, surprisingly high number of repetitions. The technology that underlies truly random number generation, such as for slot machines, is more sophisticated than humans could replicate on their own. (This human proclivity for creating patterns is, incidentally, what rock-scissor-paper champions rely on to outwit novice players in competitions.)

Our low tolerance for randomness is legendary in the fields of psychology and neuroscience; many experiments have demonstrated how much worse we are than other animals in actually coping with randomness simply because our densely wired brains work on overdrive to look for patterns. Michael Gazzaniga conducted an experiment in which flashes of light appeared randomly, but 80 percent of them were above a line and 20 percent below it. Rewards were offered for correct guesses where the light would flash after subjects had experimented with the procedure. Rats earned more rewards by always choosing the above-the-line option (and therefore were correct 80 percent of the time). Humans tended to average 67 percent correct guesses by trying to outsmart the system.[158]

Gazzaniga reports that all animals tend to do better on this task, as do children under the age of four. Older children and adults all try to override the randomness by guessing where the light will flash, even when they are told that the order of the flashes is entirely random and unpredictable. Similar experiments reveal that if we know, for example, that a red light will flash 60 percent of the time and a green light 40 percent of the time, we will guess red about 60 percent of the time, thereby reducing our chances of being right to 52 percent, compared with always guessing red and being right 60 percent of the time.

These experiments confirm "the illusion of control" described by psychologist Ellen Langer in various experiments that demonstrate how reluctant we are to surrender ourselves to randomness.[159] We cannot help ourselves: our intuition is that we have some ability to stay ahead of it, even though that is impossible. Interestingly, patients with damage in certain parts of the prefrontal cortex have been observed to be less prone to the error of overinterpreting randomness: it takes a brain lesion to get us to be as smart as a rat when it comes to dealing with randomness.

14

Strategies for Complex Systems
The Hunt for Signals

The world we live in is vastly different than the world we think we live in.
Nassim Nicholas Taleb*

Systems theory is a powerful way to interpret and respond to complexity. The challenge is how to concretize systems theory in responding to real-world problems. The focus of this chapter is the collection of strategies that help us do exactly that, something that is no easy feat. Consider an initiative to reduce the incidence of malaria in Africa:

> The Roll Back Malaria program was launched in 1998 by a number of non-profit organizations, including the World Health Organization. Many initiatives were undertaken to reduce malaria in Africa and other developing countries. One of the initiatives was the distribution of bed nets to pregnant women and children. The nets were given to villagers so they could protect themselves at night from mosquitoes — carriers of the disease — by placing the nets around their beds.
>
> In the follow-up studies that were conducted to determine the effectiveness of the program, researchers discovered that in farming communities, the nets were often used to protect crops, and, in fishing villages, the nets were commonly used to catch and dry fish. Many of the villagers who were interviewed were not convinced of the effectiveness of nets for malaria prevention, but saw great value

---

* *The Black Swan: The Impact of the Highly Improbable*, Random House, 2010.

in the nets for their livelihood. Many indicated that, given the choice between malaria and starvation, improving their fishing techniques was much more pressing for them.

Not only were the nets being diverted to other uses, but, according to Kenyan officials, the use of nets for fishing reduced the fish stocks in some of the major lakes: the nets catch juvenile fish, which are prevented from maturing and reproducing.

This cycle of unforeseen consequences is precisely what systems theory is designed to explore and, at its best, to guard against. The interactions that occur between moving parts makes complexity behave much differently than World #1 phenomena, and requires more sophisticated modelling as a result.

How can we use systems theory, with its unintuitive but powerful models, to engage complexity more effectively? By using strategies that allow us to describe and engage World #2 more productively than the intuitive, default models of our ancestral thinking.

Systems Theory Strategies

World #2 is different ⟶	1. Identify the landscape
Complexity's cues are hidden ⟶	2. Pursue missing information
Causality is easy to misinterpret ⟶	3. Dissect causal relationships
Signals need to be surfaced ⟶	4. Generate feedback
The system can be influenced ⟶	5. Look for leverage
Noise masquerades as signal ⟶	6. Beware randomness

14.1 Identify the Landscape (Pick the Right Mountain)

Evolutionary biologists and complexity scientists speak of the "fitness landscapes" that problems are located on: a conceptual three-dimensional field consisting of small bumps and mountainous hills. The best solution to a problem, including the evolution of a survival-enhancing genotype, is the peak of the tallest mountain on the landscape of possible solutions. In World #1, the landscape is straightforward; it is not particularly rugged: problems are obvious and clearly defined, and solutions are easy to identify even if they prove difficult to implement. It is not difficult to pick out which hill to climb in World #1, because the field is quite flat except for the conspicuous mountain, the peak of which represents the solution to the problem.

Not so in World #2. In the landscape where complex problems reside, the highest peak is not necessarily obvious; finding the right mountain to scale is half the battle. Many smaller hills look alluringly like high peaks until deeper exploration and a broader perspective reveal higher mountains. And even when we think that we have spotted the highest-peaked mountain, it is never a straight-line expedition up its side: it takes some creative and flexible navigation to find our way up. In World #2, we do not really have a clear picture of how high the mountain is that we are scaling until we get to the top — if we ever do. At any point on the journey, including arriving at the summit, we may come to realize that we are on the wrong mountain and need to relocate. We are often forced to climb a few mountains in order to ascertain whether better solutions are poking up in the distance (recall the bank struggling with poor customer service ratings, in 8.2).

World #1 Landscape

Easy to identify the tallest mountain

World #2 Landscape

Hard to know which mountain to scale

The worst mistake to make with complexity is to assume that we are dealing with a straightforward landscape, where the first conclusion we draw and solution we implement are obvious and definitive. Our rush to certainty encourages us to sprint up the closest mountain and, once atop it, to look down at the base of the mountain with the false confidence that we scaled the right mountain, instead of looking out for higher mountains. We are vulnerable to assuming that we have reached the highest point in the landscape, when, in fact, there is a good chance that we have scaled the wrong hill.

The first step to tackling complexity is to recognize that much of the time we are operating in a different set of conditions than World #1. Only by acknowledging that we are in a different landscape can we be open-minded and nimble enough in our expedition to invoke the following strategies.

14.2 Pursue Missing Information (Dig Below the Surface)

In World #2, there is always more to the picture than what is immediately available to us. Our brains prefer to work in greedily reductive ways, but when we compress sensory data into simple stories, we lose a lot of key information about how complexity works.

The causal chains that reside below the surface of quick interpretations are composed of multiple links that are far more elaborate than the single-cause analysis our intuitions gravitate to. We have to expand the boundary of our search for signal in both time and space, so that proximate causes are not overweighted at the expense of distant causes that may have significant influence. The missing information is contained in the multiple contributing causal factors that are not obvious and in the interrelationships between these causes, which are even less obvious.

Missing Multiple Causality

Studies by sociologist Charles Perrow and others have documented that most big accidents are an accumulation of a lot of little things going wrong: plane crashes are rarely the result of a single malfunction; they are a combination of causes, the nexus of which is tragedy.[160] Complexity is characterized by chains of causes interacting with one another.

One of the challenges of complexity is our failure to distinguish proximate from non-proximate causes. If our partner comes home, scowls at the dirty dishes on the counter and snaps, "Clean this mess up," we are inclined to attribute her grumpiness to the dirty dishes since they are the proximate trigger for her anger. More distant causal factors are usually more relevant than the immediate issue of dishes, such as her struggles at work or her larger concerns about the relationship. Decision scientists distinguish proximate causes from ultimate causes: the proximate cause of her irritation is dirty dishes, but the ultimate cause is bigger and less manageable, such as her feeling unappreciated. Ultimate causality is largely theoretical in the case of complexity, because it is impossible to fully capture every single element that makes up a complete description of causality for a complex system. But the notion of ultimate cause is more than just academically interesting: it serves a practical purpose by reminding us that the whole picture is always bigger and more nuanced than our first interpretations. Availability bias motivates us to overweight proximate causes and underweight causal contributors that are not as immediately obvious. Our intuitions lead us to believe that cause and effect are closely linked in time and space, but complex systems have causes that are distant from the immediate results we witness. It takes discipline to extend our search for multiple causes, especially when a single one lures us into thinking we have figured things out.

Missing Interactions

It is not just the multiplicity of causal factors but their interactions that give complexity its character. We cannot just examine each cause independently; the behaviours of complex systems emerge from the relationships between factors.

One of the failures of modern nutritional science, as alluded to in the opening of chapter 4, is, according to writer Michael Pollan, its inability to account for the interactions between nutrients, the food they are stored in and our bodies.[161] Pollan points to the identification of beta-carotene and vitamin E as cancer-fighting antioxidants: it turns out that, in capsule form, as they have been manufactured and marketed, they lose their effectiveness. Once separated from food, the nutrients behave differently: there is something different about their form in food than their form outside of food. The confusing, sometimes contradictory and often misleading nature of nutritional science arises, in part, from its examination of nutrients in isolation, without an understanding of how they operate within interactive systems. Nutrients interact with other chemical compounds in the foods that contain them; they interact with other foods that we ingest; and they interact with our digestive systems. All of these interactions make it difficult to draw definitive conclusions, and when single causal factors are isolated from multiple, interacting causal factors, our conclusions are at high risk of being wrong.

Uncovering and assessing interactions is tricky work. Some causes have bigger roles than others, despite signals that can suggest otherwise: our simplifying shortcuts were not designed to pick up on the nuanced difference between a direct cause and an indirect enabling one. For example, walking barefoot in the snow in your bathing suit does not cause you to catch a cold, since viruses are a necessary and direct cause. Wearing insufficient clothing can tax your immune system, though, which makes you more vulnerable to viruses getting the better of you. However, while insufficient clothing can enable a viral cold, it is neither necessary nor sufficient for getting sick. Even more nuanced are genetic predispositions for certain diseases that are not sufficient causes for developing the disease, since environmental factors can shape if and how a disease forms.

Complex thinkers dig for the underlying multidirectional interactions that define systems; they are never satisfied with the first reasonable causal relationship that surfaces, especially when it is a simple, one-way interaction that preserves the separateness of the parts and neglects the system between them. Because emergent properties arise from these interactions, complex systems have to be analysed with a view to uncovering these relationships. But these relationships have to be dissected very carefully since we are prone to misinterpreting them.

14.3 Dissect Causal Relationships (Proceed Very Carefully)

We are not only at risk of missing the key information, but also of misinterpreting it. Dissecting causality in World #2 is by far our most daunting challenge because the pull to rely on our basic intuitive causal model is so strong, yet so susceptible to getting causality wrong.

Wrong Direction

Do chemical reactions in the brain cause depression, or do pessimistic thoughts trigger chemical reactions in the brain? Neuroscientists favour the former interpretation; psychologists generally favour the latter. The correlation between thoughts and chemicals does not reveal a clear causal direction, a problem that scientists are still trying to sort out.

The Garcia effect, named after psychologist John Garcia, who discovered it, describes a taste aversion that we develop when we become sick after eating.[162] It is a good illustration of the error we are prone to when it comes to if-then thinking (the logical error known as "affirming the consequent"). We can confidently say that eating bad eggs causes stomach upset, but that does not entitle us to say that suffering a stomach upset after eating eggs proves that the eggs were bad, which is where our minds want to go.

If the food is rotten, then my stomach will ache. This is a perfectly reasonable argument, and the conclusion that eating rotten eggs causes stomach upset is both intuitive and valid (the logical structure of this argument is called *modus ponens*). We can also validly claim that if our stomach is not upset, then the eggs were not rotten (*modus tollens*), which is also an intuitive logic for us. Here is where our intuitions fail us: if, after eating eggs, our stomach is upset, we conclude that the eggs must have been bad. This conclusion is not valid: it is the fallacy of affirming the consequent, which will be revisited in 15.3. The presence of the consequent (what follows "then") is not proof of the presence of the antecedent (what follows "if"). Our intuitive mistake is assuming that since we can say, "If rotten eggs, then indigestion," we are allowed to infer the converse, "If indigestion, then rotten eggs." This argument is faulty because there are many possible reasons for an upset stomach that may have little or nothing to do with the meal that preceded it. The converse is not valid, because the presence of the consequent (upset stomach) does not necessitate the antecedent (rotten eggs): other causes may be having the flu, eating rotten lettuce that accompanied the eggs, or stress-induced stomach indigestion.

Logically valid conclusions cannot be contradicted as long as the premises that support them are true. Conclusions drawn from invalid arguments, however, can be contradicted, because alternative conclusions can be derived from the premises. Unfortunately, our struggle with if-then is pervasive; we often think and argue on the basis of incorrectly affirming the consequent:

- If waterboarding is effective, then terrorist activity will decline; terrorist activity has declined, so waterboarding must work. *It is possible that it works, but this argument does not prove it.*

- If global warming is human-created, then the earth's average temperature will rise; since average temperatures have risen, humans are clearly causing global warming. *We are no doubt contributing to global warming, but this argument does not prove it.*
- If my broker is a good stock picker, then my stocks will go up; my stocks have gone up, so my broker is a genius. *Perhaps, but do not bet on it.*
- If that mentalist can read minds, then he will be able to tell me what I am thinking; he correctly identified that I was worried about my job, so his talents are authentic. *Well, we at least can say he is entertaining.*
- If this herbal medicine is legitimate, then it will ward off colds; I have not been sick for a year, so these miracle pills are worth the $65 I spend to buy them every month. *Maybe, but how will you rationalize the next cold that you do get?*

All of the conclusions above may be valid, but they do not follow from the argument that precedes them.

Things get even more confusing when the direction of causality is irrelevant because there is no causation to begin with.

Correlation Versus Causation

The error of mistaking correlation for cause was the basis of an argument by scientist Ronald Fisher in 1958 to defend the tobacco industry.[163] He argued that smoking and lung cancer were unlikely to be related in a directly causal way, but could be indirectly related by the same underlying, genetic disposition that causes both cancer as well as a craving for smoking. This argument was not as preposterous at the time as it appears today because, in the absence of concrete scientific evidence, the logic is sound: co-occurring things are not necessarily causally connected since they could be merely correlated; both could be the effects of an underlying cause (called a "confounding variable"). Correlation can masquerade as causation if the underlying cause is hidden and all that we witness are the two effects occurring together.

A more current example is one addressed by Pollan, which relates to the correlation of vitamin supplements with overall health. Research indicates that, while there may be a correlation between taking vitamins and overall health, there is not necessarily a direct causal connection. The correlation could be related to the underlying causality of being well educated and having sufficient resources to pay for not just vitamins but also gym memberships and healthful food. Other examples abound: married people are purportedly happier than unmarried ones, but is that because of an underlying trait of extroversion that inclines certain people to be both happier in general and more likely to marry?

Returning to the Garcia effect: if you have an egg sandwich for lunch, suffer stomach upset and conclude the sandwich was the cause of the indigestion, then you may well have an aversion to chopped egg ever after. On the one hand, this taste aversion is a smart biological connection that protects us from poisonous food. Psychologist Martin Seligman has demonstrated that rats have a clear and strong predisposition to link stomach sickness to the food they are eating, as opposed to other possible causal factors: our learning appears to be biologically selective, biased toward assuming food poisoning.[164] On the other hand, an upset stomach can be caused by many variables, or by the interaction of variables (like eating anything at all when you are in the early stages of a flu that you are unaware of). Reducing the cause of stomach upset to "egg sandwiches don't agree with me" is usually an overgeneralization that attributes causality to factors that are merely correlated.

The Garcia effect is an example of a broader category of correlation–cause mistakes based on a proclivity for assuming that when one thing follows another, the first thing causes the second. "After this, therefore because of this" (known as the logical fallacy *post hoc ergo propter hoc*) is the mistaken assumption of causality based on the sequence of events. We assume that the egg caused the stomach upset because the latter followed the former. We make this mistake all the time as we interact with one another. An angry glare from someone leads us to believe that what we said caused irritation, but the glare could have been triggered by something completely unrelated, like a negative floating thought that the frowner had about something or someone else — nothing to do with our verbal exchange. In this latter case, correlation is purely coincidental: neither indicative of a causal relation nor related to an underlying, confounding variable.

Even if we are diligent about not jumping to causal conclusions when we perceive correlation, we still face difficulty in properly assessing the correlations themselves. Let us say you believe that the key to being a famous actor is getting formal theatrical training, but you want to validate your "inside intuition" with "outside data": you want to examine how strong the correlation is between an actor's training and her fame. You interview 725 actors with at least a decade of experience, all of whom claim that acting is their main career: some had formal training and others did not; some have had at least one major acting job each year since starting their career and others have not.

	Many Jobs	*Few Jobs*
Formal Training	500	130
No Formal Training	75	20

The top left box looks immediately convincing, and this is where most of us focus immediately: 500 working actors had formal training, which encourages the conclusion that there is a strong and presumably causal relationship between training and fame. The problem is that the same training produced 130 actors who do not get regular acting gigs, which suggests that training may be necessary but not sufficient for getting work. Confusingly, the absence of training produced 75 working actors, suggesting that training may not be necessary at all. Our intuitions are almost useless in sorting out information like this; in fact, they are worse than useless because most of us put far too much weight on the big number 500. In this case, the relevant information is a comparison of the percentage of trained actors who have steady work and the percentage of untrained actors who also get regular gigs. Five hundred of 630 trained actors get regular work: 79 percent. And 75 of 95 untrained actors get regular work: 79 percent. The conclusion? Training did not help actors get work: there simply is no meaningful correlation at all in this case.

One of science's biggest challenges is finding correlations that are both legitimate and indicative of causal relationships. The conclusions of scientific research, the highlights of which are often reported in the media if they are newsworthy, can be extraordinarily deceptive. An example, courtesy of professors Hillel Einhorn and Robin Hogarth, is a hypothetical experiment intended to reveal the correlation between intercourse and pregnancy.[165] This experiment is designed to illustrate the vulnerability of scientific research that is not validated by repeated experimentation. In this imaginary scenario, the observer has no theoretical concept of how babies are made. He makes observations and records his findings as follows:

	Pregnant	Not Pregnant
Intercourse	20	80
No Intercourse	5	95

This table indicates a possible correlation, but not a high one, between pregnancy and intercourse. The results suggest that pregnancy may be related to intercourse or perhaps some underlying cause that triggers both intercourse and pregnancy. But there appears to be no consistent and direct connection between them, since intercourse does not usually result in pregnancy. In five cases, pregnancy is induced without intercourse, so the latter is definitively not a necessary condition of the former. If the research were published, the headline would read, "Pregnancy unlikely to be related to intercourse."

Their example is representative of many experiments designed to uncover correlations: despite the immediate conclusions that one might be drawn to, the data are confusing and difficult to interpret reliably. The fact that five subjects purportedly did not have intercourse but did get pregnant is typical of any study: people forget, they lie and the experimenters make mistakes when tabulating the numbers. So five impossible scenarios can be easily reported as legitimate.

Correlations can be trouble in World #2 because they can be valid but not causal (attributable to an underlying confounding variable); they can be valid and causal but the causality runs in the opposite direction than we assume; and they can be completely invalid by virtue of being merely coincidental. Our intuitions are not geared toward untangling even the slightest complexity when it comes to causal relationships, which is why we must resist the rush to impose the most basic of causal relationships onto complexity in the most overconfident of ways.

14.4 Generate Feedback (Interact to Learn)

Our bounded brains have difficulty simulating the behaviour of complex systems: we apply models that are linear, reductionist and static, but World #2 is non-linear, interactive and dynamic. Because the cues we need to interpret complexity are hidden, the best and often only way to get a handle on a complex system is to interact with it directly. Interaction allows us to gather the diagnostic information that reveals the character of the system.

There is no learning without feedback, but high-quality feedback is hard to come by in World #2, because it is buried deep in noise and is rarely instantaneous. The only way to learn about a complex system is to engage with it, experiment with it, learn from its responses to our interventions. Unlike World #1, where our responses are based on applying past learning to similar situations, World #2 generates systems that are often sufficiently different that we cannot simply apply past learning and expect identical results. The distinct signals can be deciphered only by direct engagement.

Learning by interacting means that we are sometimes limited to accepting evidence in the absence of a coherent explanatory theory: we have to go with "what works," because we cannot fully explain the feedback that we are getting, even though the feedback is consistent enough to be useful. The quintessential example is evidence-based medicine, in which treatments are prescribed based on convincing scientific evidence that they work, even if a comprehensive explanation for their success is lacking (in contrast to theory-based medicine with scant evidence, like bloodletting, which was standard medicinal practice until the end of the nineteenth century). The uncomfortable paradox of dealing with complexity is that while we have to be very skeptical of our initial interpretations and early conclusions, we also have to be willing to give credence to evidence that may not yet have complete rationalistic underlying support.

The most notorious example of prioritizing theory over evidence was the dismissal of physician Ignaz Semmelweis' observations in 1847 when he noticed that fewer mothers died during childbirth when doctors sanitized their hands before assisting in deliveries (especially doctors who had just performed autopsies!). Germs had not yet been discovered, so there was no theoretical underpinning to his hypothesis, which was rejected by the medical community. It was only after Louis Pasteur discovered and developed germ theory, fifteen years later, that hand washing began to become standard practice in operating rooms.

We take evidence-seeking experimentation for granted, but the process of experimenting is only four hundred years old. For nineteen hundred years before that, the Aristotelian scientific approach reigned, which was exclusively rationalistic, its method being to theorize in an armchair. At the dawn of the scientific revolution in the early seventeenth century, Francis Bacon ("father of experimental science"), Galileo Galilei ("father of modern science") and others formalized the scientific method that serves not only science but also us in our daily lives, when we are patient and disciplined enough to apply it.

14.5 Look for Leverage (Where Influence Is Greatest)

Influencing a complex system requires finding the opportune spot for the greatest impact — the spot of greatest leverage. Understanding the landscape is crucial in pin-pointing where the leverage is: changing the dynamic of a system in a preferred way is highly dependent on identifying where the crux of the problem resides, which can be distant from the symptoms of the problem. If we uncover the pockets of order and pattern within a complex system, we increase the odds that our interventions will be effective. Otherwise, we are at high risk of applying solutions that treat the symptoms but not the underlying causal connections that define the system.

We can take a page from Napoleon Bonaparte's battle strategy of central position: when he faced a stronger foe, the tactic he favoured was to pick a point of weakness in the opposing army to divide it into two and then apply most of his force to attack one side before attacking the other. This divide-and-conquer strategy hinges on finding the key point in the system where influence can be leveraged for the greatest effect. The leverage in reducing the spread of malaria in Africa was not in the simple distribution of bedding nets; it was in educating the locals on the risk of mosquitoes to pregnant women and children. The leverage in improving education was not in the funding of smaller schools, but in subsidizing special programs in all schools.

Altering a system to create better outcomes depends on finding the spots that have the most influence on the system's dynamics. This is precisely why it is so crucial to ferret out the hidden signals and interpret the causal connections accurately. And why we have to be adept at recognizing randomness, so we can sidestep it rather than overweight its significance.

14.6 Beware Randomness (the Deceiving Patterns)

All signals are buried in noise, and in World #2, it is easy to be deceived by noise when randomness masquerades as signal. The trick is to recognize the patterns of randomness so they do not deceive because, to an untrained mind, the patterns are easily misinterpreted as a signal of something other than randomness.

How can randomness have patterns? This would appear to contradict the very definition of randomness — that it is characterized by a lack of connection between events and therefore no patterns can be detected that allow for prediction. While random events have no connection between them, there are patterns that arise among a collection of random events: when numerous random events occur sequentially over time, two patterns reveal themselves that do not relate the specific events to one another, but characterize a string of random events: streaking and reversion.

Streaking

First and foremost, randomness is much more "streaky" or "clumpy" than our intuitions allow for. This is why, when we write down what we think is a random string of numbers, we underrepresent repetition. If asked to guess the probability of flipping a coin three times and getting the same side each time (either three heads or three tails), most people underestimate the odds of a streak or clump of the same side, which is 25 percent. What if a coin is tossed six times: what are the odds of flipping five of one side? It is higher than most people guess, at 19 percent. And almost even odds of flipping either four heads or four tails in six tosses, at 47 percent.

Here are some random patterns that could arise in the six tosses; each scenario has the same probability. Try to guess what that probability is. Then guess what the combined probability is that any of the patterns (it does not matter which one) will show up.

- All six flips are the same (all heads or all tails)
- Perfect fluctuation between heads and tails (HTHTHT or THTHTH)
- Fluctuating groups of three (TTTHHH or HHHTTT)
- Fluctuating groups of two (TTHHTT or HHTTHH)
- Fluctuating on first and last only (HTTTTH or THHHHT)

While each of the above five scenarios has a 3 percent chance of happening, the combined possibility of one of the above scenarios occurring is 16 percent. If you also add in the possibility of flipping five heads or five tails in the six tosses, the odds of one of those six scenarios is 35 percent.

Here is the point: the odds of something "interesting" happening — something that looks conspicuously non-random to our untrained intuitions — are much higher than expected. There is a one-in-three chance that an interesting pattern will occur, because randomness kicks out patterns much more frequently than we expect. Predicting a *particular* surprise may be impossible, but predicting that *some* surprise will occur is almost a certainty. The problem is that we tend to conflate the difference, and react to *some* surprise as if it was a *particular* one, like treating the coincidence of any two people in a group sharing a birthday with the same surprise as someone sharing our particular birthday.

We get confused in part because we know, over the long term, that the ratio of heads and tails has to approximate 50 percent. This "law of large numbers," however, does not apply to small samples, which can vary from this ratio significantly, and in ways that seem counterintuitive. Mathematician Siméon-Denis Poisson was, in 1837, the first to statistically quantify the streaking or clumping that occurs in a random series of events.[166] "Poisson clumping" is what random systems do and is exactly what we often experience in our lives, leading to expressions like, "Bad luck comes in threes," or, "When it rains, it pours." In fact, rainfall is a good example of poisson clumping: if you measure precipitation over a month, the majority of it is not spread out evenly but accumulates in clumps over just a couple of days, and within those days, in clumps over a few key hours. Other examples abound: we get a large number of phone calls and emails in a short period rather than evenly spread throughout the week; we are inconvenienced by mishaps all within a few hours or days of each other, then enjoy periods of carefree living; most goals scored at hockey, soccer or football games are clumped in one or two periods; most customers arrive at the checkout counter or bank teller around the same time, after which there is no line-up; we find three items on our shopping list in one hour, then spend the rest of the day looking in vain for the last item; some families, like the Kennedys, suffer a disproportionate amount of tragedy within a short period; the list goes on.

The streakiness of randomness makes our intuitions vulnerable to imparting signal to noise. Understanding the streaky nature of randomness protects us from assuming that streaks are signals, especially if we are counting on them to keep repeating (which they will not, if they are just random). Streaks can be very deceptive because they can be random (e.g., luck), non-random (e.g., skill) or, most often, a combination of the two. Only the non-random streaks provide insight into the underlying structure of a complex scenario, by revealing the regularities that afford us a modicum of predictability and control. Distinguishing the noise of randomness from the signal of information is nowhere more problematic than confusing luck and skill.

Skill Streaks Versus Luck Streaks

We perpetually attribute far too much skill to the lucky movie star, athlete, executive and mutual fund manager. Virtually all success in any human endeavour depends on at least a modicum of luck; even the best chess players can benefit from the luck of an opponent having an off day. But it is very difficult, and in many cases impossible, to separate luck from skill. How do we judge if an individual is skilled or lucky? Or if a business is exceptionally well run or lucky? We tend to rely exclusively on results or outcomes, because they are easily available and measurable. More often than not, results are the only data that are available: we cannot attend a weekly management meeting to assess the quality of a company's decision-making process, nor can we observe a mutual fund manager for weeks in a row to assess her exact method of buying and selling stocks.

The problem with results is that they are combinations of luck and skill that are difficult to differentiate. The signal of skill is obscured by the noise of luck, and the degree of obscurity is based on the ratio of signal to noise — skill to luck — that the particular scenario entails. Slot machines require no skill since they are driven entirely by random luck; blackjack requires skill and luck; poker has an even higher skill-to-luck ratio. A concert pianist's recital involves some luck, like the quality of the piano on a particular day or whether she had a good sleep the night before, but the much larger component of the recital's quality depends on her skill.

Author Michael Mauboussin has analysed a variety of activities to determine where they fall on the skill–luck continuum, and concluded that chess has a very high skill–luck ratio: generally, you cannot luck into checkmate, because of the amount of strategic thinking it takes to orchestrate a win.[167] Between the necessary skill of chess and the pure luck of slots are countless other activities. Mauboussin rates tennis as having a higher skill quotient than other sports, followed closely by basketball, while football rates as one of the lower-skill-based sports. That is not to say that skill is not a significant feature in football; it is just to point out that the structure and rules of football create outcomes that are more influenced by random luck than do the structure and rules of basketball.

In a similar vein, economist Arthur de Vany discovered, after extensively analysing the success of films, that predicting the box office success of a film is virtually impossible because of the number of random factors that influence the outcome, and because a small number of these variables can have an enormous impact on word-of-mouth publicity that ultimately fuels a box-office winner.[168] Random variability makes it difficult to assess a film producer's track record.

Skill Versus Luck in Business

Various academics, including Phil Rosenzweig, Spyros Makridakis, Michael Raynor and Michael Porter,[169] have independently examined the skill–luck combination reflected in the sales and profits of companies. The research results are consistent and stunning. The so-called "great" or "excellent" companies rarely sustain their success, and the qualities that pundits point to as the underlying causal factors of success have statistically negligible explanatory power and no predictive power whatsoever. Their work overwhelmingly demonstrates that attributing a company's success to a small number of "key success factors," like a focused strategy or a culture of innovation, is highly suspect. While their collective analysis indicates that there is evidence for persistent superior performance in a minority of companies, there are important qualifications that relate to the inference of management skill, even in these cases:

- Skill is mixed with a lot of random luck, which explains why winning companies can become losers very quickly (or, as Rosenzweig notes, "Nothing recedes like success"). The list of "superior" companies that were unable to maintain their superiority is staggering.

- The underlying skill represented in company-specific strategy is a small component of the company's overall performance; the larger drivers are the industry structure the company competes within and inexplicable random factors. It has been estimated that only 10 percent of the difference in performance between companies is attributable to company-specific strategy. Rosenzweig suggests that what we think of as drivers of a company's performance are better understood as attributes of performance, not causes of it.

- When we look retrospectively at successful companies, it is easy to concoct stories that explain their success, but we are starting from a biased sample of perceived winners that we already view favourably (this is "the halo effect": our assessments are skewed in favour of the companies we have already categorized as successful). We disregard the losers that may have embraced the exact same strategies but were not successful because the stars just did not line up for them. So we end up drawing precarious conclusions from a highly skewed sample.

- There are no predictable "special causes" that can be classified as sufficient for success. We may be able to see causes more clearly in retrospect, but that does not mean that they were ever predictable or that they can be projected into the future as key ingredients for success. Deep research — the kind that does not make for bestselling titles — reveals that there are no ways to consistently generate superior performance that can be replicated by aspiring executive teams. Whatever contributing factors are uncovered are at best necessary but in no way sufficient. Many average or below-average companies have all the same elements in place, like clearly stated missions, employee-centred cultures and customer-centred strategies.

It makes for a good story (and a bestselling book) to say that the key to success in business is ... *insert magic formula here*. But these stories are all highly suspect. Management consultant Michael Raynor and his coauthors point out that one book categorized Campbell Soup as a "winner" based on its ten-year performance from 1986 to 1996, whereas a different book identified the company as a "loser" based on its ten-year performance from 1992 to 2002.[170] The superficial analysis in most business bestsellers belies the complexity of drivers that mingle with random factors to create systems that are virtually impossible to replicate and just as hard to predict as forecasting which actors who graduate from The Actors Studio will become movie stars.

Note that the research does not suggest that superior corporate performance does not exist, or that it is entirely attributable to luck. It does, however, indicate that random factors are very difficult to separate from the myriad of interacting causal factors that constitute repeatable outperformance. Even when long streaks of superior performance indicate the likelihood of management skill, the specific causal contributors are not easy to identify and certainly not easy to replicate by other companies.

One way to separate luck from skill is time, which, as Taleb notes, is the "cleanser of noise."[171] In the short run, skill can be swamped by luck, both good and bad, which is why Jack, as the less skilled basketball player, should prefer the shorter game described in 2.4. Over sufficiently long periods, the good and bad luck offset each other and what remains is the signal of skill, which would be Jill's superior playing. How much time is long enough? That depends entirely on the activity.

The performance of a mutual fund typically requires decades to reveal a strong signal of skill — longer than the lifespan of almost all funds that have the same manager. Even then, the stock-picking approach that worked in the 1980s and '90s may not suit the stock markets in the 2000s. Bill Miller of Legg Mason Capital managed to beat the U.S. stock index for fifteen years in a row, but his fund's underperformance in subsequent years eliminated most of the gains. In his fourth-quarter report in 2006, he wrote, "There was, of course, a lot of luck involved in the streak." At the end of 2008, his ten-year performance had dipped below the returns generated by the stock market index.

A skilled baseball player, on the other hand, is much easier to identify: one full season is usually a good indicator of skill, but at least a few seasons are required for the signal to be fully apparent. A chess player's skill can be reflected in a single challenging game and is clear in two or three. On the other end of the spectrum, but not quite as far out as the mutual fund manager, is the skill of a company CEO or an executive, which is typically buried in a lot of random noise, including the decisions of previous management that continue to influence the present. Complicating the assessment of a particular CEO is the lag that occurs between decision making and results. Unlike an athlete, a dentist or an accountant, whose talents are tightly linked to their results over short periods of time, a company's immediate results can be unreflective of the CEO's decisions. Many of the decisions have much longer-term impacts, like investing in new technology, cutting back investments in certain areas, discontinuing a product line or hiring new people. The same complications apply to politicians, who inherit the legacy of their predecessors and are often held accountable for the impact of these past public policy decisions while getting little credit for initiatives they undertake, the benefits of which take years to accrue.

The mistake we are prone to when assessing skill in complex activities is overweighting short-term results where the true signal is very weak and where we are champions at confabulating good stories about random streaks. The "law of large numbers" is a valid and useful statistical concept indicating that a large enough sample will reflect the underlying character of the population that it represents. But our intuitions tempt us to mistakenly conclude that tiny samples are similarly reflective, because we do not see just how small and therefore unrepresentative these samples truly are. In other words, we do not have good intuitions about just how dominant random factors can be in a CEO's five-year performance, or a mutual fund's ten-year record.

As reviewed in 8.3 regarding outcome bias, the inherent randomness within complexity means that process and outcomes are not tightly linked: good decisions can yield bad results, and bad decisions can yield good results. Of the four possible boxes below, we overweight the top left and bottom right in our assessments, and neglect the other two, which are equally possible scenarios.

	Good Outcome	Bad Outcome
Good decision making	BRILLIANCE	Unlucky Brilliance
Bad decision making	Lucky Mistake	MISTAKE

As management consultant Michael Raynor points out, "Success is very often a result of having made what *turned out to be* the right commitments (good luck), while failed strategies, which can be similar in many ways to successful ones, are based on what *turned out to be* the wrong commitments (bad luck)."[172] He cites Apple's many failings (the personal computers "Lisa" and "Newton" as well as "eWorld" online service) as having emerged from the same powerful strategic planning process that created its huge successes. Unpredictable luck plays a very significant role in whether solid, well-executed strategies end up succeeding or failing.

Given the severe limits of deriving a skill signal in so many of our activities from results alone, the fallback has to be assessing the process that underlies the results, as murky or ambiguous as that can be. What was the CEO's decision-making process? How does the fund manager decide which stocks to buy and sell? Complexity reveals its signals over sufficient time and through asking the right questions. We underestimate randomness and find it intuitively distasteful that luck should play such a large role in our affairs. Our stories centre on the talented, hardworking few who rise above the crowds; our stories are rarely about the talented, hardworking many who do not get ahead. Yet luck is all that differentiates them: luck determines *which* of the talented, hardworking individuals achieve success.

Reversion

The flipside of streaking is reversion, or, more specifically, "reversion to the mean" (sometimes referred to as "regression to the mean"): the tendency for extreme behaviour to gravitate to less extreme behaviour, reverting to the average over time. Whereas streaking generates surprising clumps of similar outcomes, especially over short cycles of repetition, reversion pulls surprises back to the average over longer cycles of repetition. If a coin is flipped one thousand times, there will be many streaks of the same side coming up repeatedly. But the overall ratio of heads to tails will get closer and closer to 50 percent as the number of coin flips rises. The streaks will eventually offset one another, such that the number of head and tail flips will approximate the long-term average.

Reversion to the mean reigns in outlying behaviour by pulling it to the average: it explains why very tall people have children who are usually shorter than they are and closer to the population average; why athletes who achieve a record-breaking performance are likely to see their performance slip in subsequent competitions; why extraordinary company results are highly likely to revert to the average over time (the best companies have been tracked at around ten years of superior performance before they lose their edge).

It is important for us to understand reversion as a pattern of randomness because it can deceive us into perceiving the wrong signal. Daniel Kahneman uses the hypothetical example of sad children being treated with an energy drink and showing significantly improved moods over a three-month period.[173] The reversion in this example is a function of an unusual condition reverting to the mean over time: the energy drink had no likely benefit whatsoever; the mental condition of the children moved naturally from an extreme state to a less extreme state over time. The same argument holds for herbal remedies that purport to make people feel better, or some forms of physiotherapy: our bodies tend to heal themselves much of the time, returning to their more normal state with or without intervention. The changes (of mood or physical condition) appear to be a signal when in fact they are just noise reverting to its average.

In fact, the absence of reversion, over longer cycles of repetition, is particularly useful in detecting skill. The larger the reversion of superior performance to the average, and the faster it occurs, the higher the likelihood that the performance was luck-based. Put another way, the longer that a streak of superior performance persists, the higher the likelihood that it is driven by genuine skill. Skill-based results are not nearly as quick to revert (unless there is a significant change in the circumstances, such as an athlete aging).

But the absence of a reverting trend is only a signal in large samples or over many cycles. In small samples or shorter cycles, reversion does not have an opportunity to work its magic: variability does not have a chance to wash out, so more extreme things can happen. Howard Wainer demonstrates how deceptive data can be without the benefit of reversion: he points to statistics on kidney cancer and auto accidents in the U.S.[174] The highest incidence of kidney cancer occurs in rural communities, compared with urban centres. It is tempting to explain the higher incidence in smaller counties by hypothesizing all kinds of explanations, but the lowest incidence of kidney cancer is also in rural counties. The higher and lower rates of cancer merely reflect the higher variability that occurs in less-populous centres: in large cities, the variability washes out because the sample size is so much bigger. The same variability shows up in auto accident rates: the ten cities with the highest rates of auto accidents are small cities, as are the ten cities with the lowest rates of accidents. Larger cities have auto accident frequencies that are not as extreme on the high or low end as those found in smaller cities.

Path Dependence: When Randomness Does Not Revert

Randomness is tricky. Reversion to the mean is a key feature of the kind of randomness that is linear; reversion is a feature of systems whose properties surface over time as the randomness cancels itself out ("ergodic" systems). Reversion, however, does not necessarily describe the kind of randomness that is non-linear and driven by power laws. In this latter form of randomness, extreme events do not necessarily gravitate to the average (in "non-ergodic" systems): the richer often get richer, and the famous often get more famous. In these cases, extreme events are fuelled by positive feedback loops that reinforce, rather than cancel out, the system's behaviour. While reversion to the mean pulls extremes to the middle by eliminating random effects over time, extreme events are sometimes amplified by feedback loops that reinforce initially random starting points. Path dependence characterizes this reinforcing behaviour. Path dependence describes the unfolding of things — such as events, belief systems, personalities — that are heavily influenced by the path they are set on, even though the paths themselves are usually initiated by some element of randomness at their beginning. Path dependence occurs when past events determine the rough boundaries of the path that future events take.

If, at an early age, a child has a random but disturbing confrontation with a stranger who wrongly accuses him of stealing, the set of beliefs the child develops about the threat that strangers pose will strongly influence the path of his future beliefs about people in general and his interactions with them. If a woman randomly bumps into an old friend who persuades her to apply for a job at his company, and she pursues the suggestion, ultimately meeting her future spouse at the company, the life events that unfold are highly dependent on the path created by the chance encounter with the old friend. The impact of the randomness in these two examples becomes magnified as future events unfold.

The example most commonly used to illustrate path dependence is the QWERTY typing keyboard; it was designed in the late nineteenth century to place commonly used letter combinations like "el" and "at" far enough apart from each other that the typewriter keys would not jam together when hit one after the other in rapid succession. Over a century later, we are still using this keyboard layout, notwithstanding how difficult it is to learn and how inefficient it is compared with alternative keyboard arrangements. A path had been set by the inventor of the first keyboard, and we have not deviated from it since.

Path dependence is important to recognize because it can mask randomness; for example, we may think particular people are unusually gifted, judging by their success, but more often than not, they are the lucky recipients of chance events that set them onto paths of fame and fortune that would not have materialized if random events had not put them on that course. Stellar careers, bestselling books and box office smashes are often the result of lucky starts that are reinforced by positive feedback, eventually hitting a tipping point of extreme success that was neither predictable nor proportional to the skill of the individual, or the quality of writing or directing.

The history of thought has been largely path dependent, with each thinker reacting to or building on the ideas of his predecessors. This is a clear trend in the history of philosophy, from Plato's influence on Aristotle, to Hume's influence on Kant, as well as in religion, from Hinduism's influence on Buddhism, to Judaism's influence on Christianity and Christianity's influence on Islam. Science is extremely path dependent, since most research and discoveries are based on topics that are currently interesting and related to previous research. Path dependence makes judging geniuses tricky: even putting random luck aside, sometimes all that is needed to trigger a breakthrough is the smallest of discoveries or ideas that is itself based on a long history of prior discoveries or ideas. Had Albert Einstein not been familiar with the work of Ernst Mach, Max Planck and others, he likely would never have produced the breakthrough thinking that he did and would not be the celebrity he became.

The specific paths are not predictable at the outset because elements of randomness are at play, but the influence of the paths on future events is clear in hindsight: the paths direct the systems that they shape by increasing the probability of certain events happening while decreasing the probability of other events. A career can have unpredictable turns and twists, but there is a great deal of path dependence in how it unfolds: police officers rarely switch paths to become philosophy professors, and physicists rarely change course to become advertising executives. Much of what happens to us, and what we choose to do, is highly dependent on our prior choices and the experiences that cumulate in our histories.

Because many of these early choices and experiences involve random influences, the paths we follow are never preordained. But once the paths are forged, they generate a force of momentum that is hard to break. Randomness has an uncomfortably large influence in our lives: We are not in full control of the outcomes of our actions. Complex systems, including those that define the paths of our individual lives, are far too intricate to be managed with complete control and predictability. We can influence the *process* that produces the outcomes, with knowledge of the *probable* path that our actions will lead to. The *final* path that emerges, however, is not exclusively in our hands.

15

Rethinking Truth
Provisional Truth Reveals "It's Not So Certain"

Truth is the kind of error without which a certain species could not live.
Friedrich Nietzsche*

Our effectiveness in interacting with the world depends on how well matched our beliefs are to how the world works; that is, on how "true" our beliefs are. But truth does not come easily, which is why it deserves a rethink. Consider the following fable that the Indian philosopher Jiddu Krishnamurti shared in a speech that he delivered in 1929.[175]

> The devil and a friend of his were walking down the street when they saw ahead of them a man stoop down and pick up something from the ground, look at it and put it in his pocket. The friend said to the devil, "What did that man pick up?" "He picked up a piece of the Truth," replied the devil. "That is very bad business for you then," said his friend. "Oh, not at all," the devil replied, "I am going to let him organize it."

The physicist Niels Bohr is said to have remarked that "the opposite of a great truth is also true."[176] Blaise Pascal had a similar insight: "Contradiction is not a sign of falsity, nor the lack of contradiction a sign of truth."[177] Yet it is "truth," as Nietzsche points out in the epigraph to this chapter, that we crave, in the form of feeling certain.

---

* *The Will to Power*, Vintage, 1968.

Our intuitive notion of truth is basic: truth is the label we attach to beliefs that are correct, that match the way the world actually is. This restrictive definition of truth is good enough for most of what we have to do in World #1. But, per Bohr, Pascal and Nietzsche, truth is a slippery concept when taken out of a straightforward context. There are, in fact, many forms of truth (a position known in philosophical circles as "pluralism"). Here is a sample:

- mathematical ($2^2 = 4$),
- semantic (chairs are for sitting),
- perceptual (this apple is red),
- subjective (I feel sad),
- scientific (evolution explains similar genotypes between different animals),
- probabilistic (there is a 60 percent chance of rain tomorrow),
- ethical (murder is wrong),
- aesthetic (a Picasso painting is better than a child's drawing),
- theological (God is omnipotent).

This list runs from the hardest form of truth to the softest, insofar as mathematical truths are self-evident and shared by everyone while moral, aesthetic and theological truths are less universally standardized. Our default notion of truth is the hard, straightforward kind that applies to the first four examples, and sometimes the fifth. The other types rely on proofs that are less cut and dried: they may have exceptions, like the justification for murdering a tyrannical despot or the brilliance of a child prodigy's artwork; or they may be revised, like a weather forecast or a scientific theory.

Our everyday notion of hard truth is productive in World #1 because of the tight link between our beliefs about the world and our direct perceptions of it. But much of World #2 operates below the surface of our immediate perceptions and beyond the domain of our basic intuitions. Just as navigating the complexity of World #2 requires scientific models that are less intuitive and more sophisticated than those that work in World #1, so too does it require a more sophisticated notion of truth. It has to be more flexible so it can be more productive in describing complexity.

Truth operates in each of the two worlds in fundamentally different ways:

World #1 Truth	World #2 Truth
Definitive	*Provisional*
Correspondence	*Coherence*
Dichotomy	*Probability*
Certainty	*Ambiguity*
High Predictability	*Low Predictability*

First and foremost, truth in World #2 diverges from truth in World #1 by being provisional — always subject to revision. Once again, Nietzsche: "The mostly strongly believed a priori 'truths' are for me provisional assumptions."[178] Provisional truth is the macro-level strategy for instilling methodological humility into how we conclude, just as systems theory is the macro-level strategy for instilling more useful models into how we interpret. Just as systems theory generates specific strategies for interpreting complexity, provisional truth generates specific strategies for drawing conclusions about complexity — strategies that will be explored in the chapter that follows this one.

15.1 Definitive Versus Provisional Truth

Complexity demands that we think of truth as interim because there is no certainty in World #2: we never have all of the information necessary to know something in the way that we know that a mathematical equation is correct or that a dropped apple will fall to the floor. Truth, in World #2, is something we approach asymptotically: we can get closer and closer to it but never reach it — or, at least, we cannot be absolutely sure that we have reached it.

Lying at the heart of provisional truth is the recognition that the gap between our interpretations and the real world is typically much bigger than we intuitively believe. Any conclusions that we draw about it have to be of the "for now, until further evidence surfaces" kind, not of the "this is the way it is forever" kind. Because our conclusions are always subject to refinement, provisional truth acknowledges that we are prone to overconfidence if we treat our beliefs about complexity as final.

Pragmatism

Provisional truth is a pillar of the philosophical movement associated with three American philosophers: Charles Sanders Peirce, William James and John Dewey. Pragmatism is sometimes misleadingly described as the philosophy that "truth is whatever works," with the implication that we can call a theory "true" if it is useful in some way. This is an oversimplification; the general approach of pragmatism is to focus on science and other fields of knowledge as ways of explaining the world that serve us until they are superseded by alternative explanations that serve us better. A belief, therefore, can be both true and subject to revision.

The pragmatist's view of truth is more flexible than the rigid, hard notion of truth that dominated for more than two hundred years after the scientific revolution of Copernicus, Galileo, Newton et al. The prior view was that truths were certain and never in need of revision. This perspective (philosophical realism) contrasts with the pragmatist's view that we ascribe "truth" to beliefs and theories when they successfully explain things. But the pragmatist does not preclude the possibility that we will discover error in our theories or come up with better theories. Truths, for the pragmatists, are explanations that work until we replace them with ones that work better.

Nietzsche, writing around the same time as Peirce, was particularly opposed to the notion that truth is fixed. He insisted that "there are no facts, only interpretations."[179] He declared that what we think of as truths are just metaphors that are customary in the particular era they arise. Some postmodern philosophers have taken the Nietzschean position further. Michel Foucault, for example, insisted that what we consider to be truths are merely social constructs imposed by those with power.[180] Richard Rorty argued that truths are determined by social consensus or solidarity on a particular viewpoint.[181] Our theories and beliefs are not accurate representations, according to Rorty; we have only a social justification of them, because we are a community of interpreting agents whose goal is consensus in our interpretations.

Science and Provisional Truth

Provisional truth does not mean anything goes; it means that the conclusions we draw are subject to revision at any time, based on new information or new ways of structuring existing information with greater explanatory and predictive power. The methodology of science is instructive because it operates on the basis of provisional truth: science revises and refines the insights that it develops. Scientific laws are just as much about the usefulness of the models as they are about reality itself. This is where science differs from mathematics: math proves, whereas science describes. Philosopher W. O. Quine described science as demonstrated but not proven.[182] This distinction is crucial because, unlike proofs, descriptions are always subject to revision. The best scientists proceed very carefully from description to proof-oriented explanations. Galileo Galilei insisted that description should always precede explanation, and Isaac Newton famously declared that his theory of gravity was a description of phenomena, not an explanation of it: hypotheses that are not directly deduced from observation have no place in science according to Newton.[183]

The provisional character of truth makes science an open system: hypotheses are continually tested to determine whether observations contradict predictions, because hypotheses can always be challenged. Karl Popper railed against closed systems, which are based on rigid beliefs that are not open to challenge through experiment. As reviewed in 7.1, Popper argued that a theory is only legitimate if it has the potential to be proven wrong. If it is not falsifiable, it resides in a closed system that cannot be challenged and is therefore not scientific.[184]

Popper's view of science was not only a reaction to unfalsifiable theories (like those of Marx and Freud), but also to the school of thought called logical positivism, which espoused verificationism: the view that, to be true, scientific claims must be verifiable by observation and experiment. For Popper, scientific statements cannot be verified; they can only be falsified. So certainty is available only through refutation; theories can only be disproved, not proved. Popper was influenced in part by how Einstein's theory of relativity overturned Newton's principle of universal gravitation, which had been accepted for two hundred years as the final truth on gravity. Scientists had been fooled too many times into believing that they had accurately captured the way things really are. Popper wanted to establish the legitimate function of science as proposing theories and then attempting to disprove them.

> ### The Progress of Science
>
> Scientists, according to Popper, should search for better theories by refuting old ones and inventing new ones; his argument was that a hypothesis can be corroborated by evidence but can never be categorized with the same forcefulness of truth as a definitive disproof can be. A falsifiable theory that seems to work and has not been disproven may be used until it is either refuted or replaced by a theory that explains and predicts more.
>
> Although Popper's views continue to be influential, science does not proceed in a purely Popperian way: while falsification is a crucial criterion of the scientific process, it is not an exclusive one, since it is too limiting. Theories are often refined and revised when evidence contradicts them, rather than rejected outright. Science typically operates on "inference to the best explanation," which is the most likely hypothesis. This type of inference relies on the convergence of evidence in favour of a particular explanation, enabling a probability estimate to be made of the likelihood that the explanation is true. (Popper believed that assigning a probability of correctness to a claim is unscientific, because it is largely a subjective judgment.)
>
> If the conclusions of science are never final, how do we know whether it has made progress or is just describing things differently than it used to? Scientist and philosopher Thomas Kuhn pointed to repeated episodes in history in which scientific theories were overturned in dramatic "paradigm shifts" as a demonstration that what qualifies as scientific knowledge at a given point is largely determined by the consensus of the scientific community and not by any reliable correspondence to the world itself.
>
> Contrary to Kuhn's view, however, most scientific theories are not simply overturned; rather, new ones are developed that explain previous ones as useful but limited to certain conditions. Newton's laws of motion work well until they are applied at high speeds: Newton was not wrong, but his math works only for masses moving

> much slower than the speed of light. Einstein's math works in both a Newtonian (slow) world and a very fast one. Each successive theory accommodates a greater number of conditions than its predecessors and this process of refinement reinforces the sense that our descriptions and explanations of how the world works are increasingly accurate, because they explain more, predict better and cohere more tightly with everything else we know.

The force of Popper's argument in favour of refutation over verification reveals how misleading our intuitive sense of logic is: we tend to rely on confirmation to a greater extent than disconfirmation, even though it is only the latter that offers us certainty. We tend to argue along these lines: "If my hypothesis is correct, then the evidence will support my claim; the evidence supports my claim; therefore my hypothesis is correct." This argument appears sound to most of us, but it is not. The best we can do with confirming evidence, as noted in 7.1, is to conclude that the hypothesis *might* be correct, but it does not preclude alternative hypotheses that also explain the evidence. This error of concluding based on confirming evidence, as noted in 14.3, is called the error of affirming the consequent: other possible causes explain the consequent (the evidence) besides the antecedent (the hypothesis).

Understanding this fallacy is a crucial element of complex thinking, and bears repeating, because avoiding it enables us to be open-minded about alternative interpretations. Similar to the examples of if-then statements in 14.3, the intuitive form of this error looks like this:

If I am right that my belief is correct, then I will be able to confirm my belief with evidence.
Shaky start, since only refutation can prove something conclusively.

Indeed, I do have confirming evidence that supports my belief.
OK, but proceed carefully to a tentative conclusion.

Therefore, I am correct.
Perhaps, but not necessarily.

These two premises and the resulting conclusion illustrate how most of us think most of the time — wrongly, because the conclusion does not follow from the preceding premises. Alternative explanations are always available for the consequent evidence, and these alternatives may have nothing to do with the antecedent belief. (Recall the Garcia effect in 14.3, of attributing stomach upset to a food that is not necessarily the cause of the problem.)

The invalidity of this intuitive line of thinking does not matter much in World #1, where everything is straightforward so there are few, if any, alternative interpretations that can explain the evidence. There are no reasonable alternative explanations for a rock falling to the ground other than gravity, or for feeling satiated other than eating food. In World #1, a bias for confirmation is a quick and efficient way to validate a belief. In World #2, running afoul of this logic yields countless errors because there are many possible alternatives based on hidden information, which is precisely why Popper insisted that we have to restrict ourselves to refutation:

If I am right, then the opposite of what I believe cannot be true.
Good start.

Evidence indicates that the opposite of what I believe is true.
OK.

Therefore, I am wrong.
Certainly.

Working a few decades before Popper, and a significant influence on him, was philosopher Charles Sanders Peirce, who suggested that theories that spring from the scientific method are the result neither of deduction through logic nor of induction through observation.[185] He argued that theories arise from "abduction," a term he coined to describe the mode of logic that conceives hypothetical explanations for observed phenomena. Abduction produces an educated best guess, one that not only fits the data but also is the simplest explanation of any alternative. Whereas the conclusions from deduction follow directly from their premises (e.g., all stars radiate energy into space, the sun is a star, therefore the sun radiates energy into space), and the conclusions from induction follow from repeated observations (e.g., the sun has always risen in the east, so it will continue to rise in the east), conclusions from abduction follow from a proposed explanation (e.g., the sun rises in the east because the earth rotates on its axis toward the east).

An important implication of abduction is "fallibilism," Peirce's term for describing the corollary of the provisional nature of scientific truth: any conclusion drawn from a process of abduction may be overturned or revised as new evidence or perspectives are introduced. Peirce was one of the first philosophers to argue that science operates based on the best available theories at a particular time, and that those theories are always subject to change because empirical knowledge, with additional observation and experimentation, can always alter them. Until Peirce, science generally was considered foundational: Newton's laws were not just "best explanations for now" but fundamental and final truths about how the world worked.

The inescapable fallibilism of science necessitates the notion of truth as provisional. Neuroscientist Antonio Damasio describes the goal of science as "improved provisional approximations."[186] For biologist Stephen Jay Gould, a fact can only be defined as "confirmed to such a degree that it would be perverse to withhold provisional assent."[187] Provisional truth, based on inference to the best explanation, skirts both Popper's challenge that certainty applies only to refutation and David Hume's challenge that no certainty can come from induction, no matter how many observations are made. But Popper's admonitions still have weight: the discipline of seeking out disconfirming evidence strengthens the confidence that we are entitled to, because the better our beliefs withstand efforts to contradict them, the greater the likelihood that they are valid, even if we cannot guarantee their correctness.

15.2 Correspondence Versus Coherence

Along with Galileo in the early seventeenth century, Francis Bacon was one of the pioneers in developing the scientific method. The only way to advance knowledge about the world, according to Bacon, was to collect data methodically and bring a thoughtful perspective to its analysis, always being diligent not to impose patterns formulated before the collection of the data. The patterns and laws of nature will reveal themselves eventually, but only to the patient observer who is rigorous in data collection and observation: the skilled scientist balances observation and theory. Bacon described the scientist who does not theorize as an ant, haphazardly collecting and using bits of information. Equally problematic, he wrote, is the scientist who relies solely on rationalistic contemplation, failing to engage the world experimentally; such a scientist is like a spider spinning a web secreted from within itself. Experimentation without theory is chaotic and not unified; but theory without experimentation is precarious and lacks connection to the real world.

Finding the right balance between theory and observation is easy in World #1 because it is much easier to assess how well our beliefs correspond to what we can observe. In this world, tests of correspondence are easy to come by: Does a stone always fall when I drop it? Does ingesting food consistently increase my energy level? In World #1, we are safe in assuming that our interpretations are "truthful" when they correspond to the way the world works, based on the predictions that arise from our current observations and past experiences.

World #2 challenges us to find the right balance between theory and observation. We cannot rely on immediate perceptions to the same extent that we do in World #1, but we have to be especially vigilant to avoid theorizing too wantonly. The farther we get from direct perception, the more difficult the test of correspondence becomes and the higher the risk that our constructions of reality are not representative of the way the world works. In World #2, we do not have the luxury of relying on a direct comparison between our beliefs and our experience of the world; we have no choice but to rely more heavily on assessing the quality of our thinking. In particular, we have to rely more heavily on coherence as a criterion of the usefulness of an explanation: how our beliefs "hang together" in the web of what we consider established knowledge. Bacon's admonition is crucial in World #2: neither raw experience nor pure intellect is sufficient; the two in combination are necessary to make sense of the world.

Philip Tetlock describes good judgment as a process that balances data-driven reasoning with theory-driven reasoning.[188] He makes the important distinction between "getting it right" (empirical correspondence) and "thinking the right way" (logical coherence). For complexity, we are forced to theorize to a greater extent, so thinking the right way is paramount.

```
Emphasis in World #1:          Emphasis in World #2:

   ( Correspondence )             ( Coherence )

   Beliefs match the world        Beliefs fit one another
       Data-driven                    Theory-driven
     Empircally based                Logically based
            ↓                              ↓
     Getting it right            Thinking the right way
```

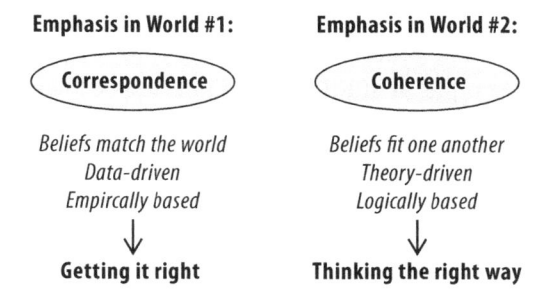

Complexity demands a greater emphasis on Type 2 thinking, because the combination of our direct perceptions with our default intuitions is often misleading. In World #2, we need a lot more Type 2 metacognition: the emphasis is on self-assessing, scrutinizing how we think about problems. Are we employing the mental models that are suited to complexity? Are we avoiding the biases and shortcuts that we otherwise rely on in World #1? Do our hypotheses cohere with the data and everything else we know?

15.3 Dichotomies Versus Probabilities

It is not that we never think probabilistically: we are born with an ability to track the relative frequency of connections between things, like how many times rain is accompanied by lightning, or how often certain animals come to a particular river to drink. We use these frequencies to establish intuitive probabilities, but our intuitive sense of probability is fairly unsophisticated: we make high-level, unquantifiable probability estimates that tend toward dichotomies, such as "very likely" or "unlikely." We are not in the habit of mindfully weighing evidence and quantifying our certainty in a rigorous way, because World #1 does not demand this of us.

One of the downsides of our simplistic notion of probabilities is that we are shockingly gullible about what we are told by sources that we consider credible; we rarely, if ever, assign a probability of correctness to such claims. When epidemiologist John Ioannidis analysed the most frequently cited research studies in the medical field, all published in the most reputable journals, he discovered that 41 percent of the studies made claims that, when retested, proved to be either patently false or grossly exaggerated.[189] That number doubled to 80 percent when he broadened the sample of studies to include less-known research and journals. He and his colleagues concluded that the majority of research culminates in claims that are more likely to be false than true.

This insight is not as peculiar as it may initially appear. Getting tenure in an academic institution depends on securing funding for research and then publishing the results of that research. Academics are funded to the extent that their work is perceived as interesting and relevant; their research is published in journals on the same basis. New claims are interesting; research that disproves the claims of others is less interesting and less urgent, especially if the refutations are targeted at reputable professors who wield clout in the academic community, including the journals. Not only is it difficult to get funding for something as uninteresting as replicating the research of others to confirm their validity, but it is also not particularly rewarding work: validating others' findings is just not that interesting. And refuting their findings does not entitle us to claim that we have discovered anything new: all we have done is prove others wrong — others who will likely put effort into discrediting our methodology in order to preserve their original findings.

The paradox is that, while replicated experiments get less attention and publication, replication of results is the backbone of the scientific method: repeating the same experiment enough times to corroborate the outcomes. Corroboration is especially important given the nature of "publication bias": a well-documented phenomenon that describes the tendency of academic journals to favour publishing positive results (experimental outcomes that prove what the researcher set out to test) over negative results, which disprove a hypothesis. Research scientist Daniele Fanelli has revealed that only 14 percent of published research focuses on negative results.[190] Because the positive conclusions get more visibility, the academic community itself, in addition to the general public, is more aware of the "interesting" (but often questionable) research, while oblivious to the myriad "uninteresting" experiments that never get published (or even undertaken in the first place).

In an analysis similar to that of Ioannidis, the physician and writer Ben Goldacre discovered that research funded by pharmaceutical companies is four times more likely to support the efficacy of a particular drug than independent studies (which are usually government funded).[191] There is considerable, documented evidence of the bias of researchers ("pharma bias") when they are contracted by pharmaceutical companies to analyse the efficacy of the newest drugs. Even research that yields contradictory conclusions about a particular medical intervention will garner attention if the intervention may be shown to have some positive results, compared with less equivocal research that cannot point to any benefits. For example, if ten studies examined the relationship between eating carrots and visual acuity, and two found a positive correlation between carrots while eight did not, the two positive ones will get much more visibility. You can be confident that if ten experiments are undertaken by a company testing the efficacy of its new pill, the two positive studies will be promoted loudly and afar, while the other eight will be discarded as "unrepresentative."

The general requirement for determining the legitimacy of an experimental result is a 5 percent or less chance of the proven claim being false (i.e., a p-value of 0.05 or better). While this may sound like an impressive threshold to meet, the more experiments someone undertakes (constituting all kinds of variations on a theme), the higher the odds of finding something that meets the minimum threshold (just as flipping a coin enough times will eventually result in some surprising patterns, via Poisson clumping). Management professor Joseph Simmons and his colleagues have demonstrated "just how unacceptably easy it is to accumulate (and report) statistically significant evidence for a false hypothesis."[192]

Unfortunately, our underdeveloped sense of probability makes us vulnerable to bold and confident claims, which are rarely marketed with a meaningful and intelligible probability estimate of their accuracy.

15.4 Certainty Versus Ambiguity

Whereas probability attempts to quantify uncertainty, there is no straightforward way to manage ambiguity, the imprecision of fact.

In World #1, truth and falsity are mutually exclusive: something cannot be both true and false at the same time. An element of inescapable ambiguity resides in World #2, where truth and falsity are not so tidily separated.

The mathematics of "fuzzy logic" is a method of applying truth-values to things where truthfulness can range from 0.0 (certainly false) to 1.0 (certainly true). For example, there is no single point at which a melting snowman becomes a pile of snow and no longer a figure: it moves through various stages of melting from clearly recognizable figure to puddle of water. Similarly, there rarely is a single point at which a dying person makes the transition from living to not living: most people expire slowly as different body systems gradually, but not synchronously, close down. At the opposite end of the life cycle, a fertilized human egg only gradually becomes a zygote, a process that unfolds over a couple of days as the sperm and egg genes slowly combine to form a new genome that eventually takes control over the cell's activity. It is equally difficult to determine when a child becomes an adult: it happens somewhere in the fuzzy zone between early teenage years when they are capable of reproducing and late teenage years when they are cognitively mature enough to be parents. Finally, there are very few people on the planet that we would be inclined to classify as "purely good" or "purely evil": most exhibit a confusing mix of ethical behaviours that varies over time and in different situations.

Reality of a complex nature cannot always be tidily disambiguated by our bounded cognitive apparatus. We are forced to tolerate a much higher level of ambiguity and uncertainty than our Type 1 thinking is programmed to accommodate.

15.5 High Versus Low Predictability

Complexity and predictability are joined at the hip: a defining feature of complexity is how much less predictable it is than straightforward situations. In World #1, explaining and predicting go hand in hand: if we can explain why something happened, it is largely predictable. We can explain why eating food is satiating, and we can trust that it will be in the future. In World #2, explaining and predicting often are far apart, and for two reasons. First, our explanations are incomplete, because we inevitably miss some information. Second, the interactions of causal factors are so intricate that we cannot rely on our understanding of one complex scenario to predict another one: no matter how similar they may appear on the surface, the slightest, imperceptible differences can make any two complex situations behave very differently.

> ### *Predictable in Theory Versus Predictable in Practice*
>
> If complex systems are unpredictable, the question arises whether they always will be, or whether better measuring equipment and analytical models might enable us to understand them well enough to make reliable predictions. How do we know if the unpredictability of complexity is an inherent ontological property of our construction of reality that we can never overcome, or if it is an epistemological limitation that we can overcome as we deepen our knowledge of how complex systems work? The answer, as in the problem of randomness, is a combination of both.
>
> With the invention of the computer, non-linear equations and fractal geometry in the past decades, we are able to model complexity more accurately and narrow the range of possible outcomes that we can predict. But at the same time, there is a fundamental limit to the predictability of complex phenomena. In the early nineteenth century, French mathematician Pierre-Simon Laplace argued that, theoretically, given Newton's laws and knowledge of the position and velocity of every particle in the universe, everything about the future could be predicted. This notion was overturned in the twentieth century by Heisenberg's uncertainty principle in quantum mechanics (which demonstrates that it is impossible to know the exact position and momentum of any particle at the same time) as well as by chaos theory (which demonstrates that, given enough time, even the slightest error can throw a forecast off by a significant factor). The best that we are capable of is continued progress in how we describe complexity, and tighter probability estimates in our predictions of it.

Good predictions depend on complete and accurate models of the past and reliable repetition of past behaviours in the future. These are two enormous hurdles. Explaining the past in sufficient detail is difficult enough, but even if we are able to fit past data into a model that does a superb job of explaining a system retrospectively, the same patterns and relationships in complex systems often change: they will not necessarily repeat in the same ways in the future. "Retrospective explanations do not travel well into the future," Tetlock notes.[193] The dynamical nature of complex systems makes them non-stationary: their properties constantly change, making it difficult to apply statistical probabilities to them. They change their forms and behaviours, and the manner in which they change also changes, making prediction even more difficult. The stock market is a good example of a non-stationary system, because it is influenced by so many shifting factors, not the least of which is the emotional volatility of investors as they swing from severe pessimism to euphoria and back, with little, if any, predictability. The economic factors that underpin stock prices are themselves no easier to predict: as David Orrell notes, "The economy is a social process that cannot be reduced to law."[194]

In addition, because the extreme outcomes of non-stationary systems can be so significant, predictive modelling can be obviated by huge surprises that a statistical analysis of the past would not have anticipated. These big anomalies (or "tail events," in the lingo of statistics, because they lie at the far ends of probability distributions) are more frequent and more significant in World #2, which makes predictive accuracy even more elusive. These rare events do not reveal themselves in the small data samples that represent our daily lives or even decades of our lifespans, so we tend to underweight or ignore them completely when we contemplate expected outcomes. We do this to our detriment, because the big surprises have such enormous impact when they happen, and in World #2, they happen more frequently because of the increasing interdependencies among all the moving parts of complex systems. Nassim Taleb observes that predictability is more difficult than ever, "since almost everything in socioeconomic life now is dominated by Black Swans."[195] (Black Swans, as reviewed in 9.1, are Taleb's label for high-impact, unpredictable events.)

Notwithstanding the shifting nature of complex systems, large-scale, predictive probability estimates are possible because the set of behaviours that complex systems generate often gravitate to attractors: patterns that stabilize systems, such that high-level ("global") forecasts are possible, even if detailed ("local") forecasts are not. In other words, the details are difficult to get right, but probabilities can be assigned to the general patterns. Think weather: The short-term weather forecast is much more reliable than the long-term forecast, but even then, short-term details, like exactly how much rain will fall or what the exact temperature will be, can only be described in probabilities.

As with weather, so with many complex systems. We can make reasonable large-scale estimations of what will happen in the immediate future, but we cannot be too specific about the details. And the farther out we go from the present, the less reliable our predictions become, because the pockets of order within the systems change or dissipate altogether. Unpredictability increases exponentially with time, as many possible paths explode into the future in different directions, from a single point in the present.

To be better decision makers in an increasingly complex world, we need to invoke more Type 2.2 thinking, monitoring and refining our thought patterns to ensure that they accommodate World #2 complexity. Our default way of thinking, with minimal Type 2.2 oversight and intervention, does not work, because our Type 1 intuitions are deceiving, and we are not accustomed to the more sophisticated Type 2.1 models and tools that suit complexity. We need specific, metacognitive strategies to resist the decision-making shortcuts that we gravitate to: strategies that temper our race to certainty.

16

Strategies for Provisional Truth
Dogma Is for Dogs

Truth is slower than fiction.

Howard Wainer*

Our beliefs about complex phenomena are not definitive, unchangeable insights into the way things work; they are highly fallible interpretations. But giving up certainty in World #2 is a constant challenge for us. Voltaire put it best: "Doubt is not a pleasant condition, but certainty is absurd."[196] Our challenge is to endure the physiological discomfort of doubt. "Our minds are like inmates, captive to our biology," Nassim Taleb writes, "unless we manage a cunning escape."[197]

The only escape route available is metacognition: Type 2.2 thinking enables us to rise above Type 1 by overriding our default way of thinking. Metacognition is the conduit for putting into practice the implications of seeing truth as provisional, which is why this chapter offers cognitive strategies that underpin truth as provisional, facilitating a more effective approach to thinking in World #2.

---

* *Picturing the Uncertain World*, Princeton University Press, 2009.

Provisional Truth Strategies

Complexity cannot be rushed ⟶	1. Slow down
Complexity resists early conclusions ⟶	2. Suspend judgment
"They" are probably wrong ⟶	3. Be skeptical
"We" are probably wrong ⟶	4. Employ humility
Disconfirming evidence offers certainty ⟶	5. Test and retest
There are always other perspectives ⟶	6. Explore alternatives
Complexity resists true-false dichotomy ⟶	7. Think probabilistically
Groups outperform individuals ⟶	8. Engage others

16.1 Slow Down (Effectiveness Over Efficiency)

Our brains are preset by World #1 to maximize efficiency: we prefer short, quick answers and solutions, since they are good enough most of the time. But "good enough" does not work with complexity, because we are not World #2 experts: what feels like "good enough" is rarely that good. We are not reliable judges of the robustness of our interpretations, and we err on the side of premature conclusions based on oversimplified explanations. World #2 demands that we reprioritize effectiveness over efficiency; that we rebalance by giving up some speed for greater accuracy.

Management professor Shane Frederick created the Cognitive Reflection Test to assess the decision-making processes of individuals.[198] According to his research, smarter people are more patient, taking longer to reflect and respond to questions. Einstein's modest self-assessment that he was not as smart as he was patient ("It's just that I stay with problems longer"[199]) should be our rallying cry in our battle with complexity.

Where we have not already developed expertise, our brains are not able to achieve conceptual depth unless we allow them time, more time than we are accustomed to taking for straightforward problems. For World #2, it is crucial to differentiate the short story (the one that we come up with that feels good enough in the early stages of our contemplation) from the long story (the one that incorporates missing information, exposes implicit assumptions and reflects alternative considerations). We survive on short stories in World #1, but they misrepresent World #2. Long stories are not as efficient, but they are more effective because they approximate complexity more closely. *Complex problems cannot be managed with maximum efficiency: the sacrifice in effectiveness, given our lack of expertise, is too significant.*

Unconscious "Incubation"

Some cognitive psychologists argue that insight comes from the "incubation" process that Type 1 thinking engages in as it mulls over a problem. This view is somewhat controversial because other psychologists argue that the benefit of taking a mental break from a problem derives from allowing conscious, Type 2 thinking to break free from whatever mental models it may be fixated on, in order to restructure the problem: to allow for "inattention" to invigorate mental creativity in a way that facilitates insight.

The conundrum of how exactly unconscious and conscious thinking interact likely will remain a mystery for a long time, but there is unanimity among scientists on the general point that novel and complex problems require sufficient time to allow the two forms of thinking to work together productively. Whether it is in order for subconscious processing to be maximized, or for conscious processing to take a break and regroup, research has convincingly demonstrated that complexity requires time — time that includes breaking from a problem and coming back to it.

Slowing down is anathema to how our brains operate. Executives market themselves as having a bias for action; management gurus intone that a wrong decision is better than no decision. While waffling analysis-paralysis can be the downfall of individual action and the bane of company strategy, our relentless tendency to err on the side of speed is the deeper and more pervasive problem. Management teams (and boards) want fast answers, and they rely on meetings with tight agendas that combine trivial updates with complicated issues, allotting too much time for the former, almost always at the expense of the latter.

A successful meeting is usually considered to be one that concludes on time, so nobody is late for the next meeting.

Wicked problems cannot be managed efficiently any more than people can be. Meaningful human relationships of every kind are fostered by time-consuming interactions, including back-and-forth adjustments and fixes for miscommunication and misunderstanding. So it is with anything complex: we may contain a problem in the short term, or even think it is solved, but if our interactions with the problem are based on short-story interpretations and the conceptual illusions that speedy thinking generates, the problems will come back with a vengeance. For a bounded brain, mindful thinking is slow and there is no way around this constraint, other than the patience and tolerance for doubt that metacognition cultivates.

16.2 Suspend Judgment (Going Against Our Grain)

The corollary of slowing down is suspending judgment. We cannot slow down if we are not prepared to hold back on concluding. Suspending judgment is not an end in itself; it is a means to slow down our thinking and ensure that we are justified in our conclusions, based on sufficient due diligence. We suspend only until we are satisfied that we have examined enough alternative explanations; this is a subjective verdict, but a better informed one than our intuitive rush for certainty promotes. It is contrary to our nature to be opinion-less. It contradicts every aspect of our physiology and brain programming to withhold a verdict, to postpone a conclusion, to back-burner a perspective. Our constitution is such that it is more pressing for us to conclude than it is to deepen our understanding.

Fighting the urge to conclude quickly not only goes against the grain of our biology but also against years of schooling. Education is based on getting the one right answer as quickly as possible (within the time allotted to a test). Kids are more often taught to memorize than to think. Math and science problems are neatly defined, with all the information ready to be transcribed into a straightforward formula. Students are taught that history is a definitive narrative with discrete cause-and-effect components that lock into place like a puzzle. They are rarely challenged with ill-defined complex problems that have insufficient relevant information mixed in with lots of irrelevant information. In other words, they are rarely presented with the kind of problems that represent real-life complexity. They are rarely taught the difference between a logical and an illogical argument, the mistakes in thinking that are common to all of us, or how to tackle problems that do not converge to single answers.

Long after we graduate, our education is reinforced by the main source of information for most of us: the media, the default approach of which is to be succinct, definitive and, wherever possible, dramatic. Information does not sell if it does not meet these criteria. We live in an artificially imposed black-and-white world from the moment we start to learn, and we escape it only with deliberate and concerted effort. The friction of time constraints and the discomfort of not knowing impede our curiosity; it takes force for us to push against this friction and maintain inquiry by suspending judgment.

Here is a version of a fable that is attributed to ancient Chinese folklore. The story has no ending, for reasons that will become obvious.

> There once was a very wise farmer who was very proud of his teenage son. The son had a beautiful horse. One day, the son rushed into the field where his father was working to tell him that the horse had bolted over the fence and run away. The son was so upset, decrying the fates who had allowed his prized possession to escape. But the father simply replied, "Who knows if it is bad?"
>
> A week went by, and the son once again came running to the field, this time with great excitement. He announced that his horse had returned, accompanied by a second, strong and beautiful horse. Now they had two horses to help till the fields! The father shrugged and responded, "Who knows if it is good?"
>
> A few days later, the son was trying to tame the new horse. Upon mounting him, the second horse abruptly reared and threw the son off, whereby the son crashed to the ground breaking his leg. He screamed for his father, who rushed from the field and carried his son to the village doctor. Upon returning home with a cast, the son wallowed in self-pity and despairingly cried that it would be weeks before he could walk again. The father calmly said, "Who knows if it is bad?"

A week later, a war broke out and all the young men from the countryside were conscripted to fight the invading army. But when the army recruiters came to the farmer's house to fetch the son, they decided he could not join them because of his broken leg. The son was relieved and celebrated his fortune for not having to fight. But his father reminded him, "Who knows if it is good?"

At no point does the father concede any conclusive view of the unfolding saga. He wisely sees that each event is neither good nor bad on its own, but a link in a long chain of never-ending events, whose significance cannot be immediately known. The wise father understands what the son does not: our perspectives are limited when we cannot see the whole — when the alternative paths of future events that unfold in front of us are hidden from our view. The son naïvely treats each episode as an endpoint, rather than as an interim and arbitrary snapshot of a moving sequence of related events. The farmer has a systems perspective of life, and he is careful not to prematurely ascribe truth or finality to what his son experiences.

The fable demonstrates how unpredictable the alternative paths are that unfold from both our decisions and from random events beyond our control. The farmer is adept at resisting the urge to draw favourable and unfavourable verdicts on discrete events, the relevance of which became meaningful only in the context of the events that unfold from them. In our daily lives, we navigate our way through countless obstacles and surprises, rendering judgment on them prematurely: we would do well to aspire to the farmer's discipline, especially keeping in mind that what first appears as unfortunate is frequently proven, as time progresses, to be fortuitous. In many cases, we will never even become aware of just how fortuitous "bad" events actually were, such as the financial catastrophe we narrowly averted when we were outbid on a house whose basement subsequently flooded.

Being mindful of the alternative paths that could unfold instills an appropriate degree of caution in our use of "because." We love the word: it feeds our addiction to certainty. Something happened "because"; someone said or did something "because." We operate on the basis that the short "because" story incorporates all relevant information into a tidy explanation. Indeed, this shortcut works just fine for us in World

#1, but in World #2, multiple causal factors are intricately woven into interrelationships that are not directly observable: unearthing causality requires a lot more digging than we usually undertake. At best, "because" masks complexity; more typically, it misrepresents it. Suspending judgment means using "because" sparingly and carefully.

Furthermore, just as "because" is *over*used in World #2, a host of expressions are *under*used: "don't know"; "could be"; "perhaps"; "I need more information"; and "I'm not in a position to render a judgment." Ludwig Wittgenstein, in his earlier writing, famously proposed, "Whereof one cannot speak, thereof one must be silent."[200] In fact, even more preferable than silence are questions — digging for more information — which is presumably what Martin Heidegger had in mind when he wrote that "questioning is the piety of thinking."[201]

Suspending judgment is also a crucial strategy for recognizing the limits of our cognitive capacities. David Hume wrote that we should limit "our enquiries to subjects as are best adapted to the narrow capacity of human understanding."[202] We need to resist the temptation that philosopher Colin McGinn warns against: in our reluctance to accept our limits (our "cognitive closure"), we opt for reductive theories that oversimplify or for magical stories that fill the gap.[203] Mathematician Gregory Chaitin argues that the complexity of nature is significantly more vast than the complexity of our combined intellect and computing power (which leads him to believe that a final unifying theory of everything in physics will elude us).[204] Suspending judgment is an imperative that both acknowledges our cognitive limits so we do not inadvertently transgress them and recognizes the bad habits our minds are vulnerable to so we can mitigate them.

Individuals vary in their ability to forego the feeling of knowing by suspending their judgment. Testing for "open-mindedness" is in its early stages, but the research thus far demonstrates differences in the efforts that people make to persevere with problems before abandoning them or concluding prematurely. Some people are more inclined toward cognitive flexibility and have a less pressing need for cognitive closure. But open-mindedness is not a "have" or "do not have" skill; it is a thinking disposition that can be improved. We can be educated on the importance and benefit of metacognition, and practice the habit of assessing our initial intuitions and resisting our early feelings of knowing. Epistemic regulation — the ability to monitor and control thinking — does not come any easier to most people than emotional regulation, but it is just as valuable a skill as learning to manage anger or depression.

16.3 Be Skeptical ("Probably Not" — Part I)

To the list of underused expressions in World #2 we can add "probably not." A significant portion of the things that we are told or believe about World #2 are not quite as true as presented or conceived; more often than not, they are just plain false.

If your default way of thinking is to rush to conclusions and resist suspending judgment to explore alternative explanations, then you will err on the side of gullibility in areas where you have not developed deep expertise. But if you accept that the ontological properties of World #2 are often mismatched with the epistemic properties of human cognition, you will realize how important it is to be skeptical. When you accept that it is not in our nature to see systems, or to understand the patterns of randomness, or to hold off on drawing conclusions, you will appreciate how cautious we must be in embracing claims of truth about complex phenomena.

> ### *A History of Skepticism*
>
> The history of skepticism is as long as recorded history, starting with the Hindu Vedas, which challenged the validity of our perceptions. The first Greek skeptical thinkers were Xenophanes and Parmenides, who, writing over a century before Plato, argued that we can never know if we have reached a final truth (Xenophanes[205]) and that our senses and reason are unreliable (Parmenides[206]).
>
> It was Pyrrho who conceptualized skepticism more formally, making it a philosophy in its own right in 330 BCE.[207] As a result of his travels as a soldier, Pyrrho was startled by how the beliefs in one culture could be so different from, and even contradict, the beliefs of other cultures. He concluded that any substantive notion of truth was fundamentally elusive. The best course of action for an individual

was to live in harmony with the cultural beliefs he happened to hold, while recognizing that these beliefs are unlikely to represent reality.

Sextus Empiricus, who wrote around 200 CE, argued that we should suspend our beliefs as a normal course of habit; we should avoid claiming truth about anything other than our own feelings and perceptions, since any claim beyond our immediate experience cannot be validated.[208] Peace of mind, for Sextus, is available to us by exploring the arguments against our beliefs, as a way of reminding ourselves that the ultimate truth about the world outside of ourselves is inaccessible. Sextus believed that skepticism was not a philosophy so much as a way of living. We must live our lives with whatever beliefs seem to work for us, while appreciating that truthfulness is not available to us about how the world works.

Plato and Aristotle had both rejected skepticism in its hardest form; however, the university that Plato founded, the Academy, was subsequently led by philosophers who subscribed to skepticism. One of these skeptics was Carneades, who, two centuries after Plato, argued eloquently in support of the Platonic conception of justice, then argued just as convincingly against the same position in a subsequent lecture, demonstrating the impossibility of truth in knowledge.[209]

David Hume, writing in the 1700s, re-engaged skepticism by building on Sextus' insights. Hume's argument, parallel to Sextus', was that anything we experience could, in theory, be different, since there is no strictly logical condition that prevents the opposite of what we experience from being true.[210] No matter how many times we see the sun rise in the east, we could surmise that it may rise in the west next time around: there is no guarantee of certainty. Factual knowledge is always conceivably false, because we have only a finite amount of information (for example, a finite number of sunrises) to base our beliefs on.

Hume was clear that his skepticism was more academic than practical; he acknowledged that we have to live our lives assuming that causal relations and other elements of reality are true, even when there is no logical necessity that guarantees the truth or universality of our beliefs. Philosophers have struggled with Hume's challenges to the concept of certainty ever since.

Just because we cannot be sure that we have arrived at a final truth does not mean that anything goes. Some beliefs are more plausible and more reliable than others because they correspond better to our experience and because they cohere better with other of our beliefs. The skeptical view is that we cannot ascertain capital-T Truth, but we can still prioritize beliefs based on their plausibility and reliability. The value of skepticism is not in rejecting all forms of truth but in keeping us cognizant that whatever we are told and whatever we read about World #2 are rarely the final verdict.

The skeptical mindset pivots on a key question: What mistake do we want to make? The idea is to start from the premise that, because our cognitive capacity is bounded, some of our beliefs are going to be wrong. In doing so, we can make one of two errors. Sometimes we will believe things that are not true; that is, we will be too gullible (Type I error, or false positive). Other times we will disbelieve things that are true; that is, we will be too skeptical (Type II error, or false negative).

Blaise Pascal used the trade-off between the two possible errors in his famous wager regarding belief in God: If you assume that God exists and behave accordingly, the upside is enormous should you be right and negligible should you be wrong, whereas if you assume that God does not exist, the potential error could have significant consequences for you in the very long term.[211] For Pascal, the downside of gullibility was far outweighed by the downside of skepticism.

Putting aside the questionable merits of his argument, this trade-off between Type I errors and Type II errors permeates the decisions we make throughout our lives whether we are aware of it or not. The terms "Type I" and "Type II" errors have a technical, mathematical meaning, relating to the two errors a scientist is vulnerable to as she attempts to validate her hypothesis (by attempting to disprove the null hypothesis that contradicts her alternative hypothesis). But Type I and Type II, and their corollaries false positive and false negative, have become standardized in more general usage.

Which mistake is worse? Naturally, the answer depends on the particular situation. However, when it comes to surviving in World #1, the second mistake, being overly skeptical, is more dangerous. It is far better to assume that a stick is a snake and be relieved if you are wrong, than to assume that a snake is a stick and die of a venomous bite if you are wrong. It is far safer to assume a rustle in a bush is a threatening stalker than to risk the outcome of assuming that it is just the wind. It is more survival enhancing to run for cover when lightning flashes than to ignore it.

Given a choice, then, it is more prudent to err on the side of gullibility in the straightforward and harsh evolutionary environment of our ancestors. It is a better strategy, in World #1, to attribute meaning to cues than to dismiss them as unimportant; better to *over*interpret than to *under*interpret; better to be inconvenienced by a Type I false alarm than to be killed by a Type II missed threat.

This evolutionary bias toward being extra-vigilant means that our subconscious processing is highly sensitive to the perception that we are under threat, no matter how small the threat may be. This "threat alert" bias translates into the well-documented fact that we tend to experience more negative emotions, and experience them more intensely, than positive emotions. This is our "negativity bias": we are more often worried, agitated, preoccupied, concerned, frazzled, upset, angry, overwhelmed, irritated, resentful, bitter, jealous, embarrassed, self-conscious, frightened and nervous than we are the opposite. Bad is stronger than good for us.

This asymmetry of our emotional responses, skewed as it is toward avoiding bad outcomes that cause us regret, is the foundational assumption of what is known as "prospect theory," the Nobel Prize–awarded work by Daniel Kahneman and psychologist Amos Tversky. Their research demonstrated that we feel the pain of loss about twice as strongly as the pleasure of gain.[212] Put another way, if we are faced with a possible gain that has the same odds as a possible loss, the gain must be twice as big as the loss to balance the scales and entice us to take the gamble.

This bias toward avoiding bad means we risk overinterpreting the possibility of bad (overreacting to false alarms) while simultaneously reducing the risk of underinterpreting (underreacting to true threats), since the two errors are traded off each other. The more gullible we are, the less skeptical we are, and vice versa.

Here is the key: *Worlds #1 and #2 have inverse risk–reward trade-offs when it comes to gullibility and skepticism.*

The Gullibility–Skepticism Trade-off

In World #1, where signals are clear and cause and effect are tightly linked, we can afford to trust our intuitions; we have developed reliable expertise in interpreting and responding to the signals. We can risk being gullible because it is a safer gamble and because the downside is limited to expending unnecessary energy on threats that did not materialize. The cost of gullibility is low compared with the life-threatening cost of skepticism.

World #2 runs on a completely different trade-off. Because we have insufficient expertise with complexity and therefore unreliable intuitions about it, we are vulnerable to locking down on initial conclusions that are less likely to be correct than they are in World #1. Our default bias for being gullible, which works in the first world, is costly in the second world. It is far more productive in World #2 to err on the side of skepticism, because signals are unclear and cause-effect relationships are buried; odds are high that our interpretations, especially our preliminary ones, are not quite accurate. Gullibility in this world generates a much higher error rate, and those errors evoke ineffective responses.

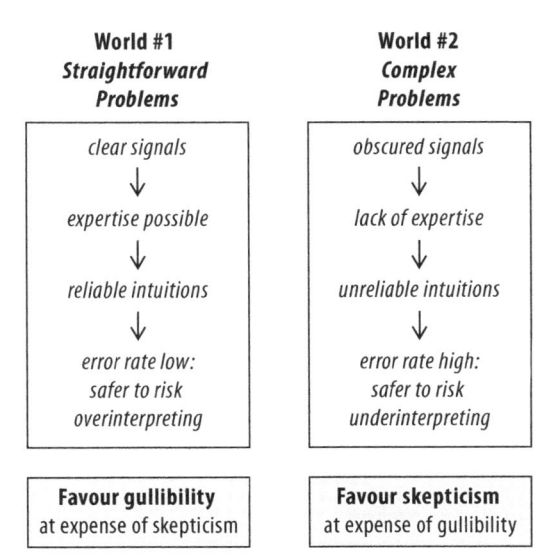

World #1 Straightforward Problems	World #2 Complex Problems
clear signals ↓ expertise possible ↓ reliable intuitions ↓ error rate low: safer to risk overinterpreting	obscured signals ↓ lack of expertise ↓ unreliable intuitions ↓ error rate high: safer to risk underinterpreting
Favour gullibility at expense of skepticism	**Favour skepticism** at expense of gullibility

With straightforward problems, the probabilities are in our favour when we risk being gullible. With complex problems, probabilities are in our favour when we risk being overly skeptical. Given the choice, gullibility is the safer bet in World #1, skepticism is the safer bet in World #2. The shaded boxes in the table below indicate which kind of extreme views are the safer bet in each world.

	Err On Side of Gullibility	Err On Side of Skepticism
Risks	Believe a falsity Overinterpret False Alarm	Disbelieve a truth Underinterpret Missed Threat
Mistake	Type I Error False Positive	Type II Error False Negative
World #1 Examples	• Interpret a stick as a snake • Alarm goes off in absence of fire • Fear all spiders, even innocuous ones • Test for cancer is positive when there is none • Innocent person convicted	• Interpret a snake as a stick • Alarm fails as fire burns • Fear no spiders, even venomous ones • Test for cancer is negative when there is cancer • Guilty person acquitted
World #2 Examples	• Stock brokers can predict the market • Lots of Vitamin C forestalls a cold • Research shows eggs are bad for you • Depression is a chemical imbalance that anti-depressants fix • Success comes down to hard work	• Stock picking is a mug's game • Vitamins are marketing gimmicks • All research is biased • Scientists do not have a handle on the origin of mental illnesses • Success depends on luck

The Trade-Off Between Type I and Type II Errors

When we draw conclusions, we try to make good decisions that accurately reflect how things work in the world. There are four possible outcomes: two ways of being right and two ways of being wrong. A good decision is one in which we pursue something useful or avoid something harmful; a bad decision leads to the two reverse outcomes: pursuing something harmful (Type I error) or avoiding something useful (Type II error). These four possible outcomes form the decision space where the objective is to maximize the odds of locating our decision in the southwest corner and the northeast corner, while minimizing the odds of being in the other two boxes.

Avoid Good (Type II error)	Pursue Good
Avoid Bad	Pursue Bad (Type I error)

This decision map (sometimes referred to as a Taylor-Russell

> diagram) is difficult to navigate, because the better we are at avoiding Type I errors, the less effective we are at avoiding Type II. The tidy, narrow ellipse depicted above, which represents the area of decision making that minimizes both types of errors, is idealized and generally unattainable in World #2. With complexity, we are always vulnerable to a much greater area occupying the northwest and southeast corners because as we reduce the area of one of those corners, we expand that area into the other: the more skeptical we are, the more likely we are to invoke Type II; the more trusting of our intuitions we are, the more likely we are to invoke Type I.
>
> Complicating the process is the troubling nature of the northwest box, because usually it is invisible for us: it is the missed opportunity where we avoid something valuable but never know that we missed it. It is the stellar employee who was not hired; the ideal mate we did not ask out; the shorter route to work that we never discovered; the mineral our body needed to fight off an infection that we neglected; the list is endless. The results of the three other boxes are usually transparent to us: we know when our decisions panned out or when they resulted in bad outcomes. But the valuable feedback that hides in the northwest corner, in the form of Type II mistakes, rarely surfaces, thereby limiting our learning in World #2 and our ability to deepen our expertise with complexity.

We can err on the side of gullibility in World #1 because we can rely on our intuitions: we can afford to "underthink" in the sense of invoking less Type 2 thinking. But in World #2, where the myriad biases and constraints that skew our thinking make us more vulnerable in what is a hostile environment, we are better off to "overthink" by invoking more Type 2 thinking and being cautious about embracing our early conclusions. Not only is it more productive to be skeptical, it is also more logical: per Popper, the only certainty available to us when we draw conclusions from experience is to prove that the contradiction of a belief is impossible. Because it is very difficult to prove that the opposite of a claim is impossible, we are automatically forced into a skeptical frame of mind. As Michael Shermer describes it, a skeptic is one who "does not believe a knowledge claim until sufficient evidence is presented to reject the null hypothesis."[213] Until the opposite of a claim is proven false, we should assume the claim to be, at best, only probable. Skepticism is a leveller: while it does not put all beliefs in the same category, it does prevent beliefs about complexity from jumping to the elite status of certainty.

"Probably not" and its more diplomatic version, "could be," are useful for reminding ourselves that early conclusions regarding complexity are provisional and rarely capture the whole story. "Probably not" also reinforces the notion that once we are satisfied that we have rigorously tested our assumptions and explored alternative explanations and are ready to move to closure of some sort, we are still restricted to statements of probability, not certainty, to darker and lighter shades of gray, not black and white. In presenting our own opinions, in lieu of, "I know that …" or even, "I believe that …" (both of which connote knowledge), we are far better to use such language as, "It appears to me that …"

All animals except us survive just fine with quick decision making that they do not second-guess. They can afford to be single-minded because they do not live in a complex environment that operates differently from the one in which they evolved. In the realm of complex problems, we need to be far more skeptical about definitive conclusions than we instinctively are. Dogma is for dogs.

16.4 Employ Humility ("Probably Not" — Part II)

Skepticism directed inward is humility.

In his poem "The Second Coming," W.B. Yeats wrote, "The best lack all conviction, while the worst are full of passionate intensity." In a similar vein, Nietzsche wrote, "Convictions are more dangerous enemies of truth than lies."[214] Both writers were keenly aware of our proclivity for overconfidence, as was Aristotle: "It is the mark of an educated mind to rest satisfied with the degree of precision that the nature of the subject admits, and not to seek exactness when only an approximation is possible."[215]

The research of Philip Tetlock on experts' predictions is revealing in many respects. First and foremost is how poorly so-called experts predict future events. Random guesses are just as reliable as their forecasts when it comes to the surprise events that dominate the news (uprisings, revolutions, wars); in fact, statistical models that extrapolate the future from the past generate consistently better forecasts. Tetlock's work also reveals a contrast in the thinking habits of people: he categorizes them into two classes based on different cognitive styles. Hedgehogs are narrow and overly confident thinkers; foxes are broader and more modest thinkers. Hedgehogs are mentally stubborn, sticking to their beliefs and predictions despite inconsistent supporting evidence. Foxes are much more willing to change their minds and update their beliefs as events unfold, and they are more inclined to integrate a variety of information sources and assumptions. While neither hedgehogs nor foxes have reliable predictive powers, hedgehogs tend to underperform more consistently.[216]

The rigidly thinking hedgehogs are much more confident in their abilities, despite making forecasts that are generally less accurate than the foxes. The more expert they perceive themselves to be, the more rigid and less accurate their forecasts. Foxes achieve far more of a balance between self-criticism and self-defence, whereas hedgehogs tend to be very defensive of their positions and predictions and more reluctant to admit mistakes and change their minds. Hedgehogs assign higher probability to being right than do foxes, while at the same time being almost three times as likely to describe a future event as either inevitable or impossible, rather than somewhere in the middle. Foxes have better-calibrated probability estimates of their forecasts: in general, their predictions are more modest; they weigh the probabilities of different scenarios in their minds rather than focusing on the unfolding of one particular scenario.

Foxes, according to Tetlock, are more "integratively complex." He summarizes the underperformance of hedgehogs by noting that their cognitive style is "less suited for tracking the trajectories of complex, evolving social systems." The irony is that hedgehogs, of course, make for great media pundits. Most of us have neither the patience nor the attention span for cautious forecasts and ambiguous explanations: foxes do not make for good TV. This is one of the reasons that the world seems straightforward and predictable to us: it is packaged and marketed to be quickly digested and delivered to us by experts who are firm in their cut-and-dried world views.

Tetlock's work dovetails with other studies on overconfidence.

Experiments run by Valerie Thompson demonstrate virtually no relationship between confidence in a judgment and its accuracy.[217] She and her colleagues asked subjects to judge both the validity of a sample of logical problems and the confidence they felt in their assessment of that validity. The subjects were also asked to reassess the problems after a period of deliberation, and to rate again their confidence levels in their responses. Interestingly, while there was no improvement in accuracy after additional deliberation, there was an increase in confidence. Not only are we extremely vulnerable to our addiction to the feeling of knowing, but that feeling is also a shockingly poor indicator of our actually being right.

World #1 shaped us into being overconfident; it trained our ancestors, just as it trains us, to disregard the difference between the world outside our heads and the one we depict inside, because in a straightforward world, they are not that far apart. In World #2, the inside and outside worlds can be very far apart, hence the gap.

Each of us produces, directs and stars in a play performed in the theatre of our mind, and we are completely oblivious to the fact that we are the one and only one audience member for this play. Others are not privy to this drama: they witness a very small portion of it, and this tiny slice is *their* version of our play, not the lines we ourselves hear or the actions we ourselves observe. Our habit is to assume that our play is the one everyone else is watching since it is the one-and-only play available for viewing as far as we are concerned. But in World #2, our play and the real thing are usually poorly correlated. Being humble means keeping this discrepancy foremost in our consciousness: the unavoidable and insurmountable difference between our internal play and the real world. Being humble means accepting this difference, and acknowledging how far the play and the real thing can diverge if we are not careful.

Because we never have the whole story, it is important for us to ask if we have *enough* of the story to make any sort of judgment. There is no way of knowing: if we do not know the whole thing, we cannot know what portion of the total picture we have a handle on. All we can be sure of is that the more complex the issue, the more digging we will need to do and the harder we will have to work to suspend judgment in the meantime. We also need to anticipate being surprised. As Taleb is fond of repeating, it is not the forecast that matters so much as the forecast error that we ascribe to it.[218] Estimating the range of our possible error is at least as important as the prediction itself: when it comes to complexity, we know that we will be wrong, so we have to consider *how* wrong we might be.

Tetlock's research demonstrates that when our confidence in our predictions about complexity is high, the odds that we are wrong are better than even. The best indicator of the quality of people's judgments, he says, is their metacognition: "soul-searching, Socratic commitment to thinking about how they think." Business professor Chris Argyris has a term for thinking about how we tackle a problem; he calls it "double-loop learning."[219] Feedback from single-loop learning provokes us to adjust our conclusions, but double-loop learning encourages us not only to alter our judgments but also to update the mental models that we are applying to the problem — to adjust how we are thinking about it. Double-loop learning is deeper than single-loop: it explores the conflicts and unspoken assumptions in our thinking, enabling us to tweak how we address a problem so we avoid getting stuck in the single-loop process of attempting to solve problems with models that do not work. World #2 is accessible only to a bounded brain if it is humble enough to engage in double-loop learning, which is a willingness to self-examine.

Taleb positions this in a characteristically indelicate way, writing that "we are a bunch of idiots who know nothing and are mistake prone, but happen to be endowed with the rare privilege of knowing it."[220] In fact, Taleb is too generous: most of us are very attuned to cognitive frailties — those of others, that is. We do not apply the same degree of scrutiny to our own thinking, which we tend to overestimate. As Taleb points out, "to be sophisticated you need to accept that you are not so."[221]

16.5 Test and Retest (and Retest Again)

"There is always a well-known solution to every human problem — neat, plausible, and wrong," according to essayist H.L. Mencken.[222] So how do we avoid tidy solutions? How do we leverage humility to deepen our understanding?

Back to Francis Bacon, originator of the scientific method, and his view that experimentation should be based on testing hypotheses that are carefully crafted from observed data, with the experimenter always on the lookout for negative instances: examples that disprove the hypothesis.[223] Bacon's approach was cautious, because he recognized our temptation to jump to premature conclusions that we then validate with a strong bias for confirmation. Testing means we have to be aggressive prosecutors of our own thoughts. Leveraging the asymmetry of induction is one of the most powerful ways for us to keep ourselves honest: rather than relying on confirmation testing, we have to seek disconfirming evidence.

Disconfirmation and Cognitive Therapy

In the late 1960s, the psychiatrist Aaron Beck used the twin strategies of disconfirming early conclusions and generating alternative explanations in his development of cognitive therapy, which is now a widely practiced form of psychotherapy for people suffering from depression and anxiety.[224] The aim of cognitive therapy is to help patients change their thought patterns in order to relieve their negative feelings, based on the premise that self-defeating thoughts are a principal cause of depression and anxiety.

For example, a person prone to depression is likely to overgeneralize in self-destructive ways, such as believing that nobody appreciates him, or that he fails at everything that he attempts, or that he will never find someone to love. Because these broad, general beliefs reinforce a depressive state, the cognitive therapist asks the patient to identify and confront these assumptions, to test whether they really stand up to scrutiny. The process is geared toward invalidating unproductive beliefs with disconfirming evidence: after articulating the evidence that supports the thought, the patient is instructed to identify the counterevidence that contradicts the thought. The final step is to generate alternative hypotheses that are more balanced and sanguine than the initial interpretations.

This process recognizes that the feeling of knowing reinforces our beliefs about ourselves and the world in ways that close us off from alternative perspectives that are often both more rational and more optimistic. It recognizes that many of our less productive thoughts are based on underlying beliefs that do not hold up to rational scrutiny, no matter how strongly we feel them to be true.

The logic of cognitive therapy is equally applicable to our thinking in general: narrow, imbalanced thinking not only puts us at risk of emotional turmoil, it also distorts how we think about complexity. The ultimate aim in both cases is to generate alternative interpretations that are more productive, by first disconfirming our initial conclusions. Even if certainty is unattainable, the judgments that we are entitled to have the greatest confidence in are those that we have attempted, unsuccessfully, to disprove.

Question Asking

Key to the process of disconfirming is asking questions. Consider this problem: A fifteen-year-old girl wants to get married right away. What should she consider, and what should she do? Psychologist Paul Baltes and his colleagues used this type of dilemma to assess the wisdom of individuals, as part of a project called "The Berlin Wisdom Paradigm," which researched the elements that constitute "the pragmatics of life," including empathy, psychological insight and tolerance for ambiguity.[225] In this model, a response to the above problem can be categorized in one of two ways: a rigid imperative ("She is far too young to be marrying") and a question-oriented response ("What are her circumstances? Is she pregnant, or did her parents recently die, or does she live in a culture that promotes early marriage?"). Wisdom, as defined by Baltes et al., accommodates complexity through questions, a process that uncovers the many subtle nuances of each situation and encourages the sensitivity required to tackle ambiguous problems that involve other people who may have differing values.

Questions get us closer to truth, which is presumably what Voltaire had in mind when he wrote, "Judge a man by his questions rather than his answers."[226] Testing involves asking the right questions in the right way: asking questions that seek disconfirmation of preliminary hypotheses, and asking them in a truth-seeking, expansive manner. Overconfidence is kept at bay with question asking: questions open up possibilities, new angles, new interpretations, new perspectives.

The paradox is that although questions move us closer to deep understanding, we are disinclined to ask questions and not very expert at asking the right kinds in the right kind of way. Once again, the cognitive therapeutic approach is instructive because it emphasizes the influence of framing on the responses we get when we do ask questions. Open-ended and constructive questions ("How can I enjoy life more?") yield very different responses than closed-ended and negative questions do ("Why is my life so rotten?"). When we engage others, how we frame our questions has a significant influence on the responses we get. Others are very adept at reading our signals: if our face, body and tone convey a desire for sympathy and alignment with our perspective, and our questions are framed in a way that invites confirmation, we will only receive validation of what others think that we want to hear. For example, if a couple has a bitter argument and each person separately canvasses a friend to confide in, each will elicit sympathy and confirmation that their partner is in the wrong. Each person frames the sequence of events and questions ("I'm not overreacting, right?") in such a way to guarantee the sought-after outcome.

For most of us, honest, probing and open-minded question asking is not our forte. We have to fight a powerful bias toward framing our questions in a way that solicits mere confirmation of our suspicions. But testing and retesting with sincere questioning of our assumptions and beliefs is the only sure-fire way to deepen our understanding of complexity.

16.6 Explore Alternatives (the Hidden Paths)

The aim of cognitive therapy is to invalidate unproductive thoughts and replace them with new, productive, optimistic interpretations. We constantly fight the headwind of our own stubborn belief systems; it takes perseverance for us to uncover alternative explanations and paths. The more rigorous we are at exploring explanations that differ from our first satisfactory ones, the closer we can get to an accurate representation of how the world works.

An alternative path is a route of causal events that could have led to a different outcome. The more open-minded we become at imagining alternative paths, the more deeply we will see and accept both the contingency of past events and the infinite possible futures that await us. Alternative paths are always present, but they are hidden in the background of our lives.

Imagine that you are driving home, taking your usual route, but traffic is jammed up at a certain point, so you decide to take one of two detours that you use when traffic is bad. You get home later than usual, which disappoints your daughter, who was hoping that you would take her to the park before dinner. You decide that the next time you are in a traffic jam on the main route, you will take the other detour, since it might be faster.

What you do not know, and never will, is that the main route cleared up right after you abandoned it (a stalled car was being towed away at the very moment you opted for the detour). If you had waited just a few minutes, you would have reached home much sooner and could have gone to the park. If you had got home on time that particular day, a rabid dog would have bitten your daughter at the park. After a hospital visit and a shot, she would be fine, but she would become fearful of dogs and ultimately veer from being the animal-loving veterinarian that she actually ended up being fifteen years after that day. (What actually happened on that day was that the rabid dog was captured by police after a neighbour spotted it before it attacked anyone.) What you also do not know is that if you had taken the second detour as an alternative route home, you would have been in a head-on collision with a truck, whose driver lost control of the wheel when he nodded off. (What actually happened is that the truck driver regained control, just narrowly missing a car that was four seconds behind the spot you would have been if you had taken that route.)

Of course there are an infinite number of paths, all different from the one that actually unfolds, all lingering behind the chain of events that we actually experience. So what value could imagining these scenarios possibly have? The contemplation of alternative paths has both prospective and retrospective benefits. Gary Klein coined the term "pre-mortem" for the useful exercise of imaging in advance the failure of a particular decision, and attempting to brainstorm the reasons for its failure, so that corrective action can be taken proactively.[227] Prospectively imagining how scenarios can play out allows us to anticipate and manage future contingencies. Just as useful is the retrospective contemplation of our experiences and decisions: how events could have unfolded differently than they did. Reflecting on alternative paths makes us humble and smart students of life, because such ruminations remind us that hindsight clarity is illusory. Events could have unfolded differently than they did, despite our best intentions and decisions.

The key to considering past alternative paths is retrospective imagination, also known as counterfactual thinking. We are pretty good at this type of thinking in particular situations, like when we suffer regret: "If only I had looked in my rear-view mirror before backing up"; "If only I hadn't had that second helping"; "If only I hadn't bought that stock." We are experts at if-only counterfactuals when it comes to disappointment (and when we are relieved by near regrets, as in, "Thank goodness I had my seat belt on, because if I hadn't …"). But we are very selective in our use of counterfactuals: typically we invoke them to contemplate a single alternative path, usually the one where things work out better for us than they actually did. We are not nearly as creative as we could be in reviewing how the past might have unfolded differently.

The method of historical analysis known as counterfactual history investigates the past by considering how events might have unfolded if certain key events had occurred differently. This type of history is a creative and effective way to isolate key events and explore their significance. How might religion have evolved if Constantine had not endorsed Christianity as the state religion of the Roman Empire? How might geopolitics have unfolded if one of the assassination attempts on Hitler had been successful? Often the paths that did not unfold are clues to help us separate the critical causal events from the merely random ones, so that we avoid interpreting the past as causally inevitable.

Contemplating alternative paths does not come naturally to us, because evolution has trained us to think about what we see, not about what we do not see or what could have been but was not. We are cognitively lazy when it comes to brainstorming possibilities, because it takes less energy to stand firm on our preliminary conclusions. Professors J. Edward Russo and Paul Shoemaker have researched the power of "prospective hindsight," or pretending to look back from the future.[228] This way of thinking reveals how miserly we are in contemplating imaginary scenarios. One group of subjects is asked to generate reasons why an employee *might* quit, and another group is asked to generate reasons why an employee *did* quit. The latter group came up with 25 percent more reasons than the former. This experiment and others like it demonstrate that we are much more prolific in imagining alternatives when we are explaining something that has occurred rather than something that has not actually happened but might. The difference between the two groups explains why Klein's pre-mortem exercise is effective: it assumes that failure has occurred in the future, which now has to be explained.

The powerful force of knowing pulls our ideas in one direction to a final destination (a single "basin of attraction," to use the language of complexity science), when what we really need to do is explore other possible resting points (other basins) without getting prematurely stuck in the first one that seems to make sense. We need to constantly ask the two key questions that are the centrepiece of Beck's cognitive therapy: "How could my interpretation be wrong?" and "What alternative explanations are possible?" We need to stretch our imaginations both into the past and into the future, to envision the alternative paths that could have arisen and could arise. What-if questions are the fuel of counterfactual thinking, the underlying assumption of which is "it could have been otherwise."

16.7 Think Probabilistically (Be a Good Bayesian)

We tend to think in either/or dichotomies. This cognitive strategy works in World #1, where conclusions and solutions are straightforward: they are either correct and work, or they are incorrect and do not work. But this strategy does not work well in World #2, because we rarely have enough information available to be as definitive. With complexity, we have to think probabilistically. Broadly defined, probability is a numerical representation of confidence in a belief. It is the principal way of quantifying the possibility of truth in World #2.

While probability theory was developed in the mid-1650s, it was not until the mid-1950s that the concept of probability as a quantifiable representation of uncertainty became part of our common language, largely through the advent of computer technology, which enabled the science of meteorology to introduce probabilistic weather forecasting to the public at large.

The problem is that our intuitions about probability are unsophisticated, because we develop them in World #1. For example, we underestimate the probability of random coincidences, such as the likelihood of two people sharing a birthday in a group of fifty people, which is 97 percent. We misinterpret predictions, like weather: if a forecast calls for a 90 percent chance of rain and the sun shines all day, we assume that the forecast was incorrect. In fact, it was not wrong at all: the forecast indicated that the prevailing set of environmental conditions would *not* generate precipitation in one out of every ten occasions, and that is exactly what happened.

And when it comes to interpreting combined probabilities correctly, we are virtually helpless. Using a version of the question in 2.4, say you are told that a particular disease, which is contracted by only 0.1 percent of the population, has shown up in your blood test. The test is quite reliable: it *wrongly* indicates, only 5 percent of the time, that a person has the disease (when the disease is not actually present). What are the odds that you have the illness? Much lower than most of us anticipate: only 2 percent. We have to combine the two probabilities (incidence of disease in the general population and likelihood of the test being accurate) in a mathematically unintuitive way (using Bayes' theorem).

Not only can elements of probability be unintuitive, but the concept itself is multifarious because it can contain varying degrees of subjectivity. The following four examples reflect increasing degrees of subjective judgment, from lowest to highest:

- A coin has a 50 percent chance of coming up heads when flipped.
- If a disease is present, the screening test has a 90 percent chance of correctly identifying it.
- Tomorrow's forecast calls for a 70 percent chance of rain.
- The producer estimated an 80 percent chance of the film being a hit.

Although there is a gradient that runs from purely objective to purely subjective, decision theorists tend to distinguish two general categories of probability, each going by three different names: objective, physical and frequentist for the first category, and subjective, evidential and Bayesian for the second. The first category is the dominant form of probability in World #1; the second category is more flexible, which is why it suits World #2.

Objective, Physical, Frequentist Suitable for World #1	Subjective, Evidential, Bayesian Required for World #2
Excludes subjective assessment.	Reflects a **subjective** expression of confidence — the strength of a belief.
The probability of a flipped coin turning up heads is 50 percent: this is an **objective** fact based on the **physical** properties of a fair coin, which determine the relative **frequency** of flipping heads.	The probability of a suspect being guilty of a crime is based on the **evidence** that supports the belief, which requires subjective judgment.
Captures the likely behaviour of repeatable, physical events that give rise to relative frequencies.	Captures the likelihood of an event that does not necessarily repeat itself, in which case there are no relative frequencies that can be relied upon. The mathematician (and minister) Thomas Bayes is credited with the mathematical (**Bayes**) formula that combines conditional probabilities.

Although they are often presented in opposition, each of the two forms of probability does a different job; in this sense, they are complementary. Frequentist probability is a reliable method of statistically summarizing the probabilities of repeatable events (particularly those that form normal distribution patterns). The frequentist version is not useful, however, when we are confronted with one-off situations, there being no history of repetition to inform us of the odds of future events. For these problems, we need the Bayesian version.

Bayesian probability can have many applications, so the difference between the two forms of probability is not quite as tidy as depicted in the above table. Bayesian probability can combine purely frequentist probabilities (such as combining the likelihood of testing positive for a disease with the likelihood of having the disease in general). It also can combine frequentist probabilities with subjectivist ones (such as combining the odds of a crime-scene DNA match with the odds that the same person has an alibi). Finally, it can combine purely subjectivist probabilities (such as combining the odds of a person liking a particular style of dress with the odds of her liking a particular clothing store).

Bayes himself did not intend the formula to be used in subjective assessments; he would be more pleased with the extensive use of his theorem for medical testing, where the probabilities are not "best guesses" but standardized numbers based on test error rates in the past and on disease incidence in large populations.[229] But the breadth and flexibility of Bayes' theorem make it an ideal way to think about all kinds of World #2 problems: Bayes allows us to combine probabilities in unintuitive but useful ways; it prevents us from illicitly inverting probabilities; and it fosters a mindset that our estimates are always subject to revision.

Combining Probabilities

Bayes is useful for predicting the likelihood of an elderly person with a good driving record getting in a car accident, a company with high management turnover going bankrupt, a romantic relationship lasting between two people with a large age gap, and countless other examples in which there is a general base rate that can be adjusted for the conditions of a particular situation.

Bayes' theorem starts with a prior probability or base rate (what we expect to occur in general, like the likelihood of any person in the population having a particular disease). Then it factors in the additional or new information (the test results, which are not 100 percent accurate), to arrive at a posterior probability (reflecting the prior probability that has been adjusted for the new information, like the odds of having the disease given the fallible test results).

Because the essence of Bayesian probability is combining related (dependent) conditions, it is sometimes referred to as conditional probability, since Bayes' objective was to expand probability theory to include events whose probabilities are conditional on one another. The trick with conditional probabilities is that they cannot be multiplied in the simple way that non-conditional (independent) probabilities can. For example, the chance of flipping a head followed by a tail are 50% x 50% = 25%: the two flips have nothing to do with each other so they can be multiplied together to determine their joint probability. Whereas if two events are related to each other, we cannot simply multiply them: we need Bayes' theorem to calculate the conditional probability.

Bayes' Theorem

The theorem (which was worked out in greater detail by Pierre-Simon Laplace after Bayes' death) calculates the probability of x happening, conditional on y also happening, which equals:

- the probability of *x*,
- divided by the probability of *y*,
- multiplied by the probability of *y* happening conditional on *x* also happening.

If your company is going to lay off 25 percent of its employees, you know that you have a one-in-four chance of losing your job. But there is additional information to factor into the odds: your boss gave you a very favourable performance review after submitting the names to the president of people she intended to let go. How do you revise the probability of losing your job? The odds are likely to be less than 25 percent, since poor-performing employees will be first out the door, but how much less? Bayes' theorem starts with the base rate probability (the prior probability) of 25 percent and then adjusts that rate to arrive at the all-in conditional (posterior) probability that reflects the additional information.

How does Bayes make the adjustment? First you estimate the probability of getting a favourable performance review in general: 75 percent of your past reviews have been "exceeds expectations." Next you estimate the probability of your boss giving you a favourable review while simultaneously planning to fire you. You estimate that at 10 percent: unlikely but not impossible. Bayes' theorem lets you combine these three probabilities — 25, 75 and 10 percent — in a way that yields an all-in probability. The conditional probability of being fired after a good review equals the unconditional probability of being fired (25 percent), multiplied by the conditional probability of getting a good review even though you are about to be fired (10 percent), divided by the unconditional probability of getting a good review independent of the layoffs (75 percent). So the revised conditional probability is 25% x 10% ÷ 75%, which equals 3%.

On the one hand, then, you can relax; on the other hand, one of the key inputs — the conditional probability of getting a good review then being fired — was purely subjective and may have been underestimated (i.e., if it is higher than 10 percent, Bayes would give you higher odds of being terminated).

Bayes helps untangle the intricacies of probability that are not always obvious, including a subtle but important distinction: the probability of being fired after a good review (3 percent) is not the

> same as the inverse probability of getting a good review even though you are about to be fired (10 percent). The former quantifies the uncertainty of being fired, whereas the latter quantifies the reliability of a performance review as an indicator of job security.

Avoiding the Temptation to Invert

Bayes not only reveals the critical insight that conditional probabilities cannot simply be multiplied together; the theorem also reveals that conditional probabilities cannot be inverted, even though our intuitions lead us to think they can. For example, the probability of having breast cancer if you test positively for a mammogram is not the same as the probability of testing positively if you have cancer. The first probability is important to know, whereas the second probability just reflects the accuracy of the test, which is an input into the first probability, but different from it.

Assuming that the two probabilities are equivalent and can be inverted is called the prosecutor's fallacy (and is related to the logical fallacy of affirming the consequent, discussed in 14.3 and 15.1. This fallacy gets its name from the faulty argument a prosecutor may invoke to demonstrate the guilt of an accused person. The error of inversing conditional probabilities may seem academic, but it is a feature of modern-day court hearings. The lawyers defending O.J. Simpson against the charge of murdering his ex-wife argued that the number of spousal abusers who actually kill their spouse is less than half of 1 percent. This argument appeared to sway the jury, but a good Bayesian knows that the relevant probability is not the odds that a spousal abuser killed his wife but the odds that a murdered woman, who had previously been the victim of abuse, was murdered by her abusive spouse. That probability has been estimated at 90 percent. As unintuitive as this may seem to most of us, each of these probabilities — 0.5 and 90 percent — represents completely different probability assessments with completely different implications.

Another common example is confusing the odds of matching DNA with the odds of being guilty when there is a DNA match. An aggressive prosecutor may argue that the one-in-one-million probability of a DNA match is proof that the charged person is guilty, but this low probability is not the same as the conditional probability that a person is both innocent and has the same DNA found at the crime. The relevant question is not, "What are the odds that any two DNA samples match?" but, "What are the odds that an innocent person would have DNA that matches the sample found at the crime?" For example, if the suspect has a credible alibi, no motive and no discernible relation to the victim, we assign a prior probability of 99 percent to his being innocent. But when a test reveals that his DNA matches that found at the crime scene, Bayes helps us to combine the 99 percent with the one-in-one-million chance of a DNA match, to arrive at a combined 50 percent probability of the suspect's being both innocent and a match. Which, of course, is not nearly as convincing as the prosecutor suggests.

Constantly Revising the Base Rate

World #1 has a limited need for sophisticated probability assessments: we develop intuitions quickly through our direct experience, like the probability of being sick after eating unripe fruit or being cold after swimming in icy water. The frequency of repeated experience is a good teacher of basic probability in World #1: our base rate beliefs – the ones that we have before introducing additional information – are reliable since there is no additional information that could be gathered that would alter our perspective in a meaningful way.

World #2 is not a good teacher of the nuances of probability, because most events in this world are different from each other in subtle ways that preclude reliable inductions. Bayes mitigates this problem by encouraging us neither to neglect nor overweight base rates but to continue adjusting them as we probe and learn more. Bayes allows us to quantify uncertainty in a more sophisticated way than our intuitions promote, reminding us how susceptible our beliefs are to revision, and how easily our base rate assumptions can be overwhelmed by enormous surprises (the Black Swans, as Taleb dubs them). Without a Bayesian perspective, we underestimate the size and frequency of these unexpected events (the "kurtosis" or "fat tail risk" that hides beneath the surface of our everyday experience), such as the terror attack that dismantles our notion of safety, the bank failure that alters our confidence in the financial system or the car accident that reshapes our outlook on life.

Probabilistic thinking fosters the kind of metacognition that stops us from locking down conclusions prematurely, because it encourages us to rise above our beliefs and think about the likelihood of their correctness as well as the likelihood of valid alternative explanations. There is perhaps no greater discipline that we can impose on our thinking than asking ourselves the probability we would assign to our conclusions about complex phenomena. When we ask ourselves, "What is the chance that …?", we open up our minds to possibilities that are not as narrow as our initial conclusions. A complex thinker is a good Bayesian, one who constantly updates and refines her beliefs as she interacts with, and learns from, complexity. A Bayesian mentality is one in which we are aware that no belief is final but is always in standby mode, waiting to be revised as more insight is accumulated.

16.8 Engage Others (the Power of Cognitive Diversity)

Back to Bacon once again, this time to consider his espousal of a collective effort for the scientific method: science should not be a lonesome activity but based on joint effort and rigorous communication between scientists. Four hundred years later, his views have been vindicated, not just in practice but also in theory. The research of Tetlock,[230] decision scientist Scott Page,[231] and others demonstrates that, when it comes to prediction and problem solving:

- groups do better than individuals working on their own, including the "best" or most expert individuals; and
- groups of cognitively diverse individuals do better than groups of like-minded, high-ability problem solvers.

Cognitive diversity is what enables the collective to outperform the individual. First, people bring ways of interpreting that are different from one another; the consolidation of distinct perspectives and problem-solving strategies enables greater creativity and insight. Second, while some biases are common to us all, we each have our own idiosyncratically skewed ways of looking at things; individual errors can be corrected by others and, even if they are not, in large enough groups, they tend to cancel one another out (for example, one person's extreme conservatism will be offset by another's extreme liberalism).

Page argues that the benefit of cognitive diversity really comes to the fore in problem solving. Unlike the simple averaging of many individuals' conclusions and predictions, an interacting group tackling a problem can reject bad ideas and build on good ones. Diverse perspectives help increase the number of possible solutions to a problem: so much so that, as Page demonstrates, a group of high-IQ individuals will tend to underperform in problem solving compared with a group of randomly selected individuals. Cognitive diversity is limited in the first group: the high-IQ people tend to share similar models and approaches to problem solving. In fact, a large enough group (of twenty or more individuals) consisting of sufficient diversity in thinking, tackling a sufficiently difficult problem, will outperform not only a homogenous group of high-IQ individuals but also the highest-IQ individual.

Groups outperform individuals by expanding the solution space beyond what individuals come up with on their own. But there are two conditions for outperformance: there must be sufficient cognitive diversity within the group (enough people and enough divergent ways of thinking), and individuals must be independent of one another to the extent that the diversity can flourish. Putting a group of diverse people together but limiting the ability of individuals to contribute hampers the potential of collective intelligence. "Groupthink" was the term coined by psychologist Irving Janis in the early 1970s to describe the tendency of individuals in discussion to move toward unanimity because the motivation to conform overpowers the motivation to critically evaluate ideas that are supported by others.[232] It is a common weakness in group decision making, because most of us are more comfortable acquiescing than challenging when confronted by a consensus point of view. All the benefits of cognitive diversity are lost when individuals do not operate independently.

Independence in groups can also be undermined by an insidious form of groupthink: path dependence. When a point of view is tabled, it can gain momentum through the support of a minority of participants, which carries it down a path to a firm conclusion without ever being sufficiently challenged. Legitimate expressions of concern can be swept away by the appearance of majority support when only a minority of views has been tabled. The same problem arises when a discussion has gone on for a period of time and participants become weary and inclined to acquiesce to a conclusion, even when the complexity of the problem may require more poking and prodding. Lack of participant independence has also been documented to foster polarization, whereby, as a result of the momentum of a single path gathering strength without interference, decisions are reached that are more extreme than any individual would have arrived at on his own.

The benefits of cognitive diversity are so substantial when tackling complex problems that Page summarizes the phenomenon by saying that diversity trumps ability. But the interactions between diverse and independent individuals must be constructive — correcting one another's errors and building on good ideas, while not undermining the independent thinking of each member. Interestingly, research by psychologist Charlan Nemeth complements the argument for diversity: her experiments reveal that the traditional form of brainstorming, where "no idea is a bad idea," generates fewer ideas than the kind where participants are encouraged to challenge and critique each another.[233] Constructive dissent both deepens thinking (uncovering weakness) and broadens it (increases creativity); it reduces the subtle but insidious limitations of path-dependent thinking.

The greater the complexity of thought that we can corral to confront complex problems, the greater our success will be in managing them. The thinking produced by a collection of minds is substantially more complex than an individual's, which is why harnessing the benefits of collective thinking is one of the most important challenges that any leader must master. Group facilitation is an undervalued leadership skill. Generating productive conversations among a diverse group of people is a talent that few leaders, no matter the field, have adequately developed.

Fostering high-quality conversation among a collection of diverse mindsets is as challenging as it is because of the complexity of individuals and their interactions, which is the focus of part V.

Part V

Brains and People
Human Complexity

17

The Complexity of Self
The Depths of Our Hypocrisy

*We are only falsehood, duplicity, contradiction;
we both conceal and disguise ourselves from ourselves.*

Blaise Pascal*

Who are we? We are self-righteous, per John Locke, who wrote that "everyone is orthodox to himself."[234] Our arrogance is ironic, given how hypocritical we can be, per Benjamin Franklin: "Mankind are very odd creatures: One half censure what they practice, the other half practice what they censure; the rest say and do as they ought."[235] Our hypocrisy is not transparent to us, because we are extraordinarily adept at deceiving ourselves, per George Orwell: "We are all capable of believing things which we know to be untrue."[236]

We are a combination of many things that do not tidily fit together. But we make them fit, because we are expert at deceiving ourselves. Consider this experiment conducted by psychologist Dan Ariely:[237]

> Participants were given a multiple-choice IQ test with the answers to the questions listed at the bottoms of the pages. Unsurprisingly, most people looked at the answers before completing the test; they rationalized that they were merely confirming the response they planned to make. But the more surprising part of this experiment came afterwards, when a second, different test was administered, one that did not include the answers. Naturally, the participants' scores in

---
* *Pensées*, Penguin, 2003.

the second test were lower. The interesting twist, however, was that for the second test, they were asked to predict what their scores would be and were told they would be paid a bonus based on how accurate their predictions were. Most test-takers predicted scores that were higher than what they ultimately achieved. If they did well in the first test, they predicted they would do just as well in the second answer-free test. So the test-takers did not perceive that their access to the answers in the first test had any influence on their scores, despite their conscious, but rationalized, cheating.

Self-deception may have evolved as a means of coping with the stream of competing impulses and motivations that constantly well up within us, which our conscious awareness struggles to make sense of and cope with. We deal with these impulses in the most efficient way that we can: in situations that require us to triage our conflicting motivations, we suppress certain desires and beliefs in order to prioritize others. We feel disgust when someone throws a wrapper out a car window but give ourselves a pass when we spit a piece of gum onto the ground when no garbage bin is in sight; we feel enraged at being betrayed, while excusing ourselves for indulging in an "unavoidable" indiscretion; we chastise our children for lying, yet constantly concoct elaborate fibs to extract ourselves from many uncomfortable predicaments.

Our self-deception allows us to be many contradictory things at once:

- We are not simple, unified, consistent personalities. Each of us is made up of multiple selves.
- The contradictions between these selves are opaque to our automatic, largely subconscious Type 1 thinking; our hypocrisy is invisible to ourselves (although rarely to others).
- This opacity can weaken our willpower and undermine our objectives.

In short, we are faced with the challenge of dealing with *two* World #2s: not just the one that resides outside us but also the one that lies within us. We are more complex than our ancestors were: we have grander ambitions for ourselves and more complicated existential conflicts to contend with, as will be explored in chapter 19. Our expectations of how fulfilling life should be are much higher than those of our ancestors a handful of generations ago when most of us were farming the land for sustenance and trade. As our base needs have been met (for sustainable food and shelter), we have moved our focus to higher-order needs (for personal meaning). Increasingly, the fear of death is not nearly as pronounced for us as the fear of not taking advantage of the life we have been given. We are more afraid of not living — of wasting the opportunity for a meaningful life — than we are of dying.

Smarter decision making requires an awareness of this extra layer — this *inner* layer — of World #2 complexity: the complexity of self.

17.1 Multiple Selves

Plato and Sigmund Freud had similar notions about the constituents of "self": how our minds (or soul, in the case of Plato) are divided into three distinct parts. Freud posited a moral super ego, a rational ego and an instinctual id,[238] a tripartite model that loosely mirrors Plato's intellectual charioteer, rational horse and irrational horse.[239] Other thinkers have been critical of the notion of a constant, fixed self. Friedrich Nietzsche insisted that "self" is not a fixed substance because individuals are always evolving.[240] Jean-Paul Sartre argued that we are the sum of our choices and therefore always in a constant process of becoming.[241] Still others were skeptical about any notion of the self. Buddha believed that any sense of self that we have is illusory, manufactured by our minds to structure our subjective experiences. And David Hume applied the same skepticism to the notion of self that he did to the concept of causality: despite having a notion of a continuous self, we cannot actually locate such a thing anywhere, he insisted, since all we can observe is a collection of fleeting thoughts, feelings and memories.[242]

These philosophers have been at least partially vindicated by recent research in cognitive psychology, which argues for an anti-Cartesian view of self. While philosopher René Descartes assumed that the self is unitary and transparent to its proprietor,[243] there is, in fact, neither a single-self-identity nor a tidy, three-part version; "self" is a constantly evolving series of *selves* that show up in different situations and over extended periods of time. As essayist Michel de Montaigne noted, "There is as much difference between us and ourselves as there is between us and others."[244] There is no single, "true self," because we adapt and alter ourselves for each situation, all the while struggling to reconcile conflicts between our many motivations.

It is not until around age five that we experience a unified and coherent sense of self as identity. But how unified and coherent is it really? Numerous experiments conducted on split-brain patients by Michael Gazzaniga and others paint a picture of self-identify as a constructed personal narrative that we create from the chaos of our perceptions, memories and behaviours. Gazzaniga suggests that there is no central command centre making final decisions; there is only a massive number of parallel series of systems that integrate in no particular place in the brain but are distributed across it.[245] We may view ourselves as a unified self, but, per Daniel Wegner, this is an illusion perpetrated by what he calls the "Great Selfini," which is the part of us that assumes we have complete freedom to decide our course of action, defying causal logic by positing a mysterious, deciding self that is completely free of external influences.[246]

As Wegner himself acknowledges, even if a unified, freely operating self is an illusion, it is a notion that will remain with us. We need a sense of a free-willed self in order to operate, especially in communities in which we hold one another responsible for our actions. The extent to which we exaggerate our sense of being unified, fixed selves, however, is surprising. Psychologist Bertram Forer administered tests to students and then delivered assessments to them that he purported were based on analysis of their tests.[247] They were told that the analysis revealed their distinct, individual personality descriptions. The catch was that everyone got the same description; for example, "At times you are extroverted and sociable, while at other times you are introverted and reserved." In what is referred to as the "Forer effect," the students were convinced that the test accurately depicted their true, distinct personalities, even though the assessments were generic. The entire horoscope industry is based on people's willingness to accept generic descriptions and predictions as uniquely determined by birth dates; believers in astrology easily disregard the anomalies and embrace the elements that appear to fit.

How well does the following description fit most of us?

You are competent and reasonably confident in your abilities, but an underlying insecurity dogs you, surfacing from time to time, especially in social settings involving large groups of people. You have a well-developed sense of humour and enjoy the company of people with whom you share a delight in the oddities of other people and situations. While you are generally satisfied with your day-to-day routine, you do not feel as fully challenged or engaged by life as you think you are capable of being. You are hard-pressed to ascertain exactly how to make your life more fulfilling, and even more so to find the time and energy to take the steps to create a richer life for yourself. While not unhappy, you sense that "there should be more." And you are fearful that, as you age, you are somehow losing the opportunity to explore and discover this "something more."

These words do not reveal a specific insight about any one individual but attempt to capture generic observations about all people. If it was a horoscope reading, it could be appropriated by any number of people with different astrological signs.

Notwithstanding the skeptical view that a persisting self-identity is somewhat of an illusory construction, most psychologists subscribe to the five factor model of human personality, also known as the big five personality traits. The traits are collectively referred to as OCEAN:

- openness (to new ideas and experiences);
- conscientiousness (about accomplishing tasks and achieving goals in a timely and efficient way);
- extroversion (the degree to which social interaction energizes or depletes);
- agreeableness (toward others);
- neuroticism (in the form of experiencing unpleasant and unstable emotions).

Differences between people within this model are considered to be about 50 percent attributable to genetic inheritance, with extroversion/introversion being the most highly heritable trait. Whereas introverts need time on their own to decompress, reflect and re-energize, extroverts need social interaction to talk through their ideas and energize themselves. While each of us tends to gravitate more strongly to one or the other, the vast majority of us show both extroverted and introverted tendencies. Few extroverts are completely dependent on others for all of their stimulation, and most introverts need a degree of social engagement to feel rounded. Most of us are not categorically just one type: despite our leanings, particular situations can influence how our traits manifest themselves.

Situation-Dependent Personalities

Experiments conducted by psychologist Kennon Sheldon demonstrate how our reactions and behaviours can be highly dependent on the roles we perceive ourselves to be playing (reflecting "cross-role variability").[248] For example, a person may have extroverted tendencies in his role as a friend, but introverted tendencies in his role as an employee. Sheldon's research does confirm that there are underlying personality dispositions that surface across similar situations, but it also confirms that there is no "true self" consistent in all situations.

Walter Mischel was one of the early psychologists who proposed that personality traits are highly influenced by particular situations, and that any consistency in a person's personality should be defined in the context of the types of situations rather than as monolithic characteristics that anchor a person across all situations.[249] A person may demonstrate conscientious tendencies in her professional life and be disorganized and chaotic in her personal life; be calm and stable in managing a crisis at work but anxious and volatile in dealing with her family. Mischel's work in 1968 built on earlier findings in the psychology of situational ethics, pioneered in the late 1920s by Hugh Hartshorne and M.A. May, psychologists who demonstrated that children will be honest in some situations but not others, and that character traits, therefore, are highly context-dependent.[250] The stability of our personalities is reflected in similar situations, but not necessarily across different situations.

Past, Present and Future Selves

Complicating the view of ourselves as multiple rather than unitary is the dichotomy, as proposed by Daniel Kahneman, between our "experiencing self" and our "remembering self."[251] Say you have a wonderful vacation but a terrible day travelling home because of a missed plane connection and lost luggage. In its assessment of the entire holiday, your remembering self will be highly influenced by that last, horrible day, notwithstanding the nice time that your experiencing self had on all the other days.

Kahneman's experiments on pain reveal that our actual experience is very different from our remembered experience: the latter skews heavily toward intensity and recency (like the bad flight home), with little emphasis on duration (like the entire week of relaxation). Kahneman concludes that the "real you" is more of your remembering self than your experiencing self, because the remembering self persists in time, even though it distorts the experiencing self's actual experience.

Psychologist Daniel Gilbert has run experiments to assess the reliability of people's predictions of their future selves.[252] The results reveal that we are very poor forecasters of what we will be doing and how we will be feeling. We underestimate how much our lives are likely to change, tending to think of the past as much more tumultuous than the future is likely to be. We also acclimatize to change far more rapidly than we expect: the thrill of winning a lottery does not persist, nor does the anguish of losing a limb. And what we think will bring us satisfaction — promotion, more money, having children, a new lover — virtually never delivers the sustained joy, sense of accomplishment or peace of mind that we anticipate.

So when we self-examine, we are confronted with the question of who we "really" are. The person whom the neighbours would describe as chatty and friendly, or the person screaming obscenities at the slow driver in front of us? Are we the extrovert who entertains everyone at the office party, or the introvert who struggles to make conversation at the park with strangers? Are we the person who actually had a great vacation, or the person who remembers it as disappointing? We typically do not perceive our multiple selves, because of a certain degree of opacity that masks the contradictions in our personality.

17.2 Cognitive Opacity

Physicist Richard Feynman noted, "You must not fool yourself and you are the easiest person to fool."[253] It is the opacity of Type 1 thinking that makes it easy to fool ourselves and easy for us to be hypocritical. Yet the inner conflict that arises from competing but hidden beliefs and motivations can wreak havoc on our psychic well-being. Pascal wrote, "The heart has reasons that reason cannot know."[254] We are pulled by motivations, the roots of which are not always clear to us. Decades before Freud wrote about the unconscious, philosopher Arthur Schopenhauer posited that the largest part of our thinking and motivation is hidden from us, repressed below the level of our conscious awareness because of its unsavoury nature.[255]

Opacity is partly what enables Type 1 thinking to be so fast; it is automatic, not slowed down by careful, conscious deliberation. Alongside Jonathan Haidt's research on people's inability to give reasons for certain scenarios that they find morally repugnant (reviewed in 11.1), the research of Philip Tetlock is also relevant. He surveyed people on their reaction to hypothetical proposals such as allowing potential adoptive parents to bid for babies of their choice.[256] The majority of subjects could not offer a rationale for their reactions, notwithstanding the strength of their reactions. Most of our moral views are deeply embedded in us and are not necessarily amenable to logical analysis; in fact, so deeply ingrained are they that the areas of the brain activated when we are morally disgusted have been identified as the same ones that are activated when we see rotting food.

The opacity of our thinking has led many cognitive scientists to assert that we are consciously aware only of what our subconscious thinking deems important enough to elevate to consciousness. As David Eagleman suggests, our subconscious processing will surface information to consciousness on a need-to-know basis only.[257] Going even further, cognitive psychologists like Timothy Wilson are convinced that we have little to no access to subconscious processing: awareness of our own cognitive processes is evolutionarily new, so we have no reason to assume that this new awareness has direct access to the old, subconscious part.[258] Wilson points to many self-perception experiments as demonstrations that we infer our internal states from our behaviour: we blindly confabulate reasons for our actions when, in fact, it is the situations themselves that evoke certain actions and behaviours. Wilson differentiates self-revelation, which we mistakenly think that we are engaged in, from self-fabrication, which is what we do most of the time. Our conscious Type 2 thinking is adept at confabulating reasons for our behaviour, creating a personal narrative about ourselves that does not necessarily reflect the actual subconscious thinking that engenders the feelings that we experience and the behaviours that we undertake as a result. We are vulnerable to the "introspection illusion": we mistakenly believe that we have direct access to our subconscious processes and motivations when all we are doing is inventing stories to explain ourselves. This form of confabulation is evident in how we justify our choices, as demonstrated in Petter Johansson's "choice blindness" experiments, outlined in 4.3.

Cognitive opacity was not a show-stopper for Freud, who built his psychoanalytic system on the theory that our subconscious is accessible, the entire purpose of psychoanalysis being to surface the repressed thoughts that cause anxiety. Freud believed mental disturbances to be the result of repressed sexual desires: the transfer of erotic interest in external objects to a below-conscious internal world. His one-time protégé, Carl Jung, disagreed with him that contact with the outside world was largely sexual. For Jung, mental illness was characterized by a disunity of the personality, a disassociation of one's sense of self among disparate elements. Just as Freud believed that the role of the psychotherapist was to surface unconscious desires, Jung advocated the importance of bringing to awareness the hidden elements lurking in the unconscious: only then could the divided self could be united. Both men viewed neurotic symptoms as signalling a psychological malfunction, just as pain signals a bodily malfunction.

While few psychotherapists currently espouse Freud's or Jung's theories and techniques in their entirety, conventional psychotherapy continues to base itself on uncovering obscured inner thoughts that cause dysfunction. Most accept the two psychoanalysts' starting assumption that a divided self is an unhappy one, and that a unified self is healthier and functions more productively. But the opacity of our subconscious thinking makes the unification of self one of the greatest challenges each of us faces.

Hypocrisy

Opacity facilitates the divisions between our multiple selves, making it effortless for us to be hypocritical. The most obvious form of our hypocrisy is the contrast between what we profess and what we do. Chris Argyris distinguishes between "espoused theory" and "theory in use": the former represents our *stated* beliefs and values, while the latter represents the theory by which we actually operate, inferred from our behaviour.[259] I may profess that I always give people a second chance if they wrong me, while my actual practice is to shun them immediately after the first perceived transgression.

The difference between "espoused" and "in use" can be invisible to us, especially in the case of deep-seated, unstated beliefs that reside below our awareness. Contemporary experiments reveal surprising insights into how our stated beliefs belie other deeply held beliefs that we are unaware of. Social psychologist Anthony Greenwald and two colleagues employed the Implicit Association Test (IAT) in experiments designed to reveal the difference between our explicit and implicit beliefs.[260] The tests demonstrated that even the most fair-minded people typically show a discrepancy between their conscious (explicit) beliefs and their subconscious (implicit) ones. When people's explicit beliefs match their implicit ones, they can complete the tests quickly, but when they do not match, they hesitate, stumble and proceed more slowly. For example, when reading a list containing names of people as well as scientific terms, they might be asked to say "hello" for all male names and all scientific terms. Some people get through this test more quickly and smoothly than when they are asked to say "hello" for all female names and scientific terms. This discrepancy reveals a closer association in their subconscious between men and science than between women and science; hence, there is a gap between their explicit thinking ("I do not discriminate") and their implicit thinking ("Scientists are men"). IAT research reveals that half of test-takers associate men with science and women with arts, even though the majority, when asked afterwards, insist that they do not believe that men are more science minded. The research also demonstrates that about 70 percent of test-takers have implicit associations between negative concepts (such as hostility or failure) and non-white people, despite the test-takers' insistence that they do not differentiate whites from non-whites in their minds.

Hypocrisy often arises, therefore, from the contradiction between various beliefs buried below the level of our awareness. Because these beliefs are subconscious, they are difficult to reconcile. These conflicts are no more obvious than in the case of our personal values — our individual sense of morality.

Moral Hypocrisy

According to philosopher Bertrand Russell, "we have two kinds of morality side by side: one which we preach but do not practice and another which we practice but seldom preach."[261] This discrepancy is partly driven by poor execution on our part: we are not able to follow through on our convictions for lack of discipline ("It is absolutely wrong to litter, but I can't find a garbage bin"). But we can be moral hypocrites without even knowing it, just as the IAT research reveals. Management professor Max Bazerman refers to this gap as "bounded ethicality," because we think that we are acting in ways that are consistent with our values but we are not.[262] We are bounded because we are unaware of how our actions differ from our belief systems. We may feel very strongly about equality of opportunity, but subconsciously favour men over women in our professional interactions, which influences those we hire, those we make eye contact with, those we listen to more carefully and so on.

Psychologist Joshua Greene's trolley experiments are designed to bring to the surface the inconsistent thinking that we engage in when we assess questions of ethical import.[263] He asks subjects to imagine a scenario in which a runaway trolley hurtles down its track toward five people who will be killed unless the subject throws a switch to divert the trolley to a different track. The alternative track has one person on it who will be killed if the switch is thrown. Most subjects indicate that it would be ethical to save five at the expense of one. Greene then asks the subjects to imagine a similar scenario, except that the only way to save the five people is to push a large man off a bridge onto the tracks: he will be killed, but his body will stop the trolley, saving the other five. Most subjects indicate that they would *not* endorse this move, citing it as morally impermissible.

Greene's thought experiments suggest that the logic that binds our moral intuitions is quite loose. Greene has demonstrated that different brain circuits are engaged when we reflect on the utilitarian body count (saving five at the expense of one), compared with when we reflect on actively killing someone (pushing someone to his death). He argues that the emotional system usually wins out over the reflective one. It appears that we may invoke different Type 1 brain systems for different aspects of moral dilemmas. His findings jibe with those of Antonio Damasio, whose work with brain-injured patients reveals that it is easier for a person to resolve moral questions if the part of her brain that generates emotional responses is not functioning properly.[264] Internal conflicts do not arise when emotional gut reactions are stymied.

It is not just anatomical brain discrepancies that may create conflict: our moral belief systems are a patchwork of distinct ethical intuitions, three in particular:

- a strong sense of what is naturally and inherently wrong alongside the duty to do the right thing (deontology);
- an equally strong sense of weighing the good and bad consequences of a particular action to determine its ultimate ethical quality (consequentialism); and
- a sense of acting out of virtue by displaying moral character over time, built by ongoing acts of moral uprightness.

Deontology, as espoused by one of its earliest and strongest proponents, Immanuel Kant, requires that a good person follow compulsory moral laws.[265] This perspective contrasts the test invoked by consequentialism, which determines whether the consequences of a particular action generate more beneficial than harmful effects. Jeremy Bentham[266] and John Stuart Mill,[267] who were the early proponents of utilitarianism — the most common form of consequentialism — argued that a moral imperative does not derive its credibility from being inherently right but from maximizing happiness and minimizing harmfulness to the greatest number of people. In contrast, Aristotle argued that it is not the discreet actions of a person that define virtue, but the quality of character developed over time through habits.[268]

These three frameworks have long historical traditions and are the main streams of ethical thinking in Western culture. They are all reflected in our intuitive sense of morality, even though they are often at loggerheads.

Following is a snapshot of the three perspectives, as articulated by their early proponents.

Aristotle: The key to morality is developing and leading a virtuous life.

Bentham: We can be much more specific: the key is how much happiness or unhappiness is being created. We can quantify the consequences of any activity on this basis.

Aristotle: Clearly, actions that are harmful to others are not virtuous, so our perspectives are not mutually exclusive. But your methodology of simply tallying the pluses and minuses of a given action is far too narrow. There is virtue in intellectual pursuit, in friendship, in balancing the extremes of our passions. A virtuous life cannot be measured by a single act or single moment because it is nurtured over a lifetime through the habits a person creates, by the choices he makes every minute.

Kant: While the utilitarian view is indeed too narrow, the concept of virtue is perhaps too broad. What both of you are neglecting is the undeniable, inherent goodness of certain activities. The real test of morality is whether an action can be universalized. Moral laws are self-evident by virtue of their ability to be generalized as a categorical imperative: the ethical good is that which you can make a universal law for everyone at all times.

Aristotle: The complexities of human life cannot be judged by simple measures any more than a jagged edge can be measured by a straight ruler. We cannot apply inflexible rules to the complexity of practical ethical dilemmas. Just as the architect needs a flexible measuring strip to bend and conform to the jagged edge of a stone, so too do we need a flexible conception of goodness that bends and conforms to the vagaries of our day-to-day lives.

While these notions are complementary in many respects, they can also lead to very different conclusions about how we gauge what is right or wrong. Aristotle, it is worth noting, was careful to acknowledge that any simple, cut-and-dried approach to segregating good from bad cannot accommodate the complexity of human dilemmas. The moral character of an individual evolves from her initial decisions that set her on a particular path in which she makes more decisions that reinforce that path. Giving in to temptation sets us on a path of further decisions in which we are likely to yield to desire, whereas practicing self-discipline leads us more directly down a path toward larger, long-term goals.

The power of path dependence, discussed in 14.6, means that one of our greatest ethical challenges as individuals is curtailing our less constructive desires, since small decisions can set us on paths that we did not plan on taking, leading us to become people that we did not plan on becoming. Self-discipline keeps us on paths toward long-term goals that we have chosen, but self-discipline does not come easily to organisms containing multiple, competing selves.

17.3 Bounded Willpower

Nowhere is the complexity of self more apparent than in the bounds of our willpower, which is presumably what Aristotle had in mind when he said that "the hardest victory is over self." *Akrasia* is the Greek word for lack of command over oneself, which Socrates, according to Plato, blamed on a lack of knowledge: his view was that if you know the right thing to do, you will automatically do it.[269] There is near-universal agreement among psychologists today that Socrates was wrong. Most of us know that we should eat better, exercise more, balance our lives better, be more present in the current moment, but that knowledge does not necessarily translate into action: knowing is not the obstacle so much as putting knowledge into practice. Why? Because we are multiple selves that do not always work in harmony.

Want-Self Versus Should-Self

These two characters are usually separated in time: my want-self is motivated by immediate rewards; my should-self is motivated by a longer-term vision of what I aspire to be. Psychologist George Ainslie uses the term "hyperbolic discounting" to describe how we weight rewards in the future: we value immediate rewards highly, but their value deteriorates quickly as the rewards are delayed for extended periods.[270] And the degree of value reduction starts to plateau as the length of time extends. So one hundred dollars gifted to us today produces much greater satisfaction than knowing we will receive a gift of one hundred and fifty dollars a year from now. And one hundred and seventy-five dollars in a year and a month from now has about the same value as the hundred and fifty in one year. After the initial deferment, any further delay in receiving the reward is negligible, because we do not discount future rewards at a constant (exponential) rate, but in an exaggerated (hyperbolic) way.

So large is the discrepancy between the value that our want-self places on what is immediately available (e.g., more dessert), compared with the value our should-self puts on longer-term rewards (e.g., a svelte physique), that the should-self struggles to compete. This "intertemporal bargaining" between selves is unfairly skewed in favour of immediate wants: even after suffering the later regret of eating too much, drinking too much, skipping a workout or watching too much TV, we find ourselves repeating these behaviours again and again when the temptation is immediate and the alternative reward is in the future.

Experiments run by psychologist Roy Baumeister indicate that willpower is a limited resource.[271] The more of it that we "use up," the weaker our self-control muscles become in the short term, until we have a chance to replenish our strength with a break from exercising our willpower. "Ego depletion" is how Baumeister describes our reduced willpower and decision-making autonomy as our internal resources are depleted by resisting temptation. If we are tired or hungry, we are even more at risk of mindlessly deferring to the strength of our short-term wants. To make matters worse, we are poor predictors of our vulnerability to ego depletion: we badly overestimate our ability to control our impulses and battle temptation. This overestimation (referred to as "restraint bias") means that we often put ourselves in precarious situations, such as all-you-can-eat buffets and turning on the TV "for just a minute before I go to bed." We habitually underestimate the temptation, because these "hot" impulses are hard to recall when we are in "cold" states. The term "empathy gap" refers to our inability to properly assess the power of our impulses when we are not experiencing them, making it difficult to get our various selves on the same page at the same time.

We know that extraordinary performance requires years of deliberate, painstaking practice. Why are only a minority of people able to make this commitment, for example, practicing a musical instrument for four hours each day, enduring the frustrating and exhausting process of perfecting their skill, while the rest of us tire easily of the tedium and discomfort? Why do some kids sit and study for a couple of hours each night, while others cannot get through fifteen minutes without becoming distracted? Where do the "stars" get their willpower?

Psychologists have identified a personality trait of persistence (sometimes referred to as "grit") to explain the star performers among us: a mindset that performance is never good enough, that it always needs relentless improvement. It is still a mystery why this trait is much more pronounced in some than others, but presumably it is highly heritable: experiments show the varying degrees of delayed gratification that different rats are capable of, even when they are raised in identical environments. Can persistence be taught? Studies by psychologist Angela Lee Duckworth and others suggest that perseverance, in the form of self-regulation, can be encouraged in children and adults, but preliminary research suggests that the range within which an individual can be coached to stretch their persistence is limited.[272]

Rationality and a Failure of Willpower

What does it mean to be rational? Rationality is yet another concept used in different ways by different disciplines; although we have a sense of what it conveys in lay conversations, there is no precise definition that is universally adhered to by economists, psychologists, sociologists and political scientists. Part of the confusion arises from the many forms that rationality can take, some of which overlap: epistemic (logical and evidence-based thinking); ecological (behaviour that suits a particular environment); normative (acting according to accepted rules); and instrumental (acting in furtherance of goals).

The various definitions do, however, differentiate rationality from IQ-type intelligence. We do not associate IQ with the hallmark of what we consider to be rational behaviour: the ability to overcome automatic thinking and motivations in order to act in accordance with beliefs, values and goals that supersede our instincts and intuitions.

> Being rational, in this sense, is no easy feat because we have multiple, often conflicting impulses to contend with. Witness Bill Clinton and Eliot Spitzer: both of them are brilliant men who took career-threatening, "irrational" risks for the sake of a sexual dalliance. Either they failed to assess the risk of being caught (which was very high for both), or they failed to override their appetites: both types of failure reflect a lack of rationality at the time of their transgressions.
>
> The very thing that advantages us over other animals — our ability to sort out multiple goals so that we can act in rational pursuit of those we prioritize — makes life more challenging for us. We have a lot more conflicting goals than other animals do, and our goals shift and change, depending on particular situations and points in our life cycle. Unlike other animals, our genetically driven goals are often in conflict with our personally created ones. Managing conflicting goals is one of our greatest personal challenges — one that relies on a heavy dose of metacognitive Type 2 thinking.

Emotional Willpower

Oversimplifying is the first grave mistake we make when confronted by complexity; overconfidence is the second. And there is a third: based on our overconfidence in our oversimplified conclusions, we overreact. We are hard-pressed to invoke the willpower *not* to be emotional, and our strong reactions are usually mismatched with the subtleties and ambiguities of complexity.

Our basic emotional response system worked very well for generations of ancestors for the obvious reason that emotions trigger the fast and decisive responses needed for meeting many challenges. Even today, our strong emotions constrain the huge number of possible responses in a given situation to a manageable few. Emotions get us into the range of likely useful responses, with a strong bias toward danger avoidance. Emotions get us to jump into action, as opposed to taking the slow cognitive route of examining each possible response available to us in detail.

This is why emotional states are not hesitant or tentative, but absolute, rigid, purposeful, conclusive, staunch, inflexible, decisive, monolithic and intensely resolute. It is also why they are more often negative than positive. As discussed in 16.3, we have evolved a strong negativity bias to our emotions to help us avoid the risks that surround us. Danger is always lurking and can never be fully eliminated, in contrast to pleasure-inducing situations, which are useful only to the point of satiation. The marginal value of eating and sex declines rapidly once we have had our fill, but the marginal value of avoiding danger never declines. When it comes to negative emotions, we rarely underreact: we are quick to invoke fear, disdain, rage and indignation at even the slightest perceived danger, injustice or offence.

> ### What Is Emotion?
>
> Emotions are neural chemical reactions that occur automatically, originating in subcortical brain structures that evolved long before the higher brain functions that allow us to regulate our reactions to a degree. Emotions are the bodily reactions that follow subconscious Type 1 thinking; they are the brain-induced physical responses that motivate us to act. Cognitive scientists differentiate emotions from feelings: feelings are the conscious awareness we have of our emotional bodily states; they are the result of our becoming consciously aware of our emotional system in action; they are the conscious icing on the subconscious emotional cake. The emotion is the racing heart, adrenalin surge and hyperawareness that accompany our perception of danger; the feeling of fear is the conscious experience of these bodily changes.
>
> Our emotional reactions are triggered by lower brain functions but moderated by higher brain functions. In other words, our emotional responses are ultimately a function of both subcortical activity (in the amygdala) that initiates the process and prefrontal cortex activity (in the OFC and VMPFC) that moderates the reaction. So we do have an ability to curtail our emotional reactions, but it is a delayed and often weak moderation. For example, frustration at work may ignite instant anger that we consciously suppress, whereas the same frustration at home, where we are less conscientious about managing our irritation, may result in a less muted response.

Positive emotions, in a mild form, win the *frequency* contest: most of the time, neutral to mildly positive emotions dominate our moods (a state of "euthymia," distinguished from a state of depression, or "dysthymia"). What is referred to as the "positivity offset" keeps us on an even keel much of the time. Negative emotions, however, win the *intensity* contest: when negative emotions arise, they are more powerful than positive ones. This imbalance between positive and negative intensities is the source of our negativity bias. Pain is deeper than pleasure, which is why prospect theory, as reviewed in 16.3, indicates that we feel loss more intensely than we feel gain. And strong negative emotions are much more persistent than strong positive ones: separate studies by psychologist Barbara Fredrickson and Roy Baumeister demonstrate that negative emotions tend to endure much longer, and strong positive feelings ("hyperthymia") tend to be fleeting.[273] Not only are negative feelings deeper, but more of them are available to us than positive ones: pain comes in many flavours, pleasure in but a few.

We share emotional imperfection with all mammals: the intensity and duration of our reactions are often overblown. But the imperfection of the human emotional system in the face of complexity is particularly problematic in two ways: first, strong emotions can dramatically reduce our effectiveness in the multitude of World #2 situations in which quick and strong reactions are not suitable; and second, the asymmetry in the intensity of our emotions, skewed as it is to negativity, can make us unnecessarily miserable.

Our negativity bias makes us extremely vulnerable to "tilting," the term used by poker players to describe emotional reactions that interfere with their ability to play the game. While the fear that motivates us to jump out of the way of a speeding car and the anger that motivates us to defend ourselves against attack are helpful emotional reactions, what about the outrage that motivates us to scream and honk at slow drivers, or the terror we feel about giving public speeches, or the debilitating anxiety we suffer writing exams, interviewing for jobs, flying in airplanes, changing jobs or moving homes? Rising blood pressure in a traffic jam does not get us to our destinations any faster, but more than likely will shorten our number of driving days. A dry mouth is not helpful when we are giving a speech. Digging our fingernails into the armrests on a plane does not reduce the already very low odds that it will crash, but it will reduce the energy available to our body to fight whatever germs are circulating in the confined space of the cabin.

Negative emotions do not lose their power as we age. A father is no less devastated at being passed over for a promotion than his daughter is at losing a favourite toy; a mother's feelings when stuck in traffic are no less intense than a five-year-old's frustration in putting a difficult puzzle together. Our emotional reactions do not dissipate any more quickly, either: the adrenalin that pumps into our bloodstream when we are triggered clears out gradually, even if the cause of the emotional reaction is eliminated (which is why spousal arguments often persist long after the initial problem is resolved — our bodies still feel ready to fight). It is not just the one-off conflicts that can plague us: our proclivity for churning negative thoughts and feelings repeatedly without resolution is referred to by psychologists as "rumination." It is a form of thinking triggered by a perception that we lack control, which subverts the confidence-inducing stop to our thinking that we otherwise rely on. We ruminate about conflicts with others, mistakes we have made and perceived injustices, as if relentless analysis will eventually lead us to insights or solutions, when, most of the time, there are none and excessive rumination just leads to anxiety or depression.

The bluntness of our emotional system is not well suited to our overpacked lives, which are riddled with every conceivable complication and frustration. Our intense and asymmetric emotional reactions are mismatched with much of World #2.

17.4 Managing Ourselves

Coping with ourselves — each of our respective (numerous) selves — can be at least as challenging as coping with other people. Much of the time, we are just as confusing and disappointing to ourselves as we find others to be. The brain–world gap is probably nowhere more obvious than in how unnecessarily miserable we make ourselves (and others) because of our failure to rein in our strong emotional reactions and because of our underdeveloped ability to self-soothe. The metacognition that Type 2 thinking furnishes us with gives us at least some awareness of our vulnerability to overreaction. But it takes substantive cognitive effort and discipline to manage our emotional negativity bias — a lot of Type 2 to moderate our automatic, Type 1 emotional responses.

The Stoic philosophers, led by Zeno of Citium (who kicked off the philosophical movement known as Stoicism around 300 BCE), espoused the view that the natural world is governed by rational principles; therefore, we must accept our fate, including our own mortality. When our emotions pull us away from the tranquil equanimity of acceptance, it is because they are based on false judgments. For example, if we are feeling greedy or jealous, it is because we mistakenly believe that wealth is worth accumulating. The Stoic solution is to make our emotions subject to our reason, reining them in with a more rationally accurate understanding of the world.

The Stoic prescription for emotional management was not dissimilar from the one espoused by Buddha, a continent away and over a century earlier. He also viewed negative emotions as arising from a misunderstanding of the true nature of reality, but he focused on the importance of recognizing our inseparability from an ever-changing, contingent universe. We are anxious and upset when we perceive ourselves as separate from the world, and when we strive to find happiness and fulfillment from it rather than simply being part of it and meditating on this truth.

A key component of managing ourselves is an insight shared by Stoicism and Buddhism: much of what we react to emotionally is noise that we mistakenly treat as signal. What we perceive as threat is often fleeting and unsubstantial, because our Type 1 thinking is programmed to err on the side of danger detection. In a busy, increasingly complex world, more and more noise taunts us; the Stoic and Buddhist recognition and control of our vulnerability to this noise is increasingly relevant to our emotional well being.

It is because negative emotions can be so damaging to us that working through them has been the focus of psychotherapy in all its various forms — Freudian, Jungian, Adlerian, Gestalt, cognitive-behavioural, interpersonal, etc. More recently, the advent of "positive psychology" has shifted focus to methods of directly creating positive emotion, to complement our efforts to curtail negative emotion. The aim of most therapies is to reveal our hidden assumptions, beliefs and values so we can integrate or reject them. Getting all the various parts of our selves coordinated and moving in the same direction is difficult, never-ending work: our multiple selves not only are in conflict much of the time but they are also malleable, shifting with circumstance and age. The operating assumption of these therapies is that our subconscious motivations are not completely opaque; they are translucent. As difficult as it is to access, understand and manage them, the effort can be rewarding, and perhaps is even necessary to leading a productive and fulfilling life. Even if the assessments of our submerged and shifting goals are only best guesses, the aim of these programs is better integration of our multiple selves as a way of managing our complexity: an incoherent, poorly reconciled self-narrative precludes the emotional peace that most of us aspire to.

Each of us will always have contradictory motivations: lose weight and eat more; work less and earn more; be more productive and watch more TV. Part of the battle to integrate is the root recognition that we have no choice but to tolerate a degree of inner conflict that cannot be eliminated. To manage this tension productively, we must mediate between our various selves so we can prioritize one over others when it matters to us: in other words, we must invoke the discipline to do what we want to do when we do not want to do it.

Our Type 1 thinking is inflexible and fast: its aim is to maximize the reproductive probability of our genes. Our Type 2 thinking is more flexible: it can prioritize our personal goals, those not necessarily or exclusively defined by our biology. Type 1 does not care about our overall, long-term happiness; so we require Type 2 to outmaneuver Type 1. Research has demonstrated that committing in advance (in our "cold states") to certain goals puts us in a stronger position to maintain our commitments when we find ourselves tempted (in our "hot states").[274] Pre-commitments have to be specific, because the vaguer the plan, the more vulnerable we are to succumbing to the exceptions encouraged by Type 1. Planning to lose ten pounds in the next month is too general and is virtually destined for failure; committing to avoid all desserts for one month makes it more difficult to cheat.

A pre-commitment is sometimes referred to as a "Ulysses contract," based on the legend of the Greek hero who instructed his ship's crew to tie him to the ship's mast so he would not be tempted by the song of the Sirens who otherwise would lure him to his death. Planning to limit your future options enables you to maximize the leverage of Type 2 thinking: you invoke 2.2 to anticipate your vulnerabilities and 2.1 to implement strategies that circumvent those vulnerabilities. Pre-commitments force us to consider in the present how our future self is likely to react in the face of temptation; they help us mediate between our current and future selves. Our present cold self has more power than our future hot self, so pre-commitments help us bargain from a position of strength.

The present–future selves dichotomy also offers us a backup plan when we find ourselves in hot situations that tempt us and for which we have not established pre-commitments. We can ask ourselves, "How will my future self feel about the decision I'm about to make?" Projecting ourselves into the future so we can look back at ourselves in the present stretches our Type 2 thinking. Understanding our competing motives and mediating between them is possible only with fully engaged Type 2.2 thinking: the highest form of metacognition that we are capable of. Our decisions in the present have enormous influence on our future selves: when we let ourselves off, by confabulating excuses for breaking our own rules, we sabotage our future selves by establishing habits that, as Arisototle admonished, eventually form the character of our future identities. Without deliberating on our choices in the present, those future selves may not match the ones we aspire to be.

18

The Complexity of Others
Our Social Shortcuts

Hell is other people.

Jean-Paul Sartre*

The question "Who are we?" in the previous chapter led to a consideration of the divided and complex state of the self. But the most challenging aspects of any primate's life are its *social* interactions, and, unsurprisingly, human beings have it toughest of all. Our never-ending quest to find common ground with others — a sometimes impossible feat — is the direct result of our complexity. If we had simpler brains and therefore simpler personalities and simpler objectives, our relationships would be more straightforward. Today, our challenge is made all the more difficult by the social complexity of World #2.

Socially, we evolved into increasingly complex groups, from bands of a few dozen individuals (most belonging to one or more extended families), to tribes of up to two hundred individuals (with everyone still knowing everyone else), to chiefdoms of thousands of individuals specializing in certain types of work (in which impersonal economic trade arose), and eventually to nation states supported by a hierarchy of administrative officials (arising only in the last couple hundred years). The extent of our social transition is unparalleled in the animal kingdom: from small, largely egalitarian communities of interdependent individuals to large urban centres that rely on law and government to protect the interests of the citizenry.

---

* *No Exit and Three Other Plays*, Vintage, 1989.

Our social instincts have not kept pace with this incredible change. We still have the basic mammalian instincts that we have always had, for both prosocial and antisocial behaviours. Survival in a harsh environment with many predators fostered intragroup co-operation. At the same time, and for most of our history, there was no government or police force to protect us from each other; a reputation for retaliation was crucial to protect yourself, your family and your food. We are extreme in both our prosocial and antisocial tendencies: we are the most co-operative of any species and the most vicious. We exhibit an outstanding level of self-control, keeping our selfish impulses in check much of the time (a form of "self-domestication" that presumably evolved over time, similar to how some wolves evolved with our training into more docile, domesticated dogs). But we simultaneously exhibit a capacity for mistreating our fellow humans that is almost unrivalled in the animal kingdom ("almost," because, for example, a group of chimps can brutally attack a lone chimp from a different group).

Our extreme co-operative and aggressive instincts represent the paradox of our social constitution. We are acutely dependent on others, while deeply distrustful and hypervigilant of them. We need others to help us not only to survive but also to thrive: numerous studies have demonstrated that our general happiness and longevity are partly the result of consistently rewarding relationships, and that being ostracized is threatening and painful for us. The fewer social relationships we have, the more vulnerable we are to infectious diseases and a shorter-than-average lifespan. Yet, like all primates, we are nervous around, and initially wary of, strangers: our stress response rises in their presence. Adding to the complexity of this dichotomy is our day-to-day selfishness. Notwithstanding our capacity for self-control and co-operation, we are not consistently empathetic: we are very stingy in expending cognitive effort to understand the perspective of others. As prosocial as we can be, there is a reason that the golden rule has been invoked and promoted by many thinkers over the course of the last twenty-five hundred years: it is not necessarily obvious or immediately intuitive for us to treat others the way that we would like to be treated ourselves.

The way that we simplify others — the shortcuts we rely on to quickly assess and respond to them — works efficiently in World #1, where snap judgments are both necessary and effective. Our shortcuts do not, however, accommodate the complexity of others; they do not enable in us the flexibility of response that is needed in World #2.

Our social shortcuts simplify others in two fundamental ways. The first and broader shortcut is tribalism. We are naturally xenophobic — distrusting of strangers — strongly preferring those who are familiar or similar to ourselves. The second shortcut is how we assess the specific personalities of others. We attribute inflexible personality traits to them, giving insufficient consideration to the influence of their particular situations.

18.1 Tribalism

"It would be nice to believe that humans were designed to love everyone unconditionally," notes Jonathan Haidt. "Nice, but rather unlikely from an evolutionary perspective."[275] An instinct for preserving our immediate community was identified by Darwin as survival-enhancing in a world of limited resources. Reproductive advantage accrues to those who co-operate with members of their own group but shun those outside the group who compete for the same resources. The exact environmental pressures that encourage cohesion among members of a group also encourage intergroup rivalry. We are generous toward our own families and other members of our tribe (within-group amity), while hostile toward those outside it (between-group enmity), which makes perfect evolutionary sense. Our altruism, therefore, is constrained: it is limited generally (although not exclusively) to those whom we identify as in-group, a bias sociologists refer to as "parochial altruism."

The depth of our tribal instinct is confirmed by countless studies, including those that have uncovered what is referred to as "own-race bias" (ORB), our proclivity for remembering the faces of people of our own race much more reliably than we do the faces of other races.[276] When we see an own-race face, we take in the entire face and remember it holistically; when we see an other-race face, we analyse distinct features, such as the shape of their nose and the size of the forehead, and remember the separate parts.

The dividing line between "in" and "out" is neither fixed nor permanent: it can move as we get to know people and determine the commonality that we might have. But our assessments of others

— especially our preliminary ones — are based on how well they fit the various tribes that we identify with (religious, national, race, etc.). As an instinctual shortcut, tribalism serves an important purpose in World #1 where we have to make immediate judgments about the likely rewards of co-operating with certain people. Our in-group instincts, however, are not unconditional. We discriminate even within our group, with a very powerful instinct for fairness and a strong instinct for status.

Fairness: Tit for Tat

Parochial altruism is not categorical: it is conditional on reciprocity ("reciprocal altruism"). In its most common manifestation, reciprocity is referred to by decision theorists as "tit for tat." We offer and expect reciprocity, but when our generosity is not returned by others, we withdraw our co-operation. Sociologists have demonstrated in experiments that co-operative behaviours within a group collapse if retaliation for cheating or freeloading is disallowed.[277] The punishment of cheaters appears to be a necessary mechanism for preserving co-operation, because if cheaters cannot be punished, cheating becomes widespread.

So consistent and pronounced is our sense of fairness and our entitlement to it that we are willing to self-sacrifice if that gives us an opportunity to punish those who treat us unfairly. Our willingness to forfeit our own gain to teach others a lesson has been demonstrated in many experiments, the most popular of which is called the Ultimatum Game. In this experiment, two players are given the task of dividing a sum of money between them. The wrinkle in the game is that one player is given all the money with the instruction that she must make an offer to the other player, of whatever portion of the money she wants to share. The other player can accept the offer and take the money, or she can reject the offer, in which case the total sum is withdrawn and neither player earns anything.

The game is designed to reveal how much money is spontaneously offered and what the second player's threshold is for acceptance. The game generates offers that are usually around 50 percent of the total, and the majority of players tend to reject offers below 30 percent of the total. It has been played numerous times in different societies all over the world; results fluctuate around these averages, depending on the size and interdependency of the communities in which the game is played, but offers rarely go above 50 percent, and accepted offers rarely fall much below 30 percent. Most of us would rather earn nothing at all than receive less than 30 percent of the total.[278] (So indignant can we become when offered an amount that we consider unfair that, in a variation of the game where the first player can keep her portion even if her offer is rejected, between 30 and 40 percent of subjects will reject a low offer anyway, just to express their dissatisfaction.)

The instinct for fairness runs deep within us: even sixteen-month-old babies have been observed to prefer cartoon characters who divide prizes equally over those who do not. And primatologist Frans de Waal has conducted experiments with chimpanzees and capuchin monkeys that respond very negatively to getting lesser rewards than their compatriots.[279] Our tribal instinct to punish non-cooperators is based on an ideal, black-and-white conception of fairness that leaves little room for the actual grayness of human nature.

An Instinct for Status

Writer Samuel Johnson proposed that "no two people can be half an hour together, but one shall acquire an evident superiority over the other."[280] We are extraordinarily status conscious.

Our instinct for status probably reflects the transition from a dominance-oriented hierarchy to a prestige-oriented one, once physical fitness became less of a distinguishing factor between people, especially men. Other primates rely on strength exclusively to signal and maintain dominance. Humans, however, not only evolved countermeasures in the form of weapons that gave weaker members a fighting chance against oppressors; we also evolved the social sophistication to outwit our competitors and woo our preferred companions (status seeking could be simply a legacy of sexual selection). Once private property became an element of our economic system, alongside the rise of agrarian societies, hierarchical social structures based on wealth became firmly embedded. (Contemporary hunter-gatherer bands are surprisingly egalitarian, leading some anthropologists to suggest that, before the advent of private property, our ancestors were better able to repress their hierarchical instincts — in both dominance and prestige forms.)

Psychologists have determined that prestige is a reliable indicator of life satisfaction: an individual's contentment appears to be less related to his actual net worth than to his perception of how his wealth compares with that of others. Given the same house size, make of car and money in our account, we are far more likely to feel good if we are among the richest minority in a modest neighbourhood than if we are among the least wealthy in an upscale one.[281] Economist Robert Frank coined the term "positional goods" for those items that we derive satisfaction from based in part on their representation of status.[282] Enjoying a sunset is a non-positional experience, but cars, houses, furniture and technological gadgets are all positional insofar as their value is determined in part by whether ours are more expensive and luxurious than others'. And our comparisons are always close to home: we are less concerned about the status of strangers than we are about those who are proximate — family, friends, colleagues, neighbours. We will attempt to maintain an image of being at least on par with them, with a strong bias for higher status, an instinct that has been transposed clumsily and unproductively into modern World #2.

Tribalism characterizes our macro in-group bias, a shortcut for easy categorization of others as friend or foe. We also have a shortcut for making a deeper assessment of an individual's personality that goes beyond just "in or out": we infer from their behaviours certain dispositional traits, and these inferences are often very shallow because they treat a critical piece of information as noise.

18.2 Assessing Others

It is virtually effortless to blame and highly effortful to self-evaluate. That is the way we are built, presumably because it was a good survival strategy to energize ourselves with anger at the slightest provocation or threat, and not particularly helpful to continually doubt ourselves. This imbalance between blame and self-evaluation is based on an operating principle embedded within our cognitive apparatus as a simplifying shortcut for assessing others: it is known as the "fundamental attribution error."

The most common version of the error reflects the asymmetry between how we assess ourselves and how we assess others: we attribute much more complexity to our own predicaments than we do to those of others. We create space for situational factors in assessing our own behaviour, attributing our actions to the intricacies of these situations (including and especially the impact of others on us). But we interpret the behaviour of others as the *exclusive* signal of their personalities, ignoring any situational factors that may influence them. What we consider noise in our view of others we consider to be a very strong signal in our view of ourselves.

"I had no choice in the matter," we say, quick to explain (and excuse) our own behaviour by referencing the dynamics of a particular situation. Or, "I only lashed out because he was intentionally provoking me," or, "If you were in my shoes, you'd do the same thing." But we are equally quick to explain the behaviour of others by attributing monolithic character traits to them as the sole explanation, disregarding any situational dynamics: "Her arrogance prevented her from being promoted," or, "He is so controlling that he can't let them make the smallest decision on their own."

We deny the complexity of others by invoking a different causal chain for them than we do for ourselves: to explain the behaviour of others, we put the spotlight on the nature of their character, which we infer from their easily observable actions; to explain our own behaviour, we focus on situational causes. Yet situation and disposition are two intimately linked causal factors that are just as difficult as nature and nurture to tidily separate. But in our rush to certainty, we do separate them, and in a way that favours ourselves. Part of the reason for this asymmetry comes back, yet again, to missing information: just as solving other people's problems is easier because we are not privy to the details of their predicaments, as outlined in 10.2, so blaming people is easier, since we are not sensitive to their circumstances unless we search out the details.

While the fundamental attribution error is one of the cornerstones of social psychology, the asymmetry in our assessment of others compared with ourselves is not limited to the situation–disposition distinction; it also extends to other aspects of our thinking about others:

- We hold others to a much higher standard of morality, judging an act more reprehensible when someone else commits it than when we have.

- We view ourselves as rational and reasonable, and others as emotional and narrow-minded.
- We acknowledge our own complexities and often view ourselves as different from how we present ourselves, whereas we tend to assume that others appear as they really are and therefore are fairly predictable.
- We attribute our success to our talents and our failure to situational factors out of our control; the success of others is attributable, at least in part, to good luck, but their failure is due to their personal shortcomings.
- We often attribute greater freedom to others than ourselves: we "had little choice in the matter, considering all the factors" whereas others "could have made better choices."
- We give ourselves the benefit of the doubt but rarely give it to others.

While scientists are reluctant to attribute a full-blown theory of mind to other animals, the irony is that our very own human ToM is quite limited, unless we expend the effort to invoke Type 2 thinking to deepen our understanding of others and their situations. The limit of our intuitive ToM is reflected in how straightforward we assume the thinking and personalities of others to be, whereas, in fact, they are no less complex and no less affected by external, situational factors than we are. While we indulge in self-pitying thoughts like, "She doesn't get me," or "He doesn't even try to understand my point of view," everyone else, without exception, is thinking the same thing about us.

18.3 Managing Others

As with all forms of complexity, the starting point in managing it is a greater degree of Type 2 thinking. The greater our reliance on unexamined Type 1 thinking, the higher the potential for interpersonal conflict. Because the intuitive starting point of Type 1 thinking is naïve realism — that the world, including other people, works exactly the way we perceive it — we set ourselves up for irreconcilable differences with others when they inevitably see the world differently.

World #2 complexity demands a diversity of thought that is not captive to tribal or oversimplifying instincts. It is a hard battle to fight our intuitions about other people; we rely on them constantly because we have limited cognitive power and many people to size up quickly. But without the benefit of a metacognitive override, these automatic inclinations can combine to alienate us from others in damaging ways, if we are not careful. One of the most crucial Type 2 overrides is listening: listening first and listening intensely.

Listening intently, without distraction, is not easy for us, in part because it is new: long before we learned to use language to communicate with one another, we assessed one another by interpreting visual cues that allowed us to infer emotional states and intentions. We are quite expert at reading visual social cues. Whereas most mammals rely to a large extent on their olfactory sense, we are highly visual in our appraisals. We have a greater number of tiny muscles in our face than any other animal, enabling myriad facial expressions that offer an enormous number of subtle cues to others, about how we are feeling, what we are thinking and what our intentions are. (A distinct part of our brain — the fusiform face area — appears to specialize in analysing these facial expressions.)

When we converse, we spend a surprising amount of time not listening to others. Not only are we busy formulating our counter-arguments and responses, but research also indicates that at least 30 percent of the time we are thinking about other things altogether.[283] Most of us are poor listeners because concentrating on someone else's talk takes a lot of energy, both to keep our minds from wandering and to understand what the other is saying, especially when that person is neither succinct nor eloquent. We are much more eager to share our own opinions than to wait patiently for others to share theirs. But by acknowledging the automatic shortcuts that we default to in categorizing and simplifying others, we can start from the humble position that we rarely have enough information to truly understand, let alone judge others. Our job, therefore, is to listen: listen by asking questions and then being silent; listen by holding off on sharing our views until we have invested enough time "doing research."

Listening intensely fosters curiosity rather than certainty, and curiosity gets us closer to understanding. Listening well means starting with the assumption that others are right, because they perceive themselves to be right. Listening to understand means repeating the other's point of view to confirm that it has been accurately interpreted. Sharing our understanding of their perspective forces us to stay in the moment with them, to focus on what they are saying, to suspend conclusion and judgment until we are certain of their view and have gathered as much information as possible.

Listening to understand is a strong antidote to our proclivity for self-righteousness. Writer Robert Wright puts the matter succinctly: "Most of us waste great quantities of time and energy indulging our indignation."[284] This is wasteful because it is usually misguided, unproductive and stress inducing. Yet our social instincts default to forms of indignation quickly and regularly: because our sense of fairness is violated; because our sense of status is compromised; because our instinct for protecting ourselves from criticism is automatic; because our frustration and anger is instant when others subvert our goals or challenge our view of things.

The deep humility that Type 2 thinking fosters helps to mitigate our inclination toward self-righteousness, by subduing it with a broader perspective. Complex thinkers attempt to extend to others the same self-forgiving bias that they offer themselves. They are willing to attribute more complex causality to the behaviour of others, based on a consideration of the intricacy of their situations, instead of restricting the causal story to fixed personality traits.

Exercising Type 2 thinking pays enormous social dividends when we use it to temper our Type 1 automatic responses: we are less inclined to oversimplify others and mismanage our relationships. Irreconcilable differences will always be with us, because of the depth and intricacy of our individual belief systems and values, many of which are not universally shared. Our only recourse is to negotiate and compromise, strategies for which Type 2 thinking is our only hope. Managing the complexity of others is one of many quintessentially human challenges — the focus of the final chapter.

19

The Complexity of Being Human
Coping With the Paradoxes

Life is one long struggle in the dark.
Lucretius*

Wherever we turn, we are confronted by the paradoxes of being human. We ask questions that have no answers. We worry about things that we cannot control. We aspire to goals that we do not have the ability or willpower to achieve. We wish for downtime to "just be" while yearning for more meaningful and active engagement. We second-guess our choices and regret foregone opportunities. We are pulled by imaginary carrots dangling in front of us that, if they materialize at all, are rarely as satisfying as anticipated (if I could just get that promotion; if I could just find the right partner; if I could just make more money; if I could just get my kids to co-operate; if I could just find some peace). One of the harshest of the difficulties that we face is how patently imperfect everything about our lives is, compared with how we envision it "should be." We expect things to go more smoothly than they typically do, from the mundane (red lights and mosquitoes) to the severe (our precarious health, difficult relationships and thwarted ambitions). And we are collectively immersed in public systems that are a constant source of disappointment to us, from our political systems to our capital markets, from our medical systems to our educational systems. To top it off, most of us struggle with the contradiction between the kind of person we want to be and the kind of person we are.

---

* *The Nature of Things*, Penguin Classic, 2007.

This final chapter considers how we deal with the fight we all face, the fight of our lives: the paradoxes of our humanity. The "darkness" that the Roman philosopher Lucretius speaks of, in the epigraph above, was not quite so dire for Blaise Pascal, who wrote, "We sail within a vast sphere, ever drifting in uncertainty, driven from end to end."[285] It is this uncertainty, or mystery, or darkness that accounts for such a significant part of our life struggle.

19.1 Being Human

Uncertainty stresses all animals, but darkness is unique to humans. Suicide in the animal kingdom is rare; many scientists are not convinced that there is any evidence that other animals deliberately take their own lives. Other animals have it easier because they:

- live in the environment for which their instincts were developed and to which they are well adapted;
- do not contend with internally conflicting goals, arising from two competing thought systems (such as the conflict between the craving for sugar and the desire for a nutritious diet);
- are not plagued by the same degree of irrational or self-destructive behaviours (while some self-injurious behaviour is observable, such as birds picking out their feathers, other animals act in ways that are more consistently aligned with their base survival instincts);
- are not burdened with the tension of reflecting on their own negative thoughts and feelings;
- are not plagued by existential questions of meaninglessness and the tension that arises from the pursuit of personal fulfillment;
- do not have brains that come into the world only 25 percent formed and highly vulnerable to postnatal influences like the kind that lead to peculiar human neuroses (such as obsessive-compulsive disorder);
- live in the present moment, without spending a disproportionate amount of time dwelling on past injustices and worrying about future problems.

Unlike other animals, our brains are not designed to focus exclusively on the present moment of our immediate perceptions, but to use our enhanced working memory (autonoetic awareness) to continually survey the past and reflect on the future. According to psychologists Matthew Killingsworth and Daniel Gilbert, our minds wander between 30 and 65 percent of the time, depending on the activity that we are engaged in (unsurprisingly, only 10 percent of the time during sex).[286] The kicker is that our thinking usually wanders to negative territory: mind wandering is biased toward unhappiness. Much of our existential distress originates from how our minds meander to worries, frustrations, anger, righteousness and sadness. Their research reveals that we are happiest when we are *not* daydreaming, and that daydreaming is what humans do 47 percent of the time on average. The paradox does not end there: the less realistic an individual is in her reflections, the less likely she is to suffer from depression. Martin Seligman has documented the correlation between reality distortion and the absence of depression: non-depressed individuals believe they have more control over their lives than they actually do, so they are more inclined toward having a positive view of pleasant events and a less negative view of unpleasant events.[287]

The contradictions that characterize humanness have been noted by many philosophers, including John Locke, who theorized that either we are purely material beings who can think and have emotions, or there are two substances in the world, matter and mind. He believed that both conceptions of human beings were unintelligible, even though only one of them could be true.[288] Pascal was not intimidated by this conundrum (presumably because of his deep religious convictions): he rejected the view that philosophers have traditionally advocated since antiquity, that there is a fundamental struggle between our bodily urges and our rational attempt to "do what is good." Insisting that the human condition is fundamentally contradictory, Pascal believed that it was impossible to reconcile the conflicts between our passions and our intellect. In his final work, *Les Pensées*, published posthumously, he explored the many inescapable contradictions that arise from the human perspective, and concluded that our only recourse is humility in the face of ignorance: "The supreme function of reason is to show man that some things are beyond reason." Pascal's approach to the human dilemma reflected the Christian theology of his late medieval times, but it also anticipated the existentialist movement by two hundred years.

19.2 Coping

Figuring out how to work our way around, or out of, the paradoxes of being human has been a topic of thought at least as far back as the early Hindu Veda texts three millennia ago. Existential philosophers, starting at the beginning of the nineteenth century, picked up the difficult task of dealing with the absurd paradoxes of our condition, and while each thinker recognized the same dilemma, each had his own take on how best to cope. The following hypothetical conversation reveals the contrasting views of a selection of influential existentialists, presented in rough chronological order.

> **Arthur Schopenhauer**: We must not be taken in by the illusion that the world is independent of us. The true, undifferentiated reality in which we are immersed is wicked and hateful. We would be better off to not exist at all. The best we can do is to turn away from the driving force, or "will," that infuses reality since everything we experience is a manifestation of evil.[289]
>
> **Friedrich Nietzsche**: On the contrary: we must not turn away from life, but affirm it! Each of us must dare to be the most we can possibly be, living in the fullest possible way by creating our own values and pushing ourselves to the limits of our capabilities. Few of us will succeed, but we must each strive to live like a superhuman, transcending the mundane herdlike morality and lifestyle that the masses gravitate to.[290]
>
> **Martin Heidegger**: We can neither turn away from the world, nor overpower it, because we are not observing subjects who are removed from a world of external objects. We are fundamentally in the world: we are coping creatures, riddled with an anxiety that has been intensified by modern technologies that push us to define ourselves by efficiency and productivity rather than by just being. There is no intrinsic meaning to life or universal rules to follow, so our only choice is to be authentic by accepting the anxiety of being human.[291]
>
> **Albert Camus**: Yes, we should accept our condition, but we must do so defiantly. Absurdity is not something to be released from (in

a Buddhist sense), nor is it something we can rise above by finding meaning within (in a Nietzschean or Sartrean sense). Absurdity is a predicament to manage, not a problem to solve. It arises at the point where reason confronts its own limitations, when purposeful beings make demands of a world that is unreasonably silent and offers no response to our needs. It is an inescapable condition that cannot be denied but must be defied, like Sisyphus of Greek mythology who is condemned to roll a boulder up a mountain until it rolls down and needs to be pushed up again. Except that we must attribute more power to Sisyphus than the traditional myth: we must imagine that he does not deny the preposterousness of his task. Rather than perceiving his fate as punishment, he accepts his condition and thereby triumphs over it and salvages his dignity. We must acknowledge and live in spite of absurdity.[292]

Jean-Paul Sartre: We have much more freedom than Sisyphus and therefore must take more responsibility for our choices than simply living in spite of absurdity. We have to create our own identities and resist the limits that others place on us: we each exist before we have essences imposed on us by other people. Our choices arise from nothingness, so not only are we free to self-determine, but, in fact, we are each condemned to be free — to live our choices. If we avoid this responsibility, we are living inauthentically, in bad faith.[293]

Viktor Frankl: Indeed, the task of each of us is to create our own individual meaning. Any kind of ultimate, non-personal meaning is beyond the intellectual capacities of human beings, so meaning cannot be defined in general terms for everyone but only in personal terms that we self-create. Our objective is neither to endure meaninglessness (as Camus would have it), nor to seek a tension-free state (as the Buddha would have it). We must create meaning-generated tension for ourselves by establishing personal goals that give us reason to live.[294]

Differences surface in how each thinker believes that we should handle the challenge. Some opt for an escape by seizing the liberties and creativity that we are endowed with (Nietzsche, Sartre, Frankl). Some opt for an attitude of resignation (Schopenhauer). Others advocate a stoic acceptance of our condition (Heidegger, Camus). Stoic acceptance is underscored in playwright Samuel Beckett's *Waiting for Godot*. When one character insists, "I can't continue," his colleague retorts, "That's what you think." Beckett returns to this idea again in his novel *The Unnamable*, closing with, "I can't go on. I'll go on." Beckett offers no solution to the problem of waiting for something meaningful to happen. He simply observes that we persist, without meaning, because we have no other choice.

The existentialist tradition advocates different forms of learning to live with angst, but most existential philosophers share the conviction that the only meaning available to us is through what each individual deliberately creates for himself or herself. This striving for personal meaning is largely a Western tradition. In fact, the Western cultural perspective of self-creation is itself fairly new: until the later Middle Ages, the highest good was the contemplation of God. The legitimacy of individual relevance and self-creation arose from the Protestant Reformation, which began in the sixteenth century and evolved into the notion that work can be purposeful and meaningful (as in the Puritan work ethic), which in turn gave rise to the modern-day notion that individual satisfaction depends on personal goal fulfillment. But the quest for personal meaning brings with it the anxiety of unfulfilled ambition: the psychotherapist Eric Maisel counsels that each of us must become "a meaning expert" while at the same time becoming "an anxiety expert," since the two go hand-in-hand.[295]

In fact, it was just this anxiety that the Buddhist tradition seeks to eliminate. The imperative to create personal meaning stands in stark contrast with this tradition's focus on releasing oneself from the empty and frustrating search for meaning. Eastern philosophy puts emphasis on relinquishing the very striving that Frankl et al. advocate, because the yearning for meaning is precisely what fuels anxiety: the attainment of our goals perpetually eludes us.

Siddhartha Gautama, the Buddha, having inherited many Hindu traditions, defined the human dilemma in the context of our misguided effort to fit the world into tidy, conceptual frameworks, including our attempt to find meaning. He viewed the clash between reality and our striving for meaning as a source of our dissatisfaction, anxiety, angst (*dukkha* — the First Noble Truth). According to the Buddha, there is nothing useful about the question of meaning since all that is real is our immediate experience of the present moment. Peace is simply in *being* (in the present); angst is in *becoming* (yearning for something different from the present moment). Cessation of angst is possible by releasing ourselves from craving and attachment. So while pain in our lives is inevitable, suffering is not if we release ourselves from craving and attachment, including, and especially, the yearning for meaning. When we see reality for what it really is — a unified, constant flux that we are an inseparable part of — all questions of meaning disappear. When we understand why we suffer, experience the cessation of our craving, and cultivate a middle way (the eightfold path) between self-indulgence and self-denial, then we can find peace by keeping our yearnings, and therefore our suffering, at bay.

Modern-day science supports much of what the Buddha taught: natural selection did not design us to *find* contentment; it designed us to *seek* it. Our biological system is geared toward the goal of activating our dopamine neurons, when the prospect of pleasure is present. Our physiology is designed to pull us away from being and toward becoming, which is precisely what Buddhism identifies as the problem. Yet most of us are not prepared to abandon our personal ambitions, because however much suffering they may entail, they provide us with the purpose that sustains our will to live, and a sense of power that energizes us. The painter Paul Klee captured this sentiment when he wrote, "He who strives will never enjoy this life peacefully."[296]

In Sartre's early work, he described a very liberating view of human freedom (which he tempered in his later years by acknowledging the constraints of social conditioning). Sartre championed the empowering view that each of us is free to invent ourselves. Nietzsche, on the other hand, believed it was a small minority of human beings who can rise to this level of full human potential; the vast majority of us are confined, by lack of initiative and creativity, to carry out Sisyphean lives of never-ending tasks and projects that amount to nothing. A version of this less optimistic perspective of human potential is aptly described by Keith Stanovich: "Most of us are cultural freeloaders — adding nothing to the collective knowledge or rationality of humanity. Instead, we benefit every day from the knowledge and rational strategies invented by others."[297] Camus did not endorse the Sartrean "completely free" view;

he espoused a "cannot escape absurdity" perspective (although he was not nearly as pessimistic as Schopenhauer). So underlying the different views of how we can best respond to the complexity of our dilemma are contrasting views of the degree to which we are constrained in our capacity to "self-overcome." Just how limited are we? How capable are we of advancing ourselves?

19.3 Contrasting Views of Human Potential

Since the beginning of life on earth, over 95 percent of all animal species have become extinct, including all human species except ours. One-third of the deaths were triggered by mass extinctions, including the "Big Five" extinction events in the last six hundred million years, the most recent of which, sixty-five million years ago, wiped out 70 percent of all species, including dinosaurs, and the biggest of which, two hundred and fifty mya, killed about 90 percent of all extant species.

Severe climatic changes would have almost certainly wiped out our species had we not evolved an extraordinary cognitive ability to adapt. Even then, there is no compelling reason that we had to evolve exactly as we did. Evolutionary biologist Stephen J. Gould suggested that if time restarted when life first emerged and another four billion years unfolded, life on earth would be extremely different than it is today.[298] The random genetic mutations that underlie natural selection likely would not recreate the same path. Gould's hypothesis is supported by what biologists refer to as "chaotic adaptation," whereby the results of natural selection are highly dependent on initial conditions, slight variations of which can direct the evolution of a species to dramatically different outcomes. Daniel Dennett disagrees with Gould's hypothesis: he cites "convergent evolution" as evidence that there are constraints within the natural world that limit the number of possible outcomes from selective pressure.[299] For example, flight has developed in distinct species separately and independently of each other, as have organs such as the eye.

No matter how contingent our development has been, there are two contrasting perspectives of the human capacity to improve ourselves. The contrast emerges in different ways and goes by different labels, but at its core is the difference between a limited view of human beings and an open-ended view. Economist Thomas Sowell and cognitive scientist Steven Pinker express this contrast in distinct but parallel ways:

	Limited View	Open-Ended View
Sowell	• "Constrained Vision" • Our potential and actuality are identical: there is no gap that can be closed • We are limited to selfish instincts and fallible thinking • Thomas Hobbes' pessimistic view of human nature	• "Unconstrained Vision" • Meaningful gap between how we operate and our ultimate potential • We are capable of moral and intellectual development • Jean-Jacques Rousseau's optimistic view of human nature
Pinker	• "Tragic Vision" • Human nature is constant and unchanging • We have limited knowledge, wisdom and virtue • We are deeply, fundamentally selfish	• "Utopian Vision" • Human nature changes with social circumstances • Our limits are artifacts that stem from social arrangements; they can be altered and improved on

Both Sowell and Pinker espouse a view of humankind that resides between these two extremes, but with a leaning toward a more limited view. Sowell cites empirical evidence of how imperfectly government and regulation work: there is no evidence to suggest that institutions and government policy can do any better for individuals than they can do for themselves without intervention.[300] Pinker points to the universality of dominance, ethnocentrism and violence in human societies, as well as the prevalence of the self-serving biases and defence mechanisms that we all share.[301] The mind, according to him, is not infinitely malleable.

While not limitless, our understanding of ourselves continues to deepen into a greater self-understanding that contradicts a completely pessimistic view of the human capacity to push our limits outward. The relentless unfolding of the evolutionary process over billions of years has not stopped in its tracks: we are still on the move, and aspects of our transition are awkward and uncomfortable. Our ape ancestors did not simply stand up one day and marvel at the view that bipedalism offered. Coming to bipedal locomotion was a slow process that involved many shifts in multiple anatomical features: the pelvis and lower-limb system was remodelled over millions of years of partial stooping, shuffling, knuckle walking and fledgling bipedalism before we managed to enjoy sustainable walking while balancing on two legs.

So it is with our cognitive powers: our ancestors did not simply wake up one morning, point to the sky and engage in conversations about the meaning of life. We continue to adapt and evolve: in our view of the universe; in our view of how to interact with one another; in our view of what is morally acceptable; and in our view of how to find fulfillment in our lives. We continue to push past the bounds of our immediate, conscious awareness and thereby exert an increasing degree of control over our destinies.

19.4 Survival of the Smartest: Pushing Our Limits

There is nothing predictable about life, except, per Buddha's First Noble Truth, that much of it is difficult. Even before we get into the world, we have to squeeze through the narrow female birth canal, a life-threatening journey that killed many of us until modern obstetrics made the trip easier. Natural selection has furnished us with cognitive tools that are up to the task of coping with many of our challenges, which is why we are still here. The increase in complexity we have created for ourselves, however, means that the same tools are not sufficient. We have to make up for this brain–world gap. We cannot rely on the slow vertical transmission of genetic adaptation that gradually empowered us with bipedalism. We have to expedite our cognitive maturity by exploiting the opportunity for horizontal transmission — cultural learning and teaching — to increase the sophistication of our thinking.

We have done remarkably well for ourselves as a species. But we could be so much better, and as our cognitive capacities mature, our excuses for not getting better wear thin. We are the only vehicles on the planet, as Richard Dawkins points out, that can rebel against the dictates of our genes: the "selfish" reproductive replicators that drive most of our behaviour.[302] In this sense we must defy our maker: not sacrilegiously, but in the spirit of overcoming some of the design specifications that hundreds of millions of years equipped us with, specifications that are not productive in World #2. We now find ourselves outside the Garden of Eden, where the challenges were straightforward and we were expert in handling them.

Enlightenment, according to Immanuel Kant, is the release from self-imposed immaturity.[303] This ought to be the exact aspiration of a continually evolving sentient species. The key to releasing ourselves, as David Bohm described it, is getting back in control of our thoughts.[304] For this is how we can close the gap between our ancestral thinking and World #2 complexity. The table in appendix 1 summarizes the contrasting cognitive strategies that work in each of the two worlds.

There is increasing urgency to closing the gap. We are facing greater ecological instability, deeper social complexity and a growing need for individual meaning that would be foreign to all but our most recent ancestors. We cannot reach the destinations we choose for ourselves if we do not alter our navigation systems to accommodate the greater complexity in our lives. In fact, closing the gap enables us to exercise our freedom to the greatest possible extent. When we fully engage Type 2 thinking, we release ourselves from the constraints of Type 1 intuition and its blunt, unproductive interpretation of, and response to, the challenges of World #2.

Ultimately, our autonomy depends on how well we ramp up the complexity of our thinking to match the complexity of our world.

EPILOGUE
Two Games for Two Worlds

Adapt or perish, now as ever, is nature's inexorable imperative.
H.G. Wells[*]

As a child ages, she will incorporate greater complexity into her thinking, but even as an adult she will continue to have greater certainty than her bounded thinking warrants. She will do this because, like all human beings, she has been programmed to jump to quick conclusions that she rarely second-guesses. She will do this because most of her waking hours will be chock-full of quick, first-satisfactory conclusions that warrant complete confidence because they are straightforward. She will do this because her brain, like that of her parents and the brains of all members of her species, has not yet evolved to easily and effectively distinguish straightforward situations from more complex ones. And this failure to distinguish the two worlds that she occupies means she is not playing her best game in World #2, where her expertise is underdeveloped.

Simon Ramo, a polymath who was a keen tennis player, recorded his insights on tennis strategy in a book published in 1970.[305] He observed that pro tennis players play a different version of the game than amateurs do. Pros win when they overpower their opponents (by outmaneuvering or forcing errors). Amateurs win when their opponents make more mistakes than they do (typically hitting easy balls into

---
[*] *The Last Books of H.G. Wells — The Happy Turning: A Dream of Life and Mind at the End of Its Tether*, Monkfish Book Publishing, 2006.

the net). The two versions were later dubbed winner's games (points are won) and loser's games (points are lost) by Charles D. Ellis, who applied Ramo's ideas to stock market investing.[306]

The key difference between a winner's and a loser's game is expertise. The expertise of professional tennis players reduces their vulnerability to the game itself, enabling them to focus on outplaying their opponents. Amateurs, on the other hand, play against their opponents as well as against the difficulties of the game itself, where their lack of expertise makes them vulnerable to mistakes. Ramo concluded that pro and amateur tennis players should have different strategies: a pro should play aggressively to score points, whereas an amateur should play conservatively to avoid mistakes. Most games, both the competitive and real-life kinds, have losing and winning versions. To be successful, we have to know which version we are playing; in other words, we have to know the depth of our expertise so that we can apply the right approach. Unlike the tennis player who, depending on his expertise, plays either a winner's or loser's game, our personal challenge is daunting because each of us is simultaneously playing two versions of "the game of life."

The first game, played in the arena of World #1, is one in which we have great expertise: our skills are perfectly suited for the challenges of this world, which is why it is a winner's game for us. We have been playing this first game for millions of years, and we are here because we know how to play it well.

The second game, played in the arena of World #2, is one in which our basic intuitions are not well adapted. It is a loser's game for us because we have not developed reliable intuitions about how complex things work, hence we oversimplify with interpretive models that are too basic, and we are too confident in our preliminary conclusions.

The challenge of invoking more and better Type 2 thinking — both 2.1 analytical and 2.2 metacognitive — is intensified by the need to unlearn as much as it is to learn. We have to unlearn the shortcuts that we rely on in World #1 if we are to operate effectively in World #2. Unlearning is complicated by the fact that we cannot abandon the shortcuts altogether since they are crucial for survival in World #1.

That World #2 requires deep humility to navigate is not obvious to us, given the urgency with which we seek certainty and the success we have in World #1 by using shortcuts that give us this certainty.

So impressive is our technique with straightforward problems that we cannot even articulate the vastness of Type 1's storehouse of practical knowledge, which is why in World #1 we know more than we can say, whereas in World #2 we know less than we think.

Playing a smart game in World #2 requires a different kind of strategy than the one that works in World #1. It requires an understanding of how our weaknesses intersect with the game's difficulty. It means slowing down in order to avoid the automatic, mindless application of the basic shortcuts that only suit straightforward problems. It means being cautious about our preliminary conclusions by retaining a skeptical bias toward them.

No matter how much expertise we develop in World #2, the exponential growth of complexity means that we may never play a winner's game: the brain–world gap is just that big. But that should not deter us from playing the best loser's game that we are capable of. The more deeply we evaluate the antiquated ways in which we tend to view complexity and the more proactively we transition from ancestral-style thinking to complex thinking, the more fully we will increase the choices that are available to us — and ultimately our power.

APPENDIX 1
Sizing Up the Gap

*The problem set of World #1 is different in form than that of World #2.
As such, each world requires a distinct cognitive approach.*

World #1
Straightforward Challenges

- Signals are differentiated from noise: cause and effect are tightly linked and easily accessible
- Patterns are consistent across situations
- Feedback is timely, direct and clean

↓

Learning is easy; prediction is reliable
Expertise is possible

What Works in World #1

Ancestral thinking: heavy reliance on Type 1 intuitive expertise
Naïve realism: one ("the") perspective
Cognitive trade-off skewed toward speed
Cognitive trade-off skewed toward gullibility (to avoid Type II error of underinterpreting)
Simplifying: reductionism • Selective attention (because missing information is irrelevant) • Linear, one-way causality • Confabulating to close gaps • Ignoring randomness
Satisficing: rushing to definitive conclusions • Addiction to certainty • Feeling of knowing • Availability, confirmation, myside biases • Reliance on observation • Dichotomies • Short "because" story • Correspondence test • Cognitive self-reliance
Emotional asymmetry skewed towards negativity
Tribalism
Personality-based behaviours
Daily routine

World #2
Complex Challenges

- Signals are buried and ambiguous: multiple, interacting causes are not immediately accessible
- Patterns vary since every situation is unique
- Feedback is delayed, indirect and dirty

↓

Learning is difficult; prediction is unreliable
Expertise is hard to develop

A Better Approach in World #2

Aspirational thinking: heavy reliance on Type 2 analytical and metacognitive cognition
Constructive realism: many ("useful") perspectives
Cognitive trade-off skewed toward accuracy
Cognitive trade-off skewed toward skepticism (to avoid Type I error of overinterpreting)
Complexity science: systems theory • Expanded spotlight (because missing information is crucial) • Multicausal, non-linear interactions • Emergence • Identifying random streaks and reversion
Cognitive science: exploring provisional truths • Alternative paths • Suspension of judgment • Testing and retesting assumptions • Requirement for interaction • Probabilities • Long "perhaps" story • Coherence test • Cognitive diversity
Emotion management
Global inclusiveness
Situation-dependent behaviours
Creating meaning

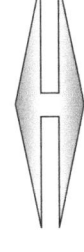

APPENDIX 2
Humans Versus Other Animals: A Dialogue

Genetic differences between animals allow us to categorize them into discrete species, but genes do not address a subject of concern to comparative psychologists: phenotypic differences — the differences in observable traits. Of particular interest are the features that are unique to humans, i.e., the "derived traits" that are not shared by our closest animal cousins.

If our cognitive abilities are extensions of what other animals are capable of, then we are not fundamentally distinct but merely endowed with a greater degree of cognitive flexibility. If, however, there are human cognitive tasks that cannot be performed by any other animal, even in very rudimentary forms, then there is a discontinuity between us and them, a difference in kind rather than degree.

Following is how this argument might play out between representatives of the two sides.

Mr. Kind: The Abrahamic religions make a clear distinction between humans and other animals. Humans are not just "more of" whatever other animals are; we are fundamentally different by virtue of having a soul and by being moral agents.

Ms. Degree: But many other religions do not make such a sharp divide, largely because they do not embrace the Abrahamic tradition as it evolved in Christianity and Islam, which emphasize the idea of individual immortal souls. In fact, Judaism, the original Abrahamic religion, does not give much prominence to the notion of soul, and immortality was not a feature of early Jewish belief systems. As for moral agency, chimpanzees display the rudimentary elements of a basic morality: a chimp will be generous with chimps who share but not with chimps who hoard. They will punish the intentional meanness of others while ignoring their accidental misdeeds. This form of elementary morality is not surprising when we consider that our human brain is made up of the same basic materials as all brains, and many animals have the necessary brain parts to experience consciousness.

Mr. Kind: But there are, in fact, some real differences in the ways that human and non-human brains are organized. Our brains have a much more pronounced "lateralization" — left and right hemispheres doing different jobs. Left-right differences are very slight in some higher primates and absent in most animals. This left-right split allows greater specialization of areas within the human brain, with less redundancy and greater cognitive efficiency. There is also distinct "frontalization" in the human brain: our prefrontal cortex is much more developed, and this is where much of our unique cognitive processing arises, particularly in "area 10," which is about twice as large as it is in the great apes. The neuronal networks of the prefrontal cortex reach deep into other brain areas, enabling a greater degree of interconnectivity between all areas of the human brain. The greater specialization of brain parts, combined with the incredible interconnectivity of these parts facilitated by our large prefrontal cortex, allows the human brain to achieve a much greater degree of cognitive flexibility than any other animal.

Ms. Degree: If our cognitive processing is fundamentally unique, then how do we explain the burgeoning field of research that is revealing the incredible ingenuity of other animals? What is a crow doing when it bends a stick with its beak to pick food out of a narrow tube? Or a western scrub jay when it hides its food in the presence of other jays, then returns to move the food to a new hiding spot after the jays have left? Or a chimpanzee when it learns to point to symbols on a chart to communicate at the level of a two-and-a-half-year-old child? Or a low-rank male baboon that threatens a top-rank male to distract him while another low-rank male has sex with one of the top rank's mates? Are animals not demonstrating humanlike thinking with their imaginative use of tools, communication and deception? It may be that a crow's extraordinary resourcefulness is simply stimulus-response behaviour, but it stretches credibility to argue that behavioural conditioning is the exclusive explanation for the practiced deception that some apes are capable of.

Mr. Kind: Bending a stick? Re-hiding food? Pointing at a chart? Those examples do not even come close to what humans are capable of: metacognition, the ability to contemplate our own thinking and to exercise control over ourselves in a way that eludes all other animals.

Ms. Degree: But, in fact, there is evidence of animal metacognition in research that focuses on examining how animals respond to uncertainty. Dolphins and rhesus macaque monkeys, when confronted by tasks that are difficult but offer high rewards, will opt for easier tasks that offer substantially lower rewards. Before deciding, they will hesitate and waver, showing signs that we associate with metacognition: thinking about what they know and are capable of. These "uncertainty monitoring experiments" reveal that the animals are monitoring and controlling their thinking when they are faced with uncertain outcomes. Furthermore, rhesus macaque monkeys are able to transfer learning from one task to another, and even ask for hints to solve problems, demonstrating a certain level of monitoring and cognitive control over the attention they apply to problem solving.

Mr. Kind: Even if those experiments indicate some sort of basic flexibility in handling uncertainty, that is not even close to the human ability to generalize learning from one activity to another. Other animals are much more restricted in their ability to apply learning beyond the single tasks for which the learning was developed. Our "domain-general" flexibility, versus the "domain-specific" adaptation of other animals, has enabled us to accumulate cultural traditions over centuries; there is absolutely no equivalent to this in other species.

Ms. Degree: But chimps have indeed been observed to pass on a small number of learned traditions.

Mr. Kind: I would not categorize anything that chimps do as "cultural." Humans accumulate innovations by adding layer upon layer of novelty, building on the creativity and cultural development of their predecessors. Accumulating knowledge comes easily to us because we are among the most prosocial animals on the planet. Our prosociality may have originated as we moved out of trees and began to rely on one another to survive in the exposed savannahs; in any case, we are one of the most co-operative breeders of all species. The protracted vulnerability of a newborn human baby requires a lot of attention and care, including the help of siblings, grandparents, cousins and even neighbours. This variety of care-giving exposes young humans to extended social learning, the kind that invokes the co-operation required for hunting, farming and ultimately trading crops and goods. There are no known human societies where individuals or individual families fend for themselves, and today there are very few societies that do not rely on trading with other societies: our co-operative nature is highly developed.

Ms. Degree: There are many animals that co-operate, and some do engage in allomaternal, or co-operative, breeding.

Mr. Kind: But our prosocial behaviour is of another kind altogether: we are the only species capable of suppressing our emotional reactions, including the facial expressions that accompany these emotions. One of the reasons we are so prosocial is that we have the unique ability to manage our emotional reactions and keep our antisocial tendencies in check most of the time. All this is possible because we can decouple our immediate sensory perceptions from our ideas in order to engage in hypothetical thinking. This "what if?" reasoning frees us from the present moment to think more strategically about our longer-term goals rather than being captive to our immediate feelings and short-term desires. The complexity of our brains gives us a kind of "command central" capability: we are not under the exclusive control of our automatic reflexes the way other animals are, no matter how many fancy tricks they can do.

Ms. Degree: Not so fast: this control over ourselves is severely limited: witness obesity and road rage, to name just two examples. More importantly, self-control is not an exclusively human talent. When chimps are rewarded with more candy if they accumulate it instead of eating it, they have been observed playing with toys as a means of distracting themselves and enforcing their willpower to not eat the candy. As more candy is introduced, their focus and engagement with the toys increases. Even more telling, chimps have been observed stockpiling and hiding rocks for future use as weapons, suggesting an ability not just to deceive but also to think outside of the present moment to a hypothetical future.

Mr. Kind: Chimps suppressing their short-term desires and contemplating hypotheticals does not come close to our ability to control our attention: we can meditate, for example, and focus our ruminations on uplifting possible outcomes. The impressive things that chimps can do only prove that there may be rudimentary elements of our cognitive skills in other animals. But that does not change the fact that the animal versions of these skills are severely limited compared with ours.

Ms. Degree: But that is precisely my point: No matter how limited their cognitive abilities, they are still versions of ours. So the difference is one of degree.

Mr. Kind: You are missing the bigger issue that our cognitive abilities put us in a separate category because we are not at the mercy of the reproductive goals of our genes in the way that all other living organisms on the planet are. Our special talent for self-reflection allows us to rebel, as Richard Dawkins describes it, against "selfish genes," which are single-mindedly concerned with replicating themselves. This ability to overcome our cravings is so much more profound than a chimp's stashing of candy. It is what led Aristotle to observe that, unlike other animals, humans are not exclusively occupied with maintaining and reproducing themselves. It prompted Kant to note that we are the only beings who set our own purpose. The inability of other animals to free themselves from their automatic cognitive processing indicates an insurmountable gap — a qualitative, not just a quantitative, one.

Ms. Degree: Let me try a different angle, starting with Darwin, who wrote, "There is a much wider interval in mental power between one of the lowest fishes ... and one of the higher apes, than between an ape and a man; yet this immense interval is filled with numberless gradations." Surely the gulf between an ant's cognitive power and a chimp's is at least as wide as that between a chimp and a human? So if there is a sharp dividing line that separates "kind," should there not be at least one more line farther down the food chain? And if you grant me that, how many categories of "kind" do we need?

Mr. Kind: But I do not grant you that. The gap between an ant and a chimp is a large one, but not one that is based on qualitative differences.

Ms. Degree: How you can say that one gap is a difference of degree and another gap is a difference of kind? There is a gray area where a sufficient difference in degree appears to be a difference in kind, but this ambiguous area is a large and indeterminate one. If you do not buy in to the ant-chimp analogy, try this one: imagine lining up all our ancestors back to the first humanlike great ape; in this lineup, there would be no ancestral creatures that demonstrate a clear dividing line between humans and non-humans. The whole notion of common ancestry nudges us toward endorsing the view that humans are nothing more than the most complex genus of the primate order in the mammalian class of animals.

Mr. Kind: Except that we are fully aware of ourselves and our place on the planet and in the universe, whereas our distant ancestors did not have a clue.

Ms. Degree: But according to your schema, the question arises of when exactly we suddenly began to "have a clue"? When did we suddenly become non-animals? The transition to human consciousness did not magically materialize one day in one of our *Homo sapiens* ancestors. Because we do not have easy access to this long line of descendants, it is tempting for us to ignore the continuum and view the difference between great ape and human cognition as a jump from one kind to another kind. But that is not how it happened. It was a very gradual transition, measured in degrees.

Mr. Kind: I do not need to look at an imaginary line of human ancestry. I can just go to a zoo and watch the animals to see that they are different from me in a fundamental way.

Ms. Degree: When I go to the zoo and watch the animals, especially when I watch human children playing in the vicinity of the monkey and orangutan cages, I know we are special animals, but animals nonetheless.

Absent a clear differentiator like "soul," the degree/kind argument comes down to how we choose to label the cognitive gap between ourselves and other animals. "Huge degree of difference" and "difference in kind," in a non-theistic context, mean the same thing.

ENDNOTES

1 Wells, H. G. (2006). *The Last Books of H. G. Wells — The Happy Turning: A Dream of Life & Mind at the End of Its Tether*. Monkfish Book Publishing.
2 Taleb, N. N. (May/June 2011). "The Black Swan of Cairo." *Foreign Affairs*, vol. 90.
3 Dunbar, R. (June 1992). "Neocortex Size as a Constraint on Group Size in Primates." *Journal of Human Evolution*, vol. 22.
4 Pascal, B. (2003). *Pensées*. Penguin Classic.
5 Simon, H. (1982). *Models of Bounded Rationality: Economic Analysis and Public Policy*. MIT Press.
6 Flynn, J. (January 1984). "The Mean IQ of Americans: Massive Gains 1932–1978." *Psychological Bulletin*, vol. 95.
7 Popper, K. (2001). *All Life Is Problem Solving*. Routledge.
8 Perkins, D. (1995). *Outsmarting IQ: The Emerging Science of Learnable Intelligence*. Free Press.
9 Campbell, J. (1984). *Grammatical Man: Information, Entropy, Language and Life*. Penguin.
10 Nagel, T. (1989). *The View From Nowhere*. Oxford University Press.
11 Thagard, P. (2010). *The Brain and the Meaning of Life*. Princeton University Press.
12 Plato (2008). *Republic*. Oxford Paperbacks.
13 Aristotle (2009). *Metaphysics*. NuVision Publications.
14 Kant, I. (2007). *Critique of Pure Reason*. Penguin Classic.
15 Descartes, R. (2013). *Meditations on First Philosophy*. Broadview Press.
16 Locke, J. (2008). *An Essay Concerning Human Understanding*. Oxford University Press.
17 Berkeley, G. (2007). *A Treatise Concerning the Principles of Human Knowledge*. Filiquarian Publishing.
18 Hume, D. (2008). *An Enquiry Concerning Human Understanding*. Oxford University Press.

19 Hegel, G. W. F. (2013). *Phenomenology of Mind*. Hardpress Publishing.
20 Sacks, O. (2011). *The Mind's Eye*. Vintage.
21 Hawking, S. and Mlodinow, L. (2012). *The Grand Design*. Bantam.
22 Dawkins, R. (2008). *The God Delusion*. Houghton Mifflin Harcourt.
23 von Uexküll, J. (2010). *A Foray Into the Worlds of Animals and Humans*. University of Minnesota Press.
24 Ryle, G. (1990). *The Concept of Mind*. Penguin UK.
25 Archer, E. (Oct. 2010). "Opinion: A Wolf in Sheep's Clothing." *The Scientist*, vol. 24.
26 Chabris, C. and Simons, D. (2011). *The Invisible Gorilla: How Our Intuitions Deceive Us*. Harmony.
27 Barrett, L. F., Mesquita, B. and Smith, E. R. (2010). "The Context Principle." In *The Mind in Context*. Guildford Press.
28 Eagleman, D. (2012). *Incognito: The Secret Lives of the Brain*. Penguin.
29 Locke, J. (2008). *An Essay Concerning Human Understanding*. Oxford University Press.
30 Spelke, E. et al. (August 1985). "Object Permanence in Five-Month-Old Infants." *Cognition*, vol. 20.
31 James, W. (2010). *The Principles of Psychology*. Bibliolife.
32 Gigerenzer, G. and Goldstein, D. (October 1996). "Reasoning the Fast and Frugal Way: Models of Bounded Rationality." *Psychological Review*, vol. 103.
33 Gazzaniga, M. S. (2011). *Who's in Charge? Free Will and the Science of the Brain*. Ecco.
34 Sacks, O. (2011). *The Mind's Eye*. Vintage.
35 Paris, J. (2000). *Myths of Childhood*. Routledge.
36 Wampold, B. (2001). *The Great Psychotherapy Debate: Models, Methods, and Findings*. Routledge.
37 Gilbert, D. (2007). *Stumbling on Happiness*. Vintage.
38 Johansson, P. and Hall, L. (April 2009). "Choice Blindness: You Don't Know What You Want." *New Scientist*, issue #2704.
39 Freud, S. (2010). *An Outline of Psychoanalysis*. Martino Fine Books.
40 Frances, A. (2013). *Saving Normal*. HarperCollins.
41 Foucault, M. (2009). *History of Madness*. Routledge.
42 Frances, A. (2013). *Saving Normal*. HarperCollins.
43 Dennett, D. C. (1996). *Darwin's Dangerous Idea: Evolution and the Meanings of Life*. Simon & Schuster.
44 Tetlock, P. (2006). *Expert Political Judgment: How Good Is It? How Can We Know?* Princeton University Press.
45 Kahneman, D. (2011). *Thinking, Fast and Slow*. Doubleday Canada.
46 Galton, F. (1970). *English Men of Science: Their Nature and Their Nurture*. Routledge.
47 Heidegger, M. (2010). *Being and Time*. State University of New York Press.
48 Wittgenstein, L. (2010). *Philosophical Investigations*. Wiley-Blackwell.
49 Shermer, M. (2012). *The Believing Brain: From Ghosts and Gods to Politics and Conspiracies — How We Construct Beliefs and Reinforce Them as Truths*. St. Martin's Griffin.

50 Nietzsche, F. (2012). *On Truth and Lies in a Nonmoral Sense*. CreateSpace.
51 Loftus, E. (November 2003). "Make-Believe Memories." *American Psychologist*, vol. 58.
52 Armstrong, K. (1994). *A History of God: The 4,000-Year-Old Question of Judaism, Christianity and Islam*. Ballantine Books. Akenson, D. H. (2002). *Saint Saul: A Skeleton Key to the Historical Jesus*. McGill-Queen's University Press.
53 Eire, C. M. N. (2002). "Pontius Pilate Spares Jesus." In R. Cowley (Ed.), *What If? II: Eminent Historians Imagine What Might Have Been*. Penguin Putnam.
54 Simon, H. (March 1956). "Rational Choice and the Structure of the Environment." *Psychological Review*, vol. 63.
55 Levin, I. P. et al. (November 2002). "A Tale of Two Pizzas: Building Up From a Basic Product Versus Scaling Down From a Fully-Loaded Product." *Marketing Letters*, vol. 13.
56 Frederick, S. (Fall 2005). "Cognitive Reflection and Decision Making." *Journal of Economic Perspectives*, vol. 19.
57 Stanovich, K. E. (2008). "Distinguishing the Reflective, Algorithmic and Autonomous Minds: Is It Time for a Tri-Process Theory?" In J. Evans and K. Frankish (Eds.), *In Two Minds: Dual Processes and Beyond*. Oxford University Press.
58 Sapolsky, R. (2004). *Why Zebras Don't Get Ulcers: The Acclaimed Guide to Stress, Stress-Related Diseases, and Coping*. Holt Paperbacks.
59 Orrell, D. (2012). *Truth or Beauty: Science and the Quest for Order*. Yale University Press.
60 Burton, R. (2009). *On Being Certain: Believing You Are Right When You're Not*. St. Martin's Griffin.
61 Kruglanski, A. and Webster, D. (October 1996). "Motivated Closing of the Mind: 'Seizing' and 'Freezing.'" *Psychological Review*, vol. 103.
62 Wason, P. (1960). "On the Failure to Eliminate Hypotheses in a Conceptual Task." *Quarterly Journal of Experimental Psychology*, vol. 12.
63 Spinoza, B. (2009). *Ethics: Demonstrated in Geometrical Order and Divided Into Five Parts*. BiblioLife.
64 Perkins, D. (1995). *Outsmarting IQ: The Emerging Science of Learnable Intelligence*. Free Press.
65 Kahneman, D. (2011). *Thinking, Fast and Slow*. Doubleday Canada.
66 Mercier, H. and Sperber, D. (2010) "Why Do Humans Reason? Arguments for an Argumentative Theory." *Behavioral and Brain Sciences*, vol. 34.
67 Wainer, H. (2009). *Picturing the Uncertain World: How to Understand, Communicate, and Control Uncertainty Through Graphical Display*. Princeton University Press.
68 Pettit, B. (2012). *Invisible Men: Mass Incarceration and the Myth of Black Progress*. Russell Sage Foundation.
69 Taleb, N. N. (2010). *The Black Swan: The Impact of the Highly Improbable*. Random House.
70 Einhorn, H. and Hogarth, R. (September 1978). "Confidence in Judgment: Persistence of the Illusion of Validity." *Psychological Review*, vol. 85.

71 Popper, K. (2002). *Conjectures and Refutations: The Growth of Scientific Knowledge*. Routledge.
72 Hume, D. (2008). *An Enquiry Concerning Human Understanding*. Oxford University Press.
73 Dunning, D. and Kruger, J. (December 1999). "Unskilled and Unaware of It: How Difficulties in Recognizing One's Own Incompetence Lead to Inflated Self-Assessments." *Journal of Personality and Social Psychology*, vol. 77.
74 Farson, R. (1997). *Management of the Absurd*. Free Press.
75 Dostoyevsky, F. (2008). *Crime and Punishment*. Oxford Paperbacks.
76 Bruner, J. and Potter, M. (April 1964). "Interference in Visual Recognition." *Science*, vol. 144.
77 Kahneman, D. (2011). *Thinking, Fast and Slow*. Doubleday Canada.
78 Burton, R. (2009). *On Being Certain: Believing You Are Right When You're Not*. St. Martin's Griffin.
79 Gladwell, M. (2008). *Outliers: The Story of Success*. Little, Brown and Company.
80 Chase, W. and Simon, H. (January 1973). "Perception in Chess." *Cognitive Psychology*, vol. 4.
81 Ericsson, K. A., Kampe, R. and Tesch-Romer, C. (July 1993). "The Role of Deliberate Practice in the Acquisition of Expert Performance." *Psychological Review*, vol. 100.
82 Simon, H. (May 1992). "What Is an 'Explanation' of Behavior?" *Psychological Science*, vol. 3.
83 Klein, G. (1999). *Sources of Power: How People Make Decisions*. MIT Press.
84 Kahneman, D. and Klein, G. (September 2009). "Conditions for Intuitive Expertise: A Failure to Disagree." *American Psychologist*, vol. 64.
85 Hogarth, R. (2010). *Educating Intuition*. University of Chicago Press.
86 Kahneman, D. (2011). *Thinking, Fast and Slow*. Doubleday Canada.
87 Tetlock, P. (2006). *Expert Political Judgment: How Good Is It? How Can We Know?* Princeton University Press.
88 Baron, J. (2007). *Thinking and Deciding*. Cambridge University Press.
89 Kahneman, D. and Frederick, S. (2002). "Representativeness Revisited: Attribute Substitution in Intuitive Judgment." In T. Gilovich, D. Griffin and D. Kahneman (Eds.), *Heuristics and Biases: The Psychology of Intuitive Judgment*. Cambridge University Press.
90 Hood, B. (2009). *Supersense: From Superstition to Religion*. Constable.
91 Drucker, P. (2001). *Management Challenges for the 21st Century*. Harper Business.
92 Dunbar, R. (1998). *Grooming, Gossip, and the Evolution of Language*. Harvard University Press.
93 Kahneman, D. and Klein, G. (September 2009). "Conditions for Intuitive Expertise: A Failure to Disagree." *American Psychologist*, vol. 64.
94 Hogarth, R. (2007). "On the Learning of Intuition." In C. Betsch, T. Betsch and H. Plessner (Eds.), *Intuition in Judgment and Decision Making*. Psychology Press.
95 Taleb, N. N. (2010). *The Black Swan: The Impact of the Highly Improbable*. Random House.
96 Schumacher, E. F. (1978). *A Guide for the Perplexed*. Harper Perennial.

97 Hammond, K. R. (1986). *Human Judgment and Social Policy: Irreducible Uncertainty, Inevitable Error, Unavoidable Injustice*. Oxford University Press.
98 Stanovich, K. E. (2011). *Rationality and the Reflective Mind*. Oxford University Press.
99 Wiener, N. (2007). *Cybernetics: Or Control and Communication in the Animal and the Machine*. Kessinger Press.
100 Shannon, C. (July and October 1948). "A Mathematical Theory of Communication." *Bell System Technical Journal*, vol. 27.
101 Kant, I. (2007). *Critique of Pure Reason*. Penguin Classic.
102 Pinker, S. (2008). *The Stuff of Thought: Language as a Window Into Human Nature*. Penguin.
103 McGinn, C. (1994). *Problems in Philosophy: The Limits of Inquiry*. Blackwell Publishers.
104 Kierkegaard, S. (1985). *Philosophical Fragments*. Princeton University Press.
105 Camus, A. (1991). *The Myth of Sisyphus and Other Essays*. Vintage.
106 Gladwell, M. (2006). *Blink: The Power of Thinking Without Thinking*. Back Bay Books.
107 Leibniz, G. (1996). *New Essays on Human Understanding*. Cambridge University Press.
108 Schopenhauer, A. (2010). *The World as Will and Representation*. Cambridge University Press.
109 Carpenter, W. B. (1998). *Principles of Mental Physiology*. Thoemmes Continuum.
110 Galton, F. (2010). *Inquiries into Human Faculty and Its Development*. Kessinger Publishing.
111 Freud, S. (2008). *The Interpretation of Dreams*. Oxford Paperbacks.
112 Stanovich, K. E. (2004). *The Robot's Rebellion: Finding Meaning in the Age of Darwin*. University of Chicago Press. Kahneman, D. (2011). *Thinking, Fast and Slow*. Doubleday Canada.
113 Lieberman, M. (2007). "Social Cognitive Neuroscience." In R. F. Baumeister and K. D. Vohs (Eds.), *Encyclopedia of Social Psychology*. Sage Publications.
114 Klein, G. (2011). *Streetlights and Shadows: Searching for the Keys to Adaptive Decision Making*. Bradford Books.
115 Stanovich, K. E. (2004). *The Robot's Rebellion: Finding Meaning in the Age of Darwin*. University of Chicago Press.
116 Descartes, R. (2013). *Meditations on First Philosophy*. Broadview Press.
117 Haidt, J. (2012). *The Righteous Mind: Why Good People Are Divided by Politics and Religion*. Pantheon Books.
118 Doidge, N. (2007). *The Brain That Changes Itself: Stories of Personal Triumph From the Frontiers of Brain Science*. Penguin Books.
119 Ridley Stroop, J. (December 1935). "Studies of Interference in Serial Verbal Reactions." *Journal of Experimental Psychology*, vol. 18.
120 Hume, D. (2008). *An Enquiry Concerning Human Understanding*. Oxford University Press.
121 Haidt, J. (2006). *The Happiness Hypothesis: Finding Modern Truth in Ancient Wisdom*. Basic Books.

122 Gazzaniga, M. S. (2011). *Who's in Charge? Free Will and the Science of the Brain.* Ecco.
123 Pinker, S. (2003). *The Blank Slate: The Modern Denial of Human Nature.* Penguin Books.
124 Flanagan, O. (1993). *Consciousness Reconsidered.* Bradford Books.
125 Wegner, D. M. (2003). *The Illusion of Conscious Will.* Bradford Books.
126 Libet, B. (2005). *Mind Time: The Temporal Factor in Consciousness.* Harvard University Press.
127 Damasio, A. (2005). *Descartes' Error: Emotion, Reason, and the Human Brain.* Penguin Books.
128 LeDoux, J. (1998). *The Emotional Brain: The Mysterious Underpinnings of Emotional Life.* Simon & Schuster.
129 Makridakis, S., Hogarth, R. and Gaba, A. (2010). *Dance With Chance: Making Luck Work for You.* Oneworld Publications.
130 Damasio, A. (2012). *Self Comes to Mind: Constructing the Conscious Brain.* Vintage.
131 Evans, J. St. B. T. (2010). *Thinking Twice: Two Minds in One Brain.* Oxford University Press.
132 Chalmers, D. J. (2003). *The Conscious Mind: In Search of a Fundamental Theory.* Oxford University Press.
133 Pinker, S. (2003). *The Blank Slate: The Modern Denial of Human Nature.* Penguin Books.
134 Schelling, T. C. (1981). *The Strategy of Conflict.* Harvard University Press.
135 Frankfurt, H. (January 1971). "Freedom of the Will and the Concept of a Person." *Journal of Philosophy*, vol. 68.
136 Darwin, C. (2004). *The Descent of Man.* Penguin Classic.
137 Bowers, K. and Natterson-Horowitz, B. (2012). *Zoobiquity: What Animals Can Teach Us About Being Human.* Doubleday Canada.
138 Stanovich, K. E. (2010). *What Intelligence Tests Miss: The Psychology of Rational Thought.* Yale University Press.
139 Stanovich, K. E. (2011). *Rationality and the Reflective Mind.* Oxford University Press.
140 Thompson, V. (2008). "Dual Process Theories: A Metacognitive Perspective." In J. Evans and K. Frankish (Eds.), *In Two Minds: Dual Processes and Beyond.* Oxford University Press.
141 Evans, J. and Stanovich, K. E. (May 2013). "Dual-Process Theories of Higher Cognition: Advancing the Debate." *Perspectives on Psychological Science*, vol. 8.
142 Kahneman, D. (2011). *Thinking, Fast and Slow.* Doubleday Canada.
143 Schwarz, N. et al. (August 1991). "Ease of Retrieval as Information: Another Look at the Availability Heuristic." *Journal of Personality and Social Psychology*, vol. 61.
144 Whittlesea, B. and Leboe, J. (2003). "Two Fluency Heuristics (and How to Tell Them Apart)." *Journal of Memory and Language*, vol. 49.
145 Winkielman, P. et al. (2006). "Prototypes Are Attractive Because They Are Easy on the Mind." *Psychological Science, vol. 17.*
146 Winkielman, P., et al. (2013). "The Hedonic Marking of Processing Fluency: Implications for Evaluative Judgment." In J. Musch and K. Klauer (Eds.), *The*

Psychology of Evaluation: Affective Processes in Cognition and Emotion. Psychology Press.

147 Burton, R. (2009). *On Being Certain: Believing You Are Right When You're Not.* St. Martin's Griffin.

148 Shynkaruk, J. and Thompson, V. (April 2006). "Confidence and Accuracy in Deductive Reasoning." *Memory and Cognition,* vol. 34.

149 Burton, R. (2009). *On Being Certain: Believing You Are Right When You're Not.* St. Martin's Griffin.

150 Senge, P. M. et al. (2005). *Presence: An Exploration of Profound Change in People, Organizations, and Society.* Crown Business.

151 Virgil. (2009). *Georgics.* Oxford University Press.

152 Hume, D. (2008). *An Enquiry Concerning Human Understanding.* Oxford University Press.

153 Hegel, G. W. F. (1979). *The Phenomenology of Spirit.* Oxford University Press.

154 Mandelbrot, B. (1977). *Fractals: Form, Chance and Dimension.* W. H. Freeman & Co.

155 Gell-Man, M. (March 2007). Beauty and Truth in Physics. TED Talk.

156 Nietzsche, F. (1974). *The Gay Science.* Vintage.

157 Taleb, N. N. (2010). *The Black Swan: The Impact of the Highly Improbable.* Random House.

158 Gazzaniga, M. S. (2011). *Who's in Charge? Free Will and the Science of the Brain.* Ecco.

159 Langer, E. (August 1975). "The Illusion of Control." *Journal of Personality and Social Psychology,* vol. 32.

160 Perrow, C. (1999). *Normal Accidents: Living With High-Risk Technologies.* Princeton University Press.

161 Pollan, M. (January 28 2007). "Unhappy Meals." *The New York Times.*

162 Garcia, J., et al. (July 1995). "Conditioned Aversion to Saccharin Resulting From Exposure to Gamma Radiation." *Science,* vol. 122.

163 Stolley, P. D. (March 1991). "When Genius Errs: R. A. Fisher and the Lung Cancer Controversy." *American Journal of Epidemiology,* vol. 133.

164 Seligman, M. (July 1971). "Phobias and Preparedness." *Behavior Therapy,* vol. 2.

165 Einhorn, H. and Hogarth, R. (January 1982). "Prediction, Diagnosis and Causal Thinking in Forecasting." *Journal of Forecasting,* vol. 1.

166 Poisson, S. D. (1837). "Researches Into the Probabilities of Judgments in Criminal and Civil Cases." (Sourced online.)

167 Mauboussin, M. (2012). *The Success Equation: Untangling Skill and Luck in Business, Sports, and Investing.* Harvard Business Review Press.

168 De Vany, A. (2003). *Hollywood Economics: How Extreme Uncertainty Shapes the Film Industry.* Routledge.

169 Rosenzweig, P. (2009). *The Halo Effect: … and the Eight Other Business Delusions That Deceive Managers.* Free Press. Makridakis, S., Hogarth, R. and Gaba, A. (2010). *Dance With Chance: Making Luck Work for You.* Oneworld Publications. Henderson, A., Raynor, M. and Ahmed, M. (April 2012). "How Long Must a Firm Be Great to Rule Out Chance? Benchmarking Sustained Superior Performance Without

Being Fooled by Randomness." *Strategic Management Journal*, vol. 33. McGahan, A. and Porter, M. (July 1997). "How Much Does Industry Matter, Really?" *Strategic Management Journal*, vol. 18.

170 Raynor, M., Ahmed, M. and Henderson, A. (2009). *A Random Search for Excellence: Why "Great" Company Research Delivers Fables and Not Facts*. Deloitte.
171 Taleb, N. N. (2012). *Antifragile: Things That Gain From Disorder*. Random House.
172 Raynor, M. (2007). *The Strategy Paradox*. Doubleday.
173 Kahneman, D. (2011). *Thinking, Fast and Slow*. Doubleday Canada.
174 Wainer, H. (2009). *Picturing the Uncertain World: How to Understand, Communicate, and Control Uncertainty Through Graphical Display*. Princeton University Press.
175 Krishnamurti, J. (2003). *Truth Is a Pathless Land*. [Audio CD]. Sounds True.
176 Delbruck, M. (1986). *Mind From Matter? An Essay on Evolutionary Epistemology*. Blackwell Scientific Publications.
177 Pascal, B. (2003). *Pensées*. Penguin Classic.
178 Nietzsche, F. (2003). *Nietzsche: Writings From the Late Notebooks*. Cambridge University Press.
179 Kaufmann, W. (1977), Ed. The Portable Nietzsche. Penguin.
180 Foucault, M. (1988). *Madness and Civilization: A History of Insanity in the Age of Reason*. Vintage.
181 Rorty, R. (1981). *Philosophy and the Mirror of Nature*. Princeton University Press.
182 Quine, W. V. O. (2013). *Word and Object*. MIT Press.
183 Newton, I. (2010). *The Principia*. Prometheus Books.
184 Popper, K. (2002). *Conjectures and Refutations: The Growth of Scientific Knowledge*. Routledge.
185 Cooke, E. (2007). *Peirce's Pragmatic Theory of Inquiry: Fallibilism and Indeterminacy*. Continuum.
186 Damasio, A. (2005). *Descartes' Error: Emotion, Reason, and the Human Brain*. Penguin Books.
187 Gould, S. J. (1984). *Hen's Teeth and Horse's Toes: Further Reflections in Natural History*. Penguin Books.
188 Tetlock, P. (2006). *Expert Political Judgment: How Good Is It? How Can We Know?* Princeton University Press.
189 Ioannidis, J. (August 2005). "*Why Most Published Research Findings Are False*." PLoS *Medicine*, vol. 2.
190 Fanelli, D. (March 2012). "Negative Results Are Disappearing From Most Disciplines and Countries." *Scientometrics*, vol. 90.
191 Goldacre, B. (2013). *Bad Pharma: How Drug Companies Mislead Doctors and Harm Patients*. Signal.
192 Simmons, J., Nelson, L. and Simonsohn, U. (November 2011). "False-Positive Psychology: Undisclosed Flexibility in Data Collection and Analysis Allows Presenting Anything as Significant." *Psychological Science*, vol. 22.
193 Tetlock, P. (2006). *Expert Political Judgment: How Good Is It? How Can We Know?* Princeton University Press.
194 Orrell, D. (2007). *Apollo's Arrow*. HarperCollins.

195 Taleb, N. H. (2012). *Antifragile: Things That Gain From Disorder*. Random House.
196 Voltaire (1770). "Letter to Frederick William, Prince of Prussia." [Sourced online.]
197 Taleb, N. N. (2010). *The Black Swan: The Impact of the Highly Improbable*. Random House.
198 Frederick, S. (Fall 2005). "Cognitive Reflection and Decision Making." *Journal of Economic Perspectives*, vol. 19.
199 Mayer, J. and Holmes, J., Eds. (1996) *Bite-Size Einstein*. St. Martin's Press.
200 Wittgenstein, L. (2001). *Tractatus Logico-Philosophicus*. Routledge.
201 Heidegger, M. (1982). *The Question Concerning Technology and Other Essays*. Harper Perennial.
202 Hume, D. (2008). *An Enquiry Concerning Human Understanding*. Oxford University Press.
203 McGinn, C. (1994). *Problems in Philosophy: The Limits of Inquiry*. Blackwell Publishers.
204 Chaitin, G. (March 2006). "The Limits of Reason." *Scientific American*, vol. 294.
205 Lesher, J. H. (1992). *Xenophanes of Colophon: Fragments: A Text and Translation With Commentary*. University of Toronto Press.
206 Parmenides. (2011). *On the Order of Nature*. Aurea Vidya.
207 Empiricus, S. (1990). *Outlines of Pyrrhonism*. Prometheus Books.
208 Empiricus, S. (1949). *Against the Professors*. Loeb Classical Library.
209 Brett, R. (Ed.) (2010). *The Cambridge Companion to Ancient Scepticism*. Cambridge University Press.
210 Hume, D. (2008). *An Enquiry Concerning Human Understanding*. Oxford University Press.
211 Pascal, B. (2003). *Pensées*. Penguin Classic.
212 Kahneman, D. and Tversky, A. (March 1979). "Prospect Theory: An Analysis of Decision Under Risk." *Econometrica*, vol. 47.
213 Shermer, M. (2012). *The Believing Brain: From Ghosts and Gods to Politics and Conspiracies — How We Construct Beliefs and Reinforce Them as Truths*. St. Martin's Griffin.
214 Nietzsche, F. (1996). *Human, All Too Human: A Book for Free Spirits*. Cambridge University Press.
215 Aristotle. (2002). *Nicomachean Ethics*. Focus Publishing.
216 Tetlock, P. (2006). *Expert Political Judgment: How Good Is It? How Can We Know?* Princeton University Press.
217 Thompson, V. and Shynkaruk, J. (April 2006). " Confidence and Accuracy in Deductive Reasoning." *Memory and Cognition*, vol. 34.
218 Taleb, N. N. (2010). *The Black Swan: The Impact of the Highly Improbable*. Random House.
219 Argyris, C. (September 1977). "Double Loop Learning in Organizations." *Harvard Business Review*, vol. 55.
220 Taleb, N. N. (2005). *Fooled by Randomness: The Hidden Role of Chance in Life and in the Markets*. Random House.
221 Taleb, N. H. (2012). *Antifragile: Things That Gain From Disorder*. Random House.

222 Mencken, H. L. (2013). *Prejudices*. Hardpress Publishing.
223 Bacon, F. (1999). *Novum Organum*. Carus Publishing.
224 Beck, Aaron. (1973). *The Diagnosis and Management of Depression*. University of Pennsylvania Press.
225 Baltes, P. and Staudinger, U. (January 2000). "Wisdom: A Metaheuristic (Pragmatic) to Orchestrate Mind and Virtue Toward Excellence." *American Psychologist*, vol. 5.
226 Voltaire. (1994). *Letters Concerning the English Nation*. Oxford University Press.
227 Klein, G. (September 2007). "Performing a Project Premortem." *Harvard Business Review*, vol. 85.
228 Russo, J. E. and Shoemaker, P. (2001). *Winning Decisions: Getting It Right the First Time*. Crown Business.
229 Bayes, Thomas (January 1763). *An Essay Towards Solving a Problem in the Doctrine of Chances*. *Philosophical Transactions*, vol. 53. [Sourced online.]
230 Tetlock, P. (2006). *Expert Political Judgment: How Good Is It? How Can We Know?* Princeton University Press.
231 Page, S. (2008). *The Difference: How the Power of Diversity Creates Better Groups, Firms, Schools, and Societies*. Princeton University Press.
232 Janis, I. (1973). *Victims of Groupthink: A Psychological Study of Foreign-Policy Decisions and Fiascoes*. Houghton Mifflin Harcourt.
233 Nemeth, C. (2011). "Minority Interest Theory." In P. A. M. van Lange, A. W. Kruglanski, E. T. Higgins (Eds.), *Handbook of Theories in Social Psychology: Collection: Volumes 1 &2*. Sage Publications.
234 Locke, J. (2011). *A Letter Concerning Toleration*. Merchant Books.
235 Franklin, B. (2000). *Wit and Wisdom From Poor Richard's Almanack*. Random House.
236 Orwell, G. (1968). "In Front of Your Nose." In Brownell Orwell, S. and Angus I. (Eds.), *The Collected Essays, Journalism and Letters of George Orwell*. Harcourt Brace Jovanovich.
237 Mazar, N. and Ariely, D. (Spring 2006). "Dishonesty in Everyday Life and its Policy Implications." *Journal of Public Policy and Marketing*, vol. 25.
238 Freud, S. (2010). *The Ego and the Id*. CreateSpace.
239 Plato, (2009). *Phaedrus*. Oxford University Press.
240 Nietzsche, F. (2010). *Beyond Good and Evil*. CreateSpace.
241 Sartre, J-P. (1993). Being and Nothingness. Washington Square Press.
242 Hume, D. (2008). *An Enquiry Concerning Human Understanding*. Oxford University Press.
243 Descartes, R. (2013). *Meditations on First Philosophy*. Broadview Press.
244 de Montaigne, M. (2004). *The Complete Essays*. Penguin Classic.
245 Gazzaniga, M. S. (July 1998). "The Split Brain Revisited." *Scientific American*.
246 Wegner, D. M. (2003). *The Illusion of Conscious Will*. Bradford Books.
247 Forer, B. (January 1949). "The Fallacy of Personal Validation: A Classroom Demonstration of Gullibility." *Journal of Abnormal and Social Psychology*, vol. 44.

248 Sheldon, K. et al. (December 1997). "Trait Self and True Self: Cross-Role Variation in the Big-Five Personality Traits and Its Relations With Psychological Authenticity and Subjective Well-Being." *Journal of Personality and Social Psychology*, vol. 73.
249 Mischel, W. (1968). *Personality and Assessment*. John Wiley and Sons.
250 Hartshorne, H. and May, M. A. (1928). *Studies in the Nature of Character*. Macmillan.
251 Kahneman, D. (2011). *Thinking, Fast and Slow*. Doubleday Canada.
252 Gilbert, D. (2007). *Stumbling on Happiness*. Vintage.
253 Feynman, R. P. (2005). "What Is and What Should Be the Role of Scientific Culture in Modern Society." *The Pleasure of Finding Things Out: The Best Short Works of Richard P. Feynman*. Basic Books.
254 Pascal, B. (2003). *Pensées*. Penguin Classic.
255 Schopenhauer, A. (2010). *The World as Will and Representation*. Cambridge University Press.
256 Tetlock, P. (2000). "Coping With Tradeoffs: Psychological Constraints and Political Implications." In A. Lupia, M. D. McCubbins and S. L. Popkin (Eds.), *Elements of Reason: Cognition, Choice, and the Bounds of Rationality*. Cambridge University Press.
257 Eagleman, D. (2012). *Incognito: The Secret Lives of the Brain*. Penguin.
258 Wilson, T. D. (2004). *Strangers to Ourselves: Discovering the Adaptive Unconscious*. Belknap Press.
259 Argyris, C. (1982). *Reasoning, Learning, and Action: Individual and Organizational*. Jossey-Bass.
260 Greenwald, A., McGhee, D. and Schwartz, J. (June 1998). "Measuring Individual Differences in Implicit Cognition: The Implicit Association Test." *Journal of Personality and Social Psychology*, vol. 74.
261 Russell, B. (1996). *Skeptical Essays*. Routledge.
262 Bazerman, M. H. and Tenbrunsel, A. E. (2012). *Blind Spots: Why We Fail to Do What's Right and What to Do About It*. Princeton University Press.
263 Greene, J. et al. (September 2001). "An fMRI Investigation of Emotional Engagement in Moral Judgment." *Science*, vol. 293.
264 Damasio, A. (2000). *The Feeling of What Happens: Body and Emotion in the Making of Consciousness*. Houghton Mifflin Harcourt.
265 Kant, I. (1997). *Critique of Practical Reason*. Cambridge University Press.
266 Bentham, J. (2013). *An Introduction to the Principles of Morals and Legislation*. Hardpress Publishing.
267 Mill, J. S. (2000). *Utilitarianism*. Broadview Press.
268 Aristotle (2009). *The Nicomachean Ethics*. Oxford University Press.
269 Plato (2008). *Republic*. Oxford Paperbacks.
270 Ainslie, G. (2001). *Breakdown of Will*. Cambridge University Press.
271 Baumeister, R. F. and Tierney, J. (2012). *Willpower: Rediscovering the Greatest Human Strength*. Penguin Books.

272 Duckworth, A. L. (2013). "Self-Regulation and School Success." In B. W. Sokol, F. M. E. Grouzet and U. Müller (Eds.), *Self-Regulation and Autonomy: Social and Developmental Dimensions of Human Conduct*. Cambridge University Press.

273 Fredrickson, B. L. (2009). *Positivity: Top-Notch Research Reveals the 3- to-1 Ratio That Will Change Your Life*. Harmony. Baumeister, R. F. et al. (December 2001). "Bad Is Stronger Than Good." *Review of General Psychology*, vol. 5.

274 Ariely, D., Wertenbroch, K. (May 2002). "Procrastination, Deadlines and Performance: Self-control by Precommitment." *Psychological Science*, vol. 13.

275 Haidt, J. (2012). *The Righteous Mind: Why Good People Are Divided by Politics and Religion*. Pantheon Books.

276 Meissner, C. A. and Brigham, J. C. (March 2001). "Thirty Years of Investigating the Own-Race Bias in Memory for Faces: A Meta-Analytic Review." *Psychology, Public Policy, and Law*, vol. 7.

277 Klaus J., Zaballa, L. (June 2010). "Co-Operative Punishment Cements Social Cohesion." *Journal of Artificial Societies and Social Simulation*, vol. 13.

278 Henrich, J. (Sept. 2009). "How Culture Drove Human Evolution: A Conversation with Joseph Henrich." *Edge* (on-line).

279 De Waal, F. (2013). *The Bonobo and the Atheist: In Search of Humanism Among the Primates*. W. W. Norton & Company.

280 Boswell, J. (2003). *The Life of Samuel Johnson*. Penguin Classic.

281 Sapolsky, R. (2004). *Why Zebras Don't Get Ulcers: The Acclaimed Guide to Stress, Stress-Related Diseases, and Coping*. Holt Paperbacks. Nobles, J., Weintraub, M. and Adler, N. (April 2013). "Subjective Socioeconomic Status and Health: Relationships Reconsidered." *Social Science and Medicine*, vol. 82.

282 Frank, R. H. (2007). *Falling Behind: How Rising Inequality Harms the Middle Class*. University of California Press.

283 Killingsworth, M. and Gilbert, D. (November 2010). "A Wandering Mind Is an Unhappy Mind." *Science*, vol. 330.

284 Wright, R. (1995). *The Moral Animal: Why We Are The Way We Are — The New Science of Evolutionary Psychology*. Vintage.

285 Pascal, B. (2003). *Pensées*. Penguin Classic.

286 Killingsworth, M. and Gilbert, D. (November 2010). "A Wandering Mind Is an Unhappy Mind." *Science*, vol. 330.

287 Seligman, M. E. P. (2004). *Authentic Happiness: Using the New Positive Psychology to Realize Your Potential for Lasting Fulfillment*. Atria Books.

288 Locke, J. (2008). *An Essay Concerning Human Understanding*. Oxford University Press.

289 Schopenhauer, A. (2010). *The World as Will and Representation*. Cambridge University Press.

290 Nietzsche, F. (2003). *Thus Spoke Zarathustra*. Penguin Classic.

291 Heidegger, M. (2010). *Being and Time*. State University of New York Press.

292 Camus, A. (1991). *The Myth of Sisyphus and Other Essays*. Vintage.

293 Sartre, J.-P. (2007). *Nausea*. New Directions.

294 Frankl, V. (2006). *Man's Search for Meaning*. Beacon Press.

295 Maisel, E. (2007). *The Van Gogh Blues: The Creative Person's Path Through Depression.* New World Library.
296 Klee, Paul. (1968). *The Diaries of Paul Klee, 1898–1918.* University of California Press.
297 Stanovich, K. E. (2004). *The Robot's Rebellion: Finding Meaning in the Age of Darwin.* University of Chicago Press.
298 Gould, S. J. (2006). *Wonderful Life: The Burgess Shale and the Nature of History.* W. W. Norton & Company.
299 Dennett, D. C. (1996). *Darwin's Dangerous Idea: Evolution and the Meanings of Life.* Simon & Schuster.
300 Sowell, Thomas. (2007). *A Conflict of Visions: Ideological Origins of Political Struggles.* Basic Books.
301 Pinker, S. (2003). *The Blank Slate: The Modern Denial of Human Nature.* Penguin Books.
302 Dawkins, R. (2000). *Unweaving the Rainbow: Science, Delusion and the Appetite for Wonder.* Houghton Mifflin Harcourt.
303 Kant, I. (2010). *An Answer to the Question: 'What Is Enlightenment?'* Penguin Books.
304 Bohm, D. (1994). *Thought as a System.* Routledge.
305 Ramo, S. (1970). *Extraordinary Tennis for the Ordinary Player.* Crown Publishers.
306 Ellis, C. (1993). *Investment Policy: How to Win the Loser's Game.* Business One Irwin.

INDEX

accuracy vs. speed, 37–38, 116, 123, 130–131, 273–277. *see also* efficiency vs. effectiveness
Ainslie, George, 318
Akenson, Donald, 86
alternative paths, 119–120, 133–135, 290–293
altruism, 330–331
ambiguity, 83, 96–97, 104, 133–135, 266
ancestral thinking, 167, 199–200, 205–206, 348
animal vs. human cognition, 7, 192–195, 355–361
Archer, Edward, 58
Argyris, Chris, 287, 314
Ariely, Dan, 305
Aristotle, 46, 284, 316, 318
Armstrong, Karen, 86
attentional spotlight, 31–32, 59–61
automatic vs. effortful thinking. *see also* Type 1/Type 2 thinking
   challenges, 25–28, 30
   and data selection, 59
   dual-process model, 171–174
   and satisficing, 88–90
   unconscious incubation, 272
availability bias, 100–101, 107–110, 130–133

awareness. *see also* theory of mind (ToM)
   boundedness, 31–32
   and cognitive opacity, 312–313
   consciousness, 18, 187–192
   human vs. animal, 7, 192–195, 355–361

Bacon, Francis, 288
Baltes, Paul, 289
Baron, Jonathan, 136
Barrett, Lisa Feldman, 61
base rate neglect, 108–111
Baumeister, Roy, 319, 323
Bayes, Thomas, 295
Bayes' theorem, 294–299
Bazerman, Max, 315
Beck, Aaron, 288
behavioural economics, 29, 31, 91
belief systems, 98–99, 103–105, 113–114
Bentham, Jeremy, 316
Berkeley, George, 48
bias
   availability, 100–101, 107–110, 130–133
   and cognitive diversity, 300–301
   and coherence, 100–105, 107–114
   confirmation, 101–103, 110–113, 259–260, 288–290
   hindsight, 118–120, 155
   myside, 103–105, 113–114
   negativity, 279–280

objective neutrality, 116
outcome, 136–137
tribalism, 330
Bohm, David, 208, 348
Bohr, Niels, 253
bounded brains, 29–37
brain physiology
    circuitry, 49, 59, 68, 176–177, 179, 308
    and consciousness, 182–183, 188
    energy needs, 20–21, 26, 179
    evolution of, 13, 15, 16, 19–22
brain vs. mind, 52–53
brain–body gap, 26
brain–brain gap, 25–26
brain–world gap, 3–4, 27, 39, 162–168
Bruner, Jerome, 120
Buddhist philosophy, 325, 343–344
Burton, Robert, 95, 121, 149, 203, 204

Campbell, Jeremy, 39
Camus, Albert, 156, 341, 345
Carneades, 278
Carpenter, William B., 172
causality
    bounded sense of, 34
    vs. correlation, 234–238, 259–260
    as mental model, 28
    and randomness, 223–225
    and resemblance, 66, 78–79
    and systems theory, 152, 209, 211–213, 231–238
certainty, need for, 96–97, 101–102, 104–105, 111–113, 135–136
Chabris, Christopher, 59
Chaitin, Gregory, 276
Chalmers, David, 188
chaos, 217–221. *see also* randomness
Chase, William, 127
cognitive dissonance, 95–96, 104
cognitive diversity, 300–302
cognitive miserliness, 90–92, 100. *see also* mental shortcuts
cognitive science, 6, 166–167, 204, 208
cognitive therapy, 288–289, 325–326. *see also* mental illness; psychotherapy
coherence, 98–100, 107–114, 261–263
complexity
    of being human, 338–348

and chaos, 219–220
and cognitive trade-off, 37–40
of self, 305–327
social, 20, 142, 328–337
and systems, 151–153, 214
and Type 2 thinking, 204–208
and World #2, 149–157, 167–168
complexity science, 6, 150–151, 166, 204, 207–208
conceptualization, 36–37, 71
confabulation, 67–72, 83–87
confidence, 88, 115–122, 135–136, 202–204, 286–287
confirmation bias, 101–103, 110–113, 259–260, 288–290
conscious vs. subconscious automatic thinking. *see* automatic vs. effortful thinking
consciousness. *see* awareness
constructive realism, 44–47, 166
control, need for, 93–94
correlation, vs. causation, 234–238
cultural evolution, 18, 144

Damasio, Antonio, 182, 187, 261, 316
Darwin, Charles, 117, 193, 330, 360
Dawkins, Richard, 51, 347
De Vany, Arthur, 244
Dennett, Daniel, 75, 345
depression. *see* mental illness
Descartes, René, 47, 178, 307
dichotomy, 79–81
Doidge, Norman, 179
Dostoyevsky, Fyodor, 118
Drucker, Peter, 141
dual process thinking. *see* automatic vs. effortful thinking
Duckworth, Angela Lee, 320
Dunbar, Robin, 142
Dunning-Kruger effect, 117

Eagleman, David, 62, 312
efficiency vs. effectiveness, 37–39, 87, 97–98, 271–274. *see also* accuracy vs. speed
Einhorn, Hillel, 110, 238
Einstein, Albert, 271
Eire, Carlos M. N., 86

Ellis, Charles D., 350
emergence, 53, 152–153, 222–223
emotion, 70–71, 321–324
Empiricus, Sextus, 278
energy conservation, 26, 88–89, 91
epigenetics, 80–81
Ericsson, K. Anders, 127
Evans, Jonathan, 187, 201
evolution
    chaotic vs. convergent, 345
    cultural, 18, 144
    imperfections, 24–25
    overview, 14–15, 16–17, 27
expertise, 127–130, 137–139, 350

fairness, 331–332
falsification, 111–112, 257, 288
Fanelli, Daniele, 264
feedback
    and complexity, 133–135, 239–240
    lack of, 77, 134
    loops, 103, 215, 250–251
Feynman, Richard, 312
Fisher, Ronald, 235
Flanagan, Owen, 181
fluency, 201–204
Flynn, James, 30
forecasting. *see* predictability
Forer, Bertram, 308
Foucault, Michel, 74, 256
Frances, Allen, 74, 75
Frank, Robert, 333
Frankfurt, Harry, 192
Frankl, Viktor, 342
Franklin, Benjamin, 305
Frederick, Shane, 91, 271
Frederickson, Barbara, 323
Freud, Sigmund, 73–74, 172, 307, 313

Galton, Francis, 79, 172
Garcia, John, 233, 236
Gazzaniga, Michael, 67, 181, 226, 308
Gell-Mann, Murrey, 223
Gigerenzer, Gerd, 65
Gilbert, Daniel, 70, 311, 340
Gladwell, Malcolm, 127, 167
Goldacre, Ben, 265
Gould, Stephen Jay, 261, 345

Greene, Joshua, 315
Greenwald, Arthur, 314
gullibility, 279–284

Haidt, Jonathan, 178, 181, 330
Hammond, Kenneth R., 146
Hartshorne, Hugh, 310
Hawking, Stephen, 51
Hegel, G. W. F., 50, 217
Heidegger, Martin, 81–82, 276, 341
heuristics, 65–67, 147, 162–163. *see also*
    mental shortcuts
hindsight bias, 118–120, 155, 292
Hogarth, Robin, 110, 130, 145, 238
Hood, Bruce, 138
human paradox, 338–348
human vs. animal cognition, 7, 192–195,
    355–361
Hume, David, 48, 112, 181, 212, 276, 278,
    307

intelligence, 30, 199, 301, 320–321
intuition, 127–139, 147–148, 158–161
Ioannidis, John, 263

James, William, 65
Janis, Irving, 301
Johansson, Petter, 71
Johnson, Samuel, 332
Jung, Carl, 313

Kahneman, Daniel, 77–78, 103, 121, 129,
    133, 137, 145, 201, 249, 280, 310
Kant, Immanuel, 46–47, 48–49, 155, 316,
    348
Kaplan, Abraham, 118
Kierkegaard, Søren, 156
Killingsworth, Matthew, 340
Klee, Paul, 344
Klein, Gary, 128, 129, 145, 175, 291
knowing, feeling of
    and cognitive dissonance, 94–95,
        104–105
    and coherence, 99
    and confidence, 115, 120–122
    and intuition, 137–139
    and Type 2 thinking, 201–202
Krishnamurti, Jiddu, 253

Kruglanski, Arie, 97
Kuhn, Thomas, 258

Langer, Ellen, 227
language, 18, 66–67, 81–83
LeDoux, Joseph, 182
Leibniz, Gottfried, 172
Libet, Benjamin, 182
Lieberman, Matthew, 175
limiting case, 54–56, 209–210
listening, 336–337
Locke, John, 48, 64, 305, 340
Loftus, Elizabeth, 84–85
Lorenz, Edward, 218
luck vs. skill, 243–248

Maisel, Eric, 343
Makridakis, Spyros, 244
Mandelbrot, Benoit, 222
Mauboussin, Michael, 244
May, M. A., 310
McGinn, Colin, 156, 276
memory gaps, 71–72, 85
Mencken, H. L., 287
mental illness, 73–75, 84, 288. *see also* cognitive therapy
mental shortcuts. *see also* cognitive miserliness; heuristics
   cognitive trade-off, 37–39
   naïve vs. constructive realism, 44–45
   and representativeness, 77–78
   and satisficing, 89
   and simplification, 58
   social shortcuts, 330–335
metacognition, 180, 196, 198
Mill, John Stuart, 316
Miller, Bill, 247
mind vs. brain, 52–53
mindfulness, 7, 198–199
Mischel, Walter, 310
models, mental, 27–29, 51–54, 62–66, 76–77, 207
Montaigne, Michel de, 308
myside bias, 103–105, 113–114

Nagel, Thomas, 45
naïve realism, 44–46, 116
Natterson-Horowitz, Barbara, 193

natural selection, 14, 21–22, 23, 345
nature vs. nurture, 79–81
negativity bias, 279–280, 323, 340
Nemeth, Charlan, 302
Newton, Isaac, 257
Nietzsche, Friedrich, 82, 223, 253, 255, 284, 307, 341, 344
nutritional science, 57–58, 164, 232

optical illusion, 28–29, 76–77. *see also* visual cues
Orrell, David, 94, 268
Orwell, George, 305
outcome bias, 136–137

Page, Scott, 300
Paris, Joel, 70
Parmenides, 277
Pascal, Blaise, 23, 253, 279, 312, 339, 340
path dependence, 113–114, 250–252, 290–293, 301–302
patterns, 62, 241–243. *see also* randomness
Peirce, Charles Sanders, 260–261
perception
   and language, 66–67, 82
   and reality construction, 44–47
Perkins, David, 39, 103
Perrow, Charles, 231
persistence, 320
Pettit, Becky, 109
Pinker, Stephen, 156, 181, 189, 346
Plato, 46, 307, 318
Poincaré, Henri, 217
Poisson, Siméon-Denis, 242
poisson clumping, 242, 265
Pollan, Michael, 232, 236
Popper, Karl, 37, 111, 257, 258
Porter, Michael, 244
Potter, Mary, 120
Pragmatism, 256
predictability, 137, 152–153, 266–269, 284–285
presentism, 100–101
probability, 32–33, 76, 263–265, 268–269, 293–300
psychotherapy, 70, 73–75. *see also* cognitive therapy; mental illness
Pyrrho, 277

Quine, W. O., 257

Ramo, Simon, 349
randomness, 33–34, 223–227, 241–252.
    see also chaos; patterns
rationality, 35, 320–321
Raynor, Michael, 244, 246, 248
reality, construction of, 44–56
reductionism, 73, 76, 87
religious evolution, 85–86
representativeness. see resemblance
resemblance, 66, 77–78
reversion to the mean, 249–250
Ridley Stroop, John, 180
Rorty, Richard, 256
Rosenzweig, Phil, 244, 245
ruminations, 84, 92–93, 324
Russell, Bertrand, 315
Russo, J. Edward, 292
Ryle, Gilbert, 52

Sacks, Oliver, 50, 69
Sapolsky, Robert, 93
Sartre, Jean-Paul, 307, 342, 344
satisficing, 89, 97–98, 123
Schelling, Thomas, 190
Schopenhauer, Arthur, 172, 312, 341
Schumacher, E. F., 146
Schwarz, Norbert, 201
scientific method, 111–112, 257–269,
    287–288, 300–302
self-deception, 305–307
    and emotion, 321–325
    and hypocrisy, 312–318
    multiplicity of, 307–311, 324–327
    and willpower, 318–321, 326
Seligman, Martin, 236, 340
sensory data. see also signal and noise;
    visual cues
    gaps, 68–69, 107–110, 130–133
    limiting case view, 55–56
    and mental models, 62
    selection of, 59–61, 67, 71
Shannon, Claude, 151
Sheldon, Kennon, 310
Shermer, Michael, 82, 283
Shoemaker, Paul, 292
signal and noise. see also sensory data
    and ambiguous feedback, 133–135
    in assessing others, 333–334
    and data selection, 59–62, 76
    and emotion, 325
    and obscured cues, 130–133
    and randomness, 150–151, 223–226,
        241–243
    skill vs. luck, 243–248
    World #1 vs. World #2, 2, 210–211
Simmons, Joseph, 265
Simon, Herbert, 29, 89, 127, 128, 181
Simons, Daniel, 59
skepticism, 277–287
skill vs. luck, 243–248
social complexity, 20, 142, 328–337
Sowell, Thomas, 346
Spelke, Elizabeth, 64
Spinoza, Baruch, 102
Stanovich, Keith, 92, 147, 175, 177, 196,
    199, 201, 344
Stoicism, 325, 343
storytelling. see confabulation
systems
    and complexity, 151–153, 213–215
    dynamics, 216–222
    and emergence, 222–223
    nonlinearity, 216
    and randomness, 223–227
    theory, 166, 209, 229

Taleb, Nassim Nicholas, 4, 110, 146, 225,
    246, 268, 270, 286, 287
Tetlock, Philip, 76, 135, 262, 268, 284–285,
    287, 300, 312
Thagard, Paul, 45
theory of mind (ToM), 20, 196–198, 335.
    see also awareness
Thompson, Valerie, 201, 203, 286
tribalism, 330–331
truth, 115–118, 253–261, 270–302
Tversky, Amos, 280
Type I/Type II error, 279–283
Type 1/Type 2 thinking. see also automatic
    vs. effortful thinking
    Type 2.1 vs. 2.2, 195–196, 198–200
    in combination, 185–187, 189–191, 202
    and complexity, 168, 200–208
    in managing multiple selves, 326–327

INDEX   381

in managing others, 335–337
strengths and weaknesses, 175–184

Virgil, 211
vision, 51, 68–69, 175–176
visual cues, 149, 159–160, 163, 336. *see also* optical illusion; sensory data
Voltaire, 270, 289
von Uexküll, Jakob, 52

Waal, Frans de, 332
Wainer, Howard, 106, 250
Wampold, Bruce, 70
Wason, Peter, 101
Wegner, Daniel, 181, 308
Wells, H. G., 4
wicked problems, 56, 145–146, 153–156, 167
Wiener, Norbert, 151
willpower, 318–321, 326

Wilson, Timothy, 312
Wittgenstein, Ludwig, 82, 276
World #1/World #2
and causality, 212–213
comparison, 2–3, 147–149, 166–168, 229–231, 353
and complexity, 149–157
and data processing, 148–149, 163–164
evolution, 140–144
"game of life," 350–351
gullibility and skepticism, 281–283
scholarly terminology, 145–147, 168
and truth, 254–255
Wright, Robert, 337

Xenophanes, 277

Yeats, W. B., 284

Zwerling, Harris, 106

About the Author

Ted Cadsby, MBA, CFA, ICD.D, is a corporate director, consultant, and a researcher, writer and speaker on complexity and decision making. As the former executive vice president of Retail Distribution at the Canadian Imperial Bank of Commerce, he led 18,000 employees in banking and wealth management services. Prior to that role, he was President and CEO of CIBC Securities Inc., Chairman of CIBC Trust Corp. and Chairman of CIBC Private Investment Counsel Inc., overseeing Canadian and international offices. In his capacity as a director on both for-profit and not-for-profit boards, he has chaired a variety of board committees.

Ted graduated as the medallist in philosophy from Queen's University, completed his MBA at the Ivey Business School at the University of Western Ontario and holds the Chartered Financial Analyst designation from the CFA Institute and the ICD.D designation from the Institute of Corporate Directors. He is the author of two books on investing and has been extensively interviewed by the national media, including CBC's The National, CTV National News and Canada AM.

More information about the author and his work is available at www.tedcadsby.com.